# ACT

## Science Prep Course

**Jeff Kolby**
**Garrett Biehle**
**Nancy Morvillo**
**Matthew Schmidt**
**Ajikumar Aryangat**

NOVA PRESS

Additional educational titles from Nova Press (available at novapress.net):

- **ACT Math Prep Course** (402 pages)
  **ACT Verbal Prep Course** (248 pages)
- **SAT Prep Course** (640 pages, includes software)
  **SAT Math Prep Course** (404 pages)
  **SAT Critical Reading and Writing Prep Course** (350 pages)
- **GRE Prep Course** (624 pages, includes software))
  **GRE Math Prep Course** (528 pages)
- **GMAT Prep Course** (624 pages, includes software)
  **GMAT Math Prep Course** (528 pages)
  **GMAT Data Sufficiency Prep Course** (422 pages)
  **Full Potential GMAT Sentence Correction Intensive** (372 pages)
- **Master The LSAT** (608 pages, includes software and 4 official LSAT exams)
- **The MCAT Physics Book** (444 pages)
  **The MCAT Biology Book** (416 pages)
  **The MCAT Chemistry Book** (428 pages)
- **Scoring Strategies for the TOEFL® iBT:** (800 pages, includes audio CD)
  **Speaking and Writing Strategies for the TOEFL® iBT:** (394 pages, includes audio CD)
  **500 Words, Phrases, and Idioms for the TOEFL® iBT:** (238 pages, includes audio CD)
  **Practice Tests for the TOEFL® iBT:** (292 pages, includes audio CD)
  **Business Idioms in America:** (220 pages)
  **Americanize Your Language and Emotionalize Your Speech!** (210 pages)
- **Postal Exam Book** (276 pages)
- **Law School Basics:** A Preview of Law School and Legal Reasoning (224 pages)
- **Vocabulary 4000:** The 4000 Words Essential for an Educated Vocabulary (160 pages)



ISBN 10: 1–944595–02–3
ISBN 13: 978-1-944595–02–9



NOVA PRESS
P. O. Box: 692023
West Hollywood, CA 90069

Phone: 1-310-275-3513
Email: info@novapress.net
Website: www.novapress.net

# ABOUT THIS BOOK

If you don't have a pencil in your hand, get one now! Don't just read this book—write on it, study it, scrutinize it! In short, for the next four weeks, this book should be a part of your life. When you have finished the book, it should be marked-up, dog-eared, tattered and torn.

Although the ACT is a difficult test, it is a *very* learnable test. This is not to say that the ACT is "beatable." There is no bag of tricks that will show you how to master it overnight. You probably have already realized this. Some books, nevertheless, offer "inside stuff" or "tricks" which they claim will enable you to beat the test. These include declaring that answer-choices B or C are more likely to be correct than choices A or D. This tactic, like most of its type, does not work. It is offered to give the student the feeling that he or she is getting the scoop on the test.

The ACT cannot be "beaten." But it can be mastered—through hard work, analytical thought, and by training yourself to think like a test writer. Many of the exercises in this book are designed to prompt you to think like a test writer.

The ACT science section is not easy—nor is this book. To improve your ACT science score, you must be willing to work; if you study hard and master the techniques in this book, your score will improve—significantly.

This book will introduce you to numerous analytic techniques that will help you immensely, not only on the ACT but in college as well. For this reason, studying for the ACT can be a rewarding and satisfying experience.

To insure that you perform at your expected level on the actual ACT, you need to develop a level of skill that is greater than what is tested on the ACT. Hence, about 10% of the science problems in this book are harder than actual ACT science problems.

# ACKNOWLEDGMENT

Behind any successful test-prep book, there is more than just the authors' efforts.

**Garrett Biehle**

The author wishes to thank James Aldridge for his comments on the manuscript.

The author wishes to thank Michelle Haller for the many hours she spent editing the book's prose.

The author especially appreciates Andrew and Judy Cordell for their critical reading of the manuscript, alerting the author to subtleties in the science, helping the author approach difficult topics, and pointing out unnecessary detail.

**Nancy Morvillo and Matthew Schmidt**

Many thanks to Michael J. Maddalena for his help with illustrations. We also wish to thank our students for their insight and critical reading of this manuscript, and for being the ultimate teachers. And to Mike Morvillo and Donna Parent, for all their love, support, and, especially, patience.

**Ajikumar Aryangat**

To my parents, Vasu and Nalini Aryangat, for their guidance and encouragement for the pursuit of educational excellence.

# CONTENTS


**ORIENTATION** 9

**Part One: SCIENCE REVIEW** 11

| | | |
|---|---|---|
| Physics Review 1 | Introduction | 13 |
| Physics Review 2 | Language of Motion | 31 |
| Physics Review 3 | Laws of Motion | 61 |
| Physics Review 4 | Gravitation | 79 |
| Physics Review 5 | Planes and Circles | 101 |
| Biology Review 1 | Introduction | 125 |
| Biology Review 2 | Biological Molecules | 131 |
| Biology Review 3 | Enzymes and Energy | 139 |
| Biology Review 4 | Cellular Metabolism | 147 |
| Biology Review 5 | DNA Structure and Function | 157 |
| Chemistry Review 1 | Introduction | 167 |
| Chemistry Review 2 | Chemical Reactions | 181 |
| Chemistry Review 3 | Electronic Structure | 197 |
| Chemistry Review 4 | Organic Chemistry | 211 |
| Chemistry Review 5 | Alkanes | 219 |

**Part Two: THE TESTS** 239

Test 1 ... 241
Test 2 ... 261
Test 3 ... 283
Test 4 ... 303
Test 5 ... 323
Test 6 ... 345

# ORIENTATION

## Format of the Science Section

The science section typically includes 40 multiple-choice questions based on 7 passages and is 35 minutes long. The test measures analytical and problem solving skills associated with the basics of science.

You will not be tested on advanced concepts in science, but you will need to know the basic concepts and nomenclature of science. The test assumes that you are in the process of taking three or more years of science courses and have complete two years of study.

| Section | Type | Time |
|---|---|---|
| Science | 40 Multiple-choice Questions | 35 minutes |

Advanced mathematical skills are not needed for the science section, but you do need basics computational skills. So, you don't need any knowledge of Calculus, Trigonometry, or even much Algebra.

The science section is the fourth section of the test.

The science passages are typically taken from Biology, Chemistry, Physics, or closely related fields of science.

## Scoring the ACT

The total score is reported on a scale from 1 to 36, with 36 being the highest score possible. The average total score is about 20. Do not become discouraged if you "blow" a few questions. The average ACT student misses about half of the questions on the science section.

In addition to the scaled score, you will be assigned a percentile ranking, which gives the percentage of students with scores below yours. For instance, if you score in the 80th percentile, then you will have scored better than 80 out of every 100 test takers.

## Calculators

Although you may use a calculator on the math section of the test, oddly enough you cannot use it on the science section. So, when studying, avoid using a calculator when doing calculations. Otherwise, you may become dependent on it, and then struggle when you no longer have access to the calculator.

For the math section, you can use any standard scientific or graphing calculator. You cannot use advanced calculators, such as ones with built-in computer algebra systems (for example, TI-89). You also cannot use the calculator in a cell phone or any other electronic communication device.

## Skipping and Guessing

Some questions on the ACT are rather hard, and it is a time-pressured test. So, you may not be able to finish the section.

Often students become obsessed with a particular problem and waste valuable time trying to solve it. To get a top score, learn to cut your losses and move on because all questions are worth the same number of points, regardless of difficulty level. So, often it is best to skip the hardest questions and concentrate on the easy and medium ones.

There is no guessing penalty on the ACT. So, if you skip a question or do not finish the section in time, be sure to mark an answer for every question. By the laws of probability, you should get about 1 out of every 4 guesses correct.

## The Structure of this Book

Frist, a through review of the basics of Physics, Biology, and Chemistry is presented. Then 6 full-length Science Tests are presented so that you can test your skills and knowledge of science.

Part One

# SCIENCE REVIEW

# Physics Review 1
# Introduction

## A. Philosophy of the Physics Review Section

This review section is about the concepts of physics, with the goal to prepare you, the reader, to pass the science section of the ACT. This section of the ACT consist mainly of a series of passages, each with several questions or problems. Often the passages involve unfamiliar situations and, rather than numbers, explanations, relationships among various quantities, and extrapolations to new situations.

How do you prepare for such a thing?

The short answer is: by thinking and doing physics.

This review section actually has several goals. One is to give you a working knowledge of the basic concepts of physics. Although you will not need a battery of specialized equations, you should remember enough equations to understand the ideas.

Another goal is to teach you how to solve problems in science. But you cannot learn to solve problems by simply reading about physics. The way to learn is to solve problems. Then you can solve future problems, for example, on the ACT, by thinking in the same way as when you solved problems before. For this reason each chapter contains problems in the text with full explanations, as well as problems for you to solve at the end. The solutions tell you how to think about the problems, which clues to look for, and what methods to apply. The goal is for you to learn how to approach new problems.

In each chapter the initial problems are simple, in order to help you to practice your understanding of the concepts in that chapter. These problems may be easy questions or single problems involving some calculations. Although they are not a close approximation of ACT questions, you will have difficulty on the exam unless you learn to do these first. Gradually, the problems in a chapter become more difficult, and at the end of every chapter there are ACT-style passages.

## B. Reading the Physics Review Section

Reading about physics is completely different from reading a novel. First of all, you should be at a desk and have paper and a pen or pencil. You should write down every equation, making sure you understand all the symbols. You should reproduce every diagram, working to understand, for example, why certain forces are there and no others, and so on. You may even try working the examples before reading the solutions and work the solutions out along side the text.

It is especially important that you keep an open mind and visualize what you read. In biology one can actually see organelles with an electron microscope. Understanding the operation of enzymes requires a little bit more imagination. In physics, you must rely on imagination even more, but it is not too different from imagining the working of enzymes. If you view physics as a mere collection of facts and equations to memorize, you will find it frustrating. Alternatively, if you approach physics looking for new concepts, themes and a new worldview, then your efforts will be better rewarded.

## C. Units

A widely held belief is that unit analysis is the least interesting activity of the physical sciences. Indeed, carefully carrying units through a difficult formula is sometimes about as interesting as painting a barn. But there are several good reasons to pay attention to units.

You can lose valuable points if you drop units, substitute into a formula, and forget to convert cm to m or the like. One way to guard against this type of error is to automatically convert any number to MKS (meters, kilograms, and seconds) as you read the passage, or at least flag the units that are nonstandard (i.e., not meters, kilograms, and seconds). Another way is to keep track of the units any time the units in the problem are nonstandard.

Another reason to pay attention to units is that they can alert you if you have written an equation the wrong way. For example, you may remember that flow rate $f$ is the volume ($m^3$) flowing past a point per unit time (s) and that it is related to the velocity $v$ and cross-sectional area $A$ of the pipe. But how do you relate $f$ [$m^3$/s], $v$ [m/s], and $A$ [$m^2$]?

The only way to correctly obtain the units is to write something like

$$f = Av$$

that is

$$\left[\frac{m^3}{s}\right] = \left[m^2\right]\left[\frac{m}{s}\right]$$

where we may have left out a proportionality constant. In this case the formula is correct as written. Units may bring back to mind an equation you would have forgotten, counting for valuable points.

A third reason for keeping track of units is that they sometimes guide you to an answer without your having to use a formula or do much work, as the next example shows.

**Example:** How much volume does 0.4 kg of oxygen gas take up at $T = 27°$ C and $P = 12$ atm? (Use the gas constant $R = 0.0821$ L atm/K mol.)

**Solution:** Well, to the question, "How much oxygen?", we can answer either in kilograms or in liters. The problem gives kilograms and asks for liters, so this is a complicated units conversion problem. We will essentially construct the ideal gas equation using the units of the elements in the problem. We start with 0.4 kg.

$$(\text{amount of } O_2) = 0.4 \text{ kg } O_2$$

In order to apply the ideal gas equation we need to convert to moles. We can do this by including the factors:

$$\left(\frac{10^3 gO_2}{1 kgO_2}\right)\left(\frac{1 mole O_2}{32 gO_2}\right)$$

Both are equivalent to 1, but the units cancel, leaving us with moles.

$$(\text{amount of } O_2) = 0.4 \cancel{\text{kg}O_2}\left(\frac{10^3 \cancel{gO_2}}{1 \cancel{kgO_2}}\right)\left(\frac{1 \text{ mole } O_2}{32 \cancel{gO_2}}\right)$$

Now we include a factor of *R* because it has liters in the numerator and moles in the denominator. We obtain

$$(\text{amount of } O_2) = 0.4\,\cancel{\text{kg}\,O_2}\left(\frac{10^3\,\cancel{\text{g}\,O_2}}{1\,\cancel{\text{kg}\,O_2}}\right)\left(\frac{1\,\cancel{\text{mole}}\,O_2}{32\,\cancel{\text{g}\,O_2}}\right)\left(\frac{0.0821\,\text{L atm}}{\text{K}\,\cancel{\text{mol}}}\right)$$

This leaves us with units of atm and K which we want to get rid of. In order to cancel them, we can just put them in. This may seem strange, but it works. (Recall 27° C = 300 K.) Thus we obtain

$$\begin{aligned}(\text{amount of } O_2) &= 0.4\,\cancel{\text{kg}\,O_2}\left(\frac{10^3\,\cancel{\text{g}\,O_2}}{1\,\cancel{\text{kg}\,O_2}}\right)\left(\frac{1\,\cancel{\text{mole}}\,O_2}{32\,\cancel{\text{g}\,O_2}}\right)\left(\frac{0.0821\,\text{L}\,\cancel{\text{atm}}}{\cancel{\text{K}}\,\cancel{\text{mol}}}\right)300\,\cancel{\text{K}}\,\frac{1}{12\,\cancel{\text{atm}}} \\ &= \frac{0.4\cdot 1000\cdot 0.0821\cdot 300}{32\cdot 12}\,\text{L}\end{aligned}$$

For ACT problems, we generally work to one digit of accuracy, so we replace 0.0821 with 0.08, so that we have

$$\begin{aligned}(\text{amount of } O_2) &= \frac{0.4\cdot 1000\cdot 0.0821\cdot 300}{32\cdot 12} \\ &= \frac{4\cdot 8\cdot 300}{8\cdot 5} = 24\text{ L}\end{aligned}$$

It is generally safe to round to one significant digit. If it happens that two choices are close, then you can always go back and gain more accuracy.

This example involved more arithmetic than most ACT problems, but its purpose was to point out that attention to units can speed up the solution to a problem. If this is the way you normally do such a problem, good. Most readers, however, would take longer working through this type of problem, using up valuable seconds on the ACT. Remember that seconds can add up to points.

## D. Equations

Students generally have one of three attitudes toward equations:

1 sheer hatred (enough said),
2. cold pragmatism (plug in numbers and get an answer), and
3. warm fondness.

Try adopting the last attitude. Many students do not realize that equations are merely a way to contain useful information in a short form. They are sentences in the concise language of mathematics. You should not have to memorize most equations in the text, because by the time you learn each chapter, the equations should feel natural to you. They should feel like natural relationships among familiar quantities.

For example, consider one of the first equations you ever encountered, distance equals rate times time, that is

$$\Delta x = v\Delta t \qquad (1)$$

It makes sense that, in a given time, we can go twice as far if we go twice as fast. Thus $\Delta x$ is proportional to $v$. On the other hand, for a given speed, we can go twice as far if we travel twice as long a time. Thus $\Delta x$ is proportional to $\Delta t$. We would never be tempted to write

$$v = \Delta x\Delta t$$

or

$$\Delta t = v\Delta x$$

because these equations give relationships among the quantities that we know to be wrong. Note also that the units work out correctly only in equation (1).

Another example is the second law of motion, which we will encounter in Section 3.B. If an object has a single force on it, then its acceleration is proportional to the magnitude of the force and inversely proportional to its mass. Instead of words, we simply write

$$a = F/m \qquad (2)$$

Now let's think about the equation. What would we do if we forgot it? If we stop to think about it, we could figure it out. First, we know that force, mass, and acceleration are connected somehow. If we have two objects of the same mass, and we apply three times as much force to the second object as to the first, then we have a picture like that in Figure 1-1. The greater force causes the greater acceleration, so we can guess that they are proportional. We write

$$a \sim F$$

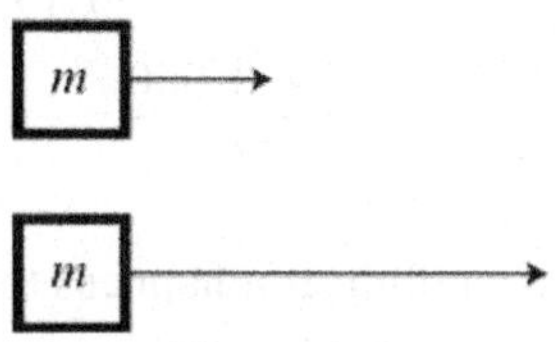


Figure 1-1

If we apply the same force to two objects of different masses, then we expect the smaller object to accelerate more (Figure 1-2). Thus we can guess that the acceleration is *inversely* proportional to the mass, so we write

$$a \sim 1/m$$

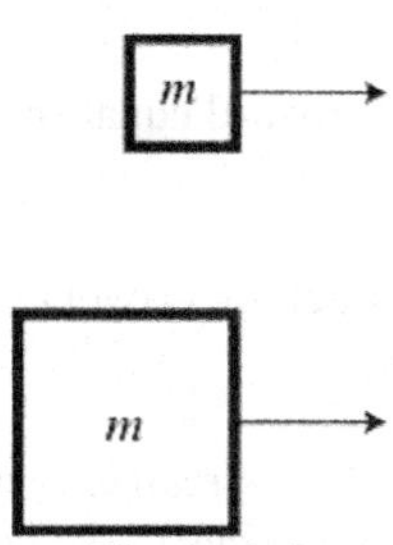


Figure 1-2

Combining these two proportions we get

$$a = F/m$$

as in equation (2).

When you take the ACT, you really should have the equation $F = ma$ in your head, but if you train yourself to think this way, it will be easier to keep the formulas in your head. This will make it possible to recover the formula if you forget it. And you will understand physics better. Most importantly, you will be better able to apply the concept behind the equation.

Some equations are a little more complicated. An example is Newton's law of gravity, which gives the force of gravity between two objects:

$$F_{grav} = \frac{Gm_1m_2}{r^2} \qquad (3)$$

where $G$ is a constant, $m_1$ and $m_2$ are masses of objects, and $r$ is the distance between them. How would you ever remember this equation?

Well, start with the idea that objects with more mass have a greater force of gravity between them, so write

$$F_{grav} \sim m_1m_2$$

Also, if objects are far apart, the force of gravity between them is less, so write

$$F_{grav} \sim \frac{m_1m_2}{r}$$

There is a constant, so write

$$F_{grav} \sim \frac{Gm_1m_2}{r}$$

The only part that needs to be memorized is the "square" in the denominator, so that we have

$$F_{grav} = \frac{Gm_1m_2}{r^2}$$

That's why we call gravity an inverse-square force.

The ACT will not ask you to substitute into an equation like equation (3), but it may ask a question like, "What happens to the gravitational force between two objects if the distance between the objects is increased by a factor of four?"

We can tell from equation (3) that an increase in distance results in a decrease in force, because $r$ is in the denominator. Because the $r$ is squared, a factor of 4 in $r$ will result in a factor of $4^2 = 16$ in $F_{grav}$. The answer is that the gravitational force decreases by a factor of 16.

If this last point seems opaque to you, try some numbers on a more familiar equation, such as that for the area of a circle

$$A = \pi r^2 \qquad (4)$$

What happens to the area when the radius increases by a factor of 3? (Answer: It increases by a factor of 9.) Try it with $r_1 = 4$ m and $r_2 = 12$ m, or with some other numbers.

Another equation is that for the surface area of a sphere

$$A = 4\pi r^2 \qquad (5)$$

What happens to the surface area of a sphere when the radius increases by a factor of 3? (Answer: It increases by a factor of 9. Surprised? What about the factor of 4? Try it with $r_1 = 4$ m and $r_2 = 12$ m.) The surface area of a sphere is an equation that you just have to memorize. It is difficult to get an intuitive grasp why the $4\pi$ should be there. On the other hand, the $r^2$ is natural in this equation. Why? (Think about units.)

Another example concerns the volume of a sphere

$$V = \frac{4\pi}{3} r^3 \tag{6}$$

What happens to the volume when the radius is doubled?

In this chapter we discussed the importance of units in solving problems. If a problem involves only simple proportionalities and there are no unitless proportionality constants, then we can obtain a quick solution simply by keeping track of units. The example in the text demonstrates all the techniques involved.

We also looked at equations as the language of physics. If you read equations as sentences containing information for you to understand, then the equations will seem less foreign than if you look at them as abstract collections of symbols. Each time you encounter a boxed equation in this section, you should spend some time thinking about what the equation means.

# Physics Review 1 Problems

In any of the following problems you may want to use one of the constants

$N_A = 6.02 \times 10^{23}$
$R = 0.0821$ L atm/K mol

1. In a certain assay, a number of microbes is measured by determining the mass of the sample. It is known that the average mass of a microbe (of this species) is $6.0 \times 10^{-16}$ g. How many microbes are in a sample of mass $1.1 \times 10^{-12}$ g?
A. 1800
B. 5500
C. $6.6 \times 10^{4}$
D. $6.6 \times 10^{28}$

2. A certain substance has a density 8.4 $\mu$g/mL. What is the mass of 422.4 mL?
A. 0.020 mg.
B. 3.55 mg.
C. 350.3 mg.
D. 3550 mg.

3. What is the mass of a water molecule?
A. $9.2 \times 10^{-26}$ g.
B. $1.7 \times 10^{-24}$ g.
C. $3.0 \times 10^{-23}$ g.
D. 0.018 g.

4. An electrical resistor is installed in a container of water to heat it. The resistor dissipates heat at a rate of 2.0 W, and the container holds 10 kg of water. How long would it take to raise the temperature of the water 5° C? (Note: The specific heat of water is $4.2 \times 10^{3}$ J/kg °C, and 1 W is 1 J/s.)

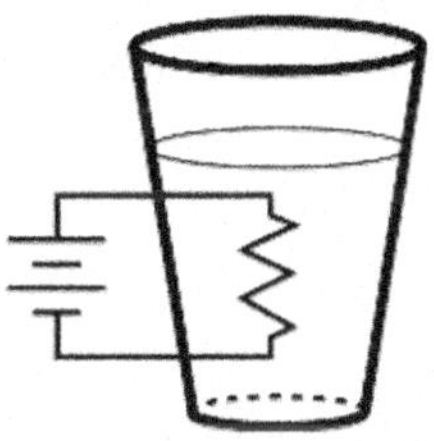

A. $2.4 \times 10^{-2}$ seconds.
B. $4.2 \times 10^{3}$ seconds.
C. $1.05 \times 10^{5}$ seconds.
D. $4.2 \times 10^{5}$ seconds.

5. Two liters of argon gas are at 10 atm of pressure. If the sample is 16 g, what is the temperature?
A. 16,000 K
B. 610 K
C. 6 K
D. 4 K

*Use the following information for questions 6–10:*

For a circle we have the formula for the circumference

$$C = 2\pi r$$

and the area

$$A = \pi r^2$$

where $r$ is the radius. For a sphere, the surface area is

$$A_{surf} = 4\pi r^2$$

and the volume is

$$V = 4/3\ \pi r^3$$

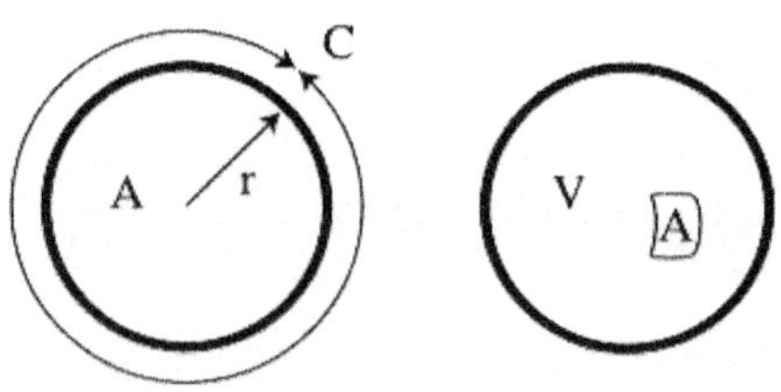


6. If the diameter of a circle is increased by a factor of 4, what happens to the circumference?
   A. It increases by a factor of 2.
   B. It increases by a factor of 4.
   C. It increases by a factor of 8.
   D. It increases by a factor of 16.

7. If the radius of a circle increases by a factor of 4, what happens to its area?
   A. It increases by a factor of 2.
   B. It increases by a factor of 4.
   C. It increases by a factor of 16.
   D. It increases by a factor of 64.

8. If the radius of a sphere increases by a factor of 4, what happens to its volume?
   A. It increases by a factor of 2.
   B. It increases by a factor of 4.
   C. It increases by a factor of 16.
   D. It increases by a factor of 64.

9. If the volume of a sphere decreases by a factor of 27, what happens to its diameter?
   A. It decreases by a factor of 9.
   B. It decreases by a factor of 3.
   C. It decreases by a factor of 4.5.
   D. It decreases by a factor of 1.5.

10. If the radius of a circle is increased by 30%, how does the area change?
    A. It increases by 30%.
    B. It increases by 60%.
    C. It increases by 69%.
    D. It increases by 75%.

*Use the following information for questions 11–13:*

A pendulum is a mass connected to a light string or rod which is connected to the ceiling. The period is the amount of time it takes the bob (as the mass is called) to swing from one side to the other and back. It is given by

$$T = 2\pi\sqrt{\frac{l}{g}}$$

where $T$ is in s, $l$ is the length of the string or rod (in m), and $g$ is the acceleration due to gravity (m/s$^2$). (See Figure.)

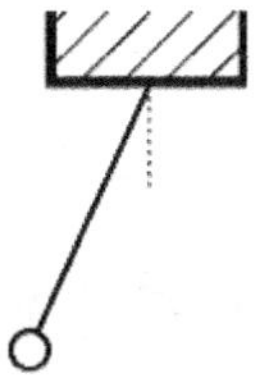

11. If the length of the string of a pendulum is increased by a factor of 4, how does the period change?
    A. It decreases by a factor of 16.
    B. It increases by a factor of 2.
    C. It increases by a factor of 4.
    D. It increases by a factor of 16.

12. The length of the rod of a certain pendulum is decreased, and the period then decreases by 20%. By how much was the rod length decreased?
    A. 20%
    B. 36%
    C. 40%
    D. 44%

13. If a pendulum is transported to the Moon, where the acceleration due to gravity is six times less than that here on Earth, how would the period of the pendulum change?
    A. It would decrease by a factor of 36.
    B. It would increase by a factor of 2.4.
    C. It would increase by a factor of 6.
    D. It would increase by a factor of 36.

*Use the following information for questions 14–16:*

A spring is characterized by a spring constant $k$ (in N/m) which gives the stiffness of the spring, or how hard you have to pull to stretch it. If you connect a mass $m$ on one end, and connect the other end to a fixed wall or ceiling, then the resulting system will vibrate. This vibration has period $T$ given by

$$T = 2\pi\sqrt{\frac{m}{k}}$$

The frequency of the vibration is defined as

$$f = 1/T$$

14. If a mass of 60 g is connected to a certain spring, the frequency is 30 Hz. If a mass of 240 g is connected to the same spring, what is the frequency?
A. 7.5 Hz
B. 15 Hz
C. 60 Hz
D. 120 Hz

15. In two trials, two masses are attached to a spring and the periods recorded. Mass P resulted in a period 36 times larger than the period of mass Q. What can be concluded?
A. Mass P is 1296 times larger than mass Q.
B. Mass P is 6 times larger than mass Q.
C. Mass P is 6 times smaller than mass Q.
D. Mass P is 1296 times smaller than mass Q.

16. If the period increases by 50%, how does the frequency change?
A. It decreases by 50%.
B. It decreases by 40%.
C. It decreases by 33%.
D. It increases by 33%.

*Use the following information for questions 17–19:*

The volume of a pyramid with a square base is given by

$$V = \frac{1}{3}s^2h$$

where $s$ is the length of a side, and $h$ is the perpendicular height.

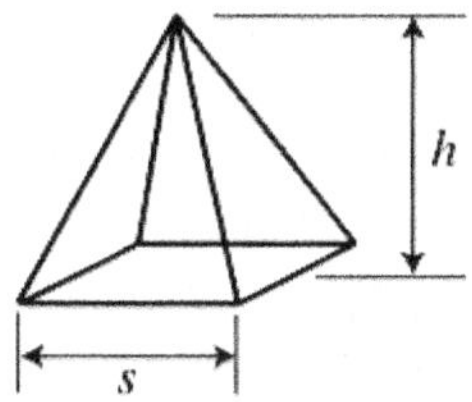


17. How does the volume of a square pyramid change if the base side length is increased by a factor of 9 and the height is unchanged?
A. It increases by a factor of 3.
B. It increases by a factor of 9.
C. It increases by a factor of 27.
D. It increases by a factor of 81.

18. How does the volume of square pyramid change if the height is increased by a factor of 12 and the base side length is unchanged?
A. It increases by a factor of 4.
B. It increases by a factor of 12.
C. It increases by a factor of 36.
D. It increases by a factor of 72.

19. If every linear dimension of a square pyramid were increased by a factor of 3, how would the volume change?
A. It would increase by a factor of 3.
B. It would increase by a factor of 9.
C. It would increase by a factor of 27.
D. It would increase by a factor of 81.

## Passage 1

In a certain experiment, we are investigating the retarding force that a fluid exerts on an object moving through it. We guess that the size of the object is a factor, so we include $A$, the cross-sectional area, in an equation. The relative velocity between the object and the fluid is a factor $v$, as well as the density of the fluid $\rho$. So we guess

$$F = k\rho^m A^n v^p$$

where $k$ is a proportionality constant with some appropriate units. Before we run the experiment, we do not know the values of the exponents $m$, $n$, and $p$.

The chart gives the data for a certain fluid.

| Experiment | object | $A$ ($cm^2$) | $v$ (m/s) | $F$ (N) |
|---|---|---|---|---|
| 1 | cork ball | 1.5 | 7.0 | 0.020 |
| 2 | cork ball | 1.5 | 3.5 | 0.005 |
| 3 | steel ball | 1.5 | 3.5 | 0.005 |
| 4 | steel ball | 3.0 | 3.5 | 0.010 |
| 5 | steel ball | 4.5 | 3.5 | 0.015 |
| 6 | steel ball | 3.0 | 14.0 | 0.160 |

20. Which pair of experiments could be used to determine $n$?
    A. 2 and 3
    B. 3 and 4
    C. 4 and 6
    D. 5 and 6

21. What is the approximate value of $p$?
    A. –1
    B. 0
    C. 1
    D. 2

22. Which pair of experiments indicates that retarding force does not depend on the density of the object?
    A. 1 and 2
    B. 2 and 3
    C. 1 and 6
    D. 4 and 6

23. Let us say $m$ and $n$ are known. What combination of experiments would be considered a minimum set for determining $p$ and $k$?
    A. 1 and 2
    B. 1, 2, and 3
    C. 3, 4, and 5
    D. 3 only

## Passage 2

The amount of energy a car expends against air resistance is approximately given by

$$E = 0.2\rho_{air} ADv^2$$

where $E$ is measured in Joules, $\rho_{air}$ is the density of air (1.2 kg/m$^3$), $A$ is the cross-sectional area of the car viewed from the front (in m$^2$), $D$ is the distance traveled (in m), and $v$ is the speed of the car (in m/s). Julie wants to drive from Tucson to Phoenix and get good gas mileage. For the following questions, assume that the energy loss is due solely to air resistance, and there is no wind.

24. If Julie increases her speed from 30 mph to 60 mph, how does the energy required to travel from Tucson to Phoenix change?
A. It increases by a factor of 2.
B. It increases by a factor of 4.
C. It increases by a factor of 8.
D. It increases by a factor of 16.

25. Julie usually drives at a certain speed. How much more energy will she use if she drives 20% faster?
A. 20% more energy.
B. 40% more energy.
C. 44% more energy.
D. 80% more energy.

26. Scott drives a very large 50s style car, and Laura drives a small 90s style car, so that every linear dimension of Scott's car is double that of Laura's car. *On the basis of energy loss due to air resistance alone*, how much more energy would you expect Scott's car to expend getting from Tucson to Phoenix than Laura's car?
A. Twice as much energy.
B. Four times as much energy.
C. Eight times as much energy.
D. Sixteen times as much energy.

27. How does Julie's energy usage change if she changes from driving 50 mph to 55 mph?
A. It increases by 10%.
B. It increases by 20%.
C. It increases by 21%.
D. It increases by 40%.

28. Julie modifies her car, so that the effective cross-sectional area is reduced by 20%. How much further can she drive and still use the same amount of energy?
A. 10% further.
B. 20% further.
C. 25% further.
D. 44% further.

# Solutions

1. A.
We start with the information $1.1\text{x}10^{-12}$ g. Unit analysis takes us to the answer:

$$1.1\times10^{-12}g\left(\frac{1\text{ microbe}}{6\times10^{-16}g}\right)=\frac{1.1}{6}\times10^{-12+16}\text{ microbes}$$
$$\approx 0.2\times10^{4}\text{ microbes}$$
$$\approx 2000\text{ microbes}$$

Here we estimate (1.1)/6 to be about 0.2, which is close enough to give us an answer. Remember, all you need is an answer. You do not need to calculate a second digit.

2. B.
The question again is, "How much?" We have information in mL, and we want it in mg. Thus we write

$$422.4\text{mL}\left(\frac{8.4\mu\text{g}}{1\text{mL}}\right)\left(\frac{1\text{mg}}{10^{3}\mu\text{g}}\right)\approx\frac{400\cdot8}{1000}\text{mg}$$
$$\approx\frac{3000}{1000}\text{mg}$$
$$=3\text{ mg}$$

Here we estimated 4 times 8 = 30. This is close enough to yield the answer B.

3. C.
Here we start with 1 molecule $H_2O$. Then we can convert to mol and g:

$$1\text{ molec}\left(\frac{1\text{ mol}}{6.02\times10^{23}\text{ molec}}\right)\left(\frac{18\text{ g}}{1\text{ mol}}\right)\approx\frac{18}{6}\times10^{-23}\text{g}$$
$$\approx3\times10^{-23}\text{g}$$

4. C.
We can guess that the amount of time required is proportional to the temperature change desired, so let's start with that. Next, we want to cancel the units °C, so we add a factor of $4.2\text{x}10^3$J/kg°C, giving

$$5°\text{C}\frac{4.2\times10^{3}\text{J}}{\text{kg°C}}$$

We want to cancel kg, so we can just place a factor of 10 kg in the numerator, giving

$$5°\text{C}\frac{4.2\times10^{3}\text{J}}{\text{kg°C}}\frac{10\text{kg}}{1}$$

To cancel J, we can use the J from the power the resistor dissipates 2 W = 2 J/s. Thus

$$5°\text{C}\frac{4.2\times10^{3}\text{J}}{\text{kg°C}}\frac{10\text{kg}}{1}\frac{\text{s}}{2\text{J}}$$

At this point we rejoice, because we have seconds in the numerator. Now we cut everything down to one digit and quickly multiply. This gives

$$\Delta t=\frac{5\cdot4\cdot10^{3}\cdot10}{2}\text{s}=\frac{20}{2}\cdot10^{4}\text{s}=10^{5}\text{s}$$

This is close to choice C.

Does this method always work? Clearly you cannot answer every conceivable question by looking at units. But it is surprising how many questions can be answered this way, specifically any question in which all the formulas involved are simple proportionalities without unitless proportionality constants.

5. B.
We know this involves the ideal gas equation, but how do we start a unit analysis? Since we know we want to end up with K in the numerator, we can just place it there at the outset:

$$\frac{\text{K mol}}{0.0821\ \text{L atm}}$$

Now we cancel L and atm, giving

$$\frac{\text{K mol}}{0.0821\ \text{L atm}}\ \frac{2\ \text{L}}{1}\ \frac{10\ \text{atm}}{1}$$

We can cancel mol if we think of a connection between mol and the 16 g Ar, so that we write

$$\begin{aligned}&\frac{\text{K mol}}{0.0821\ \text{L atm}}\ \frac{2\ \text{L}}{1}\ \frac{10\ \text{atm}}{1}\ \frac{40\ \text{g}}{\text{mol}}\ \frac{1}{16\ \text{g Ar}}\\ &\approx \frac{2\cdot 10\cdot 40}{0.08\cdot 16}\text{K}\\ &\approx \frac{100\cdot 2\cdot 10\cdot 40}{8\cdot 16}\text{K}\\ &\approx \frac{10{,}000}{16}\text{K}\\ &\approx 700\text{K}\end{aligned}$$

In the second step above, we multiplied numerator and denominator by 100. In the third step we calculated $2 \cdot 40/8 = 10$. Most calculations go pretty quickly if you look for these shortcuts. Our answer is close enough for us to realize the correct answer is B.

6. B.
If the diameter of a circle increases by a factor of 4, the radius increases by a factor of 4 also. (If you do not believe this, try it with a few numbers.) The circumference increases by a factor of 4 as well.

7. C.
Clearly the area increases, but since the radius is squared in the formula, the area increases by a factor of $4^2 = 16$. The $\pi$ in the formula does not make any difference.

8. D.
This time there is an $r^3$ in the formula, so the volume increases by a factor of $4^3 = 64$.

9. B.
If the volume of a sphere decreases by a factor of 27, then its radius must decrease by a factor of 3, since $3^3 = 27$. If the radius decreases by a factor of 3, then the diameter decreases by a factor of 3 as well. Again, you should try this method with some numbers if it does not make sense to you.

10. C.
If the radius increases by 30%, that is the same as increasing by a factor of 1.3, since $(1 + 30/100) = 1.3$. If the radius increases by a factor of 1.3, then the area increases by a factor of $1.3^2 = 1.69$. An increase by such a factor is an increase by 69%, so the answer is C. Keep in mind that you need to know how to manipulate numbers like this quickly.

11. B.
Since $l$ is in the numerator, $T$ increases if $l$ does. If $l$ increases by a factor of 4, then T increases by a factor of

$$\sqrt{4} = 2$$

12. B
A problem like this is easier if we solve for $l$, giving

$$l = g(2\pi T)^2$$

$$l = 4\pi^2 g T^2$$

A decrease by 20% in the period is equivalent to multiplying the period by (1 – 20/100) = 0.8. If the period is multiplied by 0.8, then the length is multiplied by $0.8^2 = 0.64$. We can rewrite 0.64 = (1 – 36/100). Thus the length decreases by 36%.

13. B.
Since $g$ is in the denominator, a decrease in $g$ results in an increase in $T$. If $g$ decreases by a factor of 6, then $T$ increases by a factor of

$$\sqrt{6} = 2.4$$

Of course, you do not have to work out the square root. A glance at the answers indicates that B is correct.

14. B.
Since $m$ is in the numerator, an increase in $m$ will increase $T$. This results in a decrease in frequency $f$. Since $m$ changes by a factor of 4, the period changes by a factor of

$$\sqrt{4} = 2$$

The frequency changes by a factor of 2, so the answer is B. This approach is the most straightforward way to do the problem. Any time spent calculating the value of $k$ is wasted time.

15. A.
First, if the period is larger for mass P, then mass P must be larger. Next, for there to be a change of 36 in the period, there must be a change of 1296 inside the square root. Thus the answer is A.

16. C.
Since we have (1 + 50/100) = 1.5, the period is multiplied by 1.5. Thus the frequency is multiplied by $(1.5)^{-1}$ = 0.67. We rewrite 0.67 = (1 – 33/100), so the frequency decreases by 33%.

17. D.
If $s$ increases by a factor of 9, then $V$ increases by a factor of 81. Do not let the 1/3 in the formula throw you off. If it did, try the problem with numbers to see why the 1/3 does not matter.

18. B.
This is a simple proportionality.

19. C.
In this case, both $s$ and $h$ increase by a factor of 3. The increase in $s$ causes $V$ to increase by a factor of 9, and another factor of 3 comes from the $h$.

## Passage 1

20. B.
In order to determine $n$, we need two experiments where everything stays the same except for the area, so that we can investigate the results of a change in area. As for choice A, the object changes and the area does not, so that is out. In choice B, area changes from 1.5 to 3.0 cm$^2$, and nothing else changes, so B is correct. Choice C is incorrect because the velocity is the only thing that changes. In choice D, both $A$ and $v$ change, so we would not be able to tell how much change in $F$ is due to $A$ and how much is due to $v$.

21. D.
In the previous solution, we realized that experiments 4 and 6 have the property that all the input variables stay the same except for velocity $v$, which increases by a factor of 4. The force increases by a factor of 16, which means that $p$ must be 2. That is, if $p$ is 2, then an increase by a factor of 4 in $v$ results in an increase by a factor of $4^2$ in the force.

22. B.
In choice A, experiments 1 and 2 both use a cork ball. For choice C, many input variables are altered between experiments 1 and 6, so it is impossible to isolate the effect of object density. As for choice D, experiments 4 and 6 both use a steel ball. Choice B involves two experiments in which only the density of the object changes.

23. A.
As for choice A, experiments 1 and 2 could be used to determine $p$, since the velocity changes and nothing else does. Once $p$ is determined, $k$ can be determined by substituting in values from either experiment 1 or 2. Thus experiments 1 and 2 are sufficient. We can exclude choices B and C (not minimum sets). As for choice D, there is not enough information to obtain $p$ or $k$.

## Passage 2

24. B.
If $v$ increases by a factor of 2, then the required energy increases by a factor of $2^2 = 4$.

25. C.
If Julie increases her speed by 20%, then she multiplies her speed by 1.2. Thus the required energy is multiplied by $(1.2)^2 = 1.44$, which is an increase of 44%.

26. B.
Comparing Scott's car to Laura's, all the linear dimensions are increased by a factor of 2 (see figure). The cross-sectional area $A$ is width times height ($A = hw$), so if both $h$ and $w$ increase by a factor of 2, then $A$ increases by a factor of 4. Thus the required energy increases by a factor of 4. The increase in length does not matter.

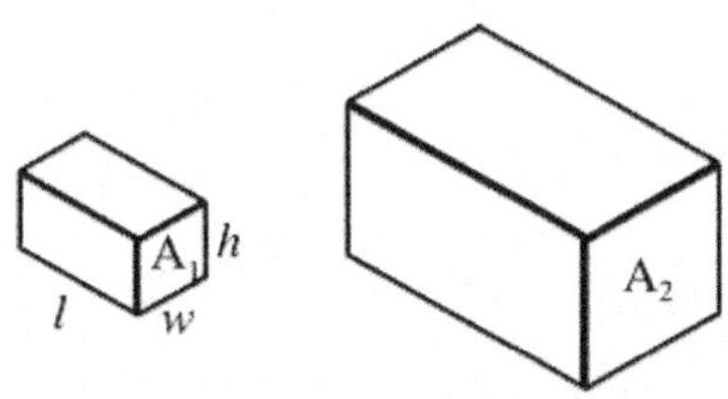


27. C.
Julie increases her speed by a factor of 55/50 = 1.1, so the energy increases by a factor of $1.1^2 = 1.21$. This is an increase of 21%.

28. C
The easiest way to do this is to solve for D, giving

$$D = \frac{5E}{\rho_{air} A v^2}$$

If *A* is reduced by 20%, then *A* is multiplied by 0.8. According to the above equation, *D* is multiplied by $(0.8)^{-1} = 1.25$, representing an increase of 25%.

# Physics Review 2
# The Language of Motion

## A. Introduction

*Mechanics* is about the motion of things. Before we can talk about motion in depth, we need to be able to describe motion and the things which affect it. Objects move, and we talk of how fast they go, that is, velocity. Their velocity changes, so we talk of acceleration. We can think of changes in acceleration, but it turns out (happily) that we rarely need to. Mechanics is concerned mainly with changes in velocity.

In this chapter we look at the fundamental elements of mechanics: force, mass, distance, velocity, and acceleration. Comparatively this chapter has a lot of equations (six that you should memorize) and the least interesting physics. It is an unpleasant way to begin, but it must be done.

## B. Force

A *force* is a push or pull, and the units for force are [Newtons = N]. (Some countries continue to use an archaic unit called the "pound".) A Newton is approximately the amount of force that you would exert on an apple near the Earth's surface to keep it from falling.

Examples of forces include the force of a horse pulling a cart, the force of a spring pushing the chassis of a car, the force of gravity pulling you down, and the force on your head due to pressure when you are at the bottom of a pool.

## C. Mass

We can think about mass in several ways. First, the *mass* of an object is a measure of the total amount of material (or stuff) in the object. The amount of stuff in an object is a fundamental property of the object. It doesn't change if you move the object to a new place, like a mountaintop or to Mars.

There is another way to think of mass. The mass of an object is a measure of how difficult it is to get it moving at a certain velocity if it starts from rest. For example, if John wants to set a car, initially at rest, to moving at 1 m/s, he has to push hard for a little while. We are assuming the car's motion has no friction. If John and the car were on the Moon, his task would be equally difficult. The fundamental concept here is the mass of the car, not the astronomical body the car is on. (See Figure 2-1.)

*A car has the same mass on the Moon as it has on Earth. It takes just as much force and time to get a car moving at a given velocity on the Earth as on the Moon.*
Figure 2-1

Saying this another way, the mass of an object is a measure of how much it hurts if your stub you toe on it. Stubbing your toe on a bowling ball is a painful proposition, even on the Moon.

There is a *wrong* way to think about mass. Many people think the mass of an object is a measure of how difficult it is to pick it up. But that definition depends on where you are. It is easy to pick up a bowling ball on the Moon, but nearly impossible on the surface of Jupiter. The difficulty in picking up an object is a matter of weight, which is a force. And weight *does* depend on the astronomical body near by.

## D. Vectors

In physics we often need to describe direction as well as size. For example, two forces $F_1$ and $F_2$ may both be 100 N and acting on a crocodile, but the crocodile's experience will be very different depending on whether the forces are both pointing north or one north and one south (Figure 2-2). In the former case he gets stretched, and in the latter case he goes flying. To describe forces we need to specify size and direction. That is, we need to use vectors. Force is a vector.

We denote vectors in diagrams by arrows, the length of the arrow showing the size of the vector and the direction of the arrow showing its direction.

We can add vectors by the tip-to-tail method. We leave the first vector fixed, and move the second vector so its tail is at the first vector's tip. If there are other vectors, then each vector gets added to the previous tip. The sum is the vector pointing from the first tail to the last tip.

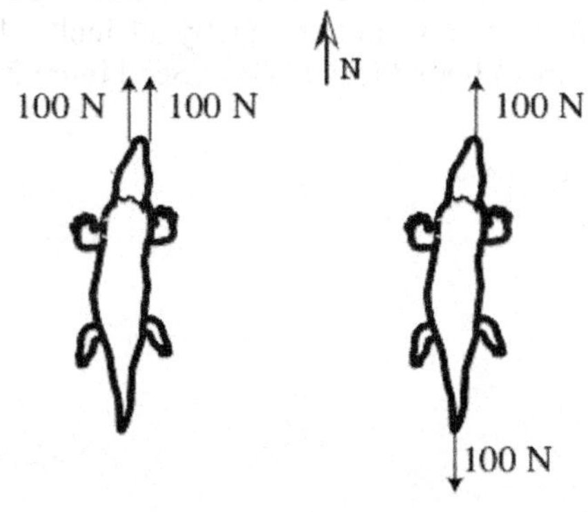


*Two forces on an object may add in different ways, depending on the relative directions of the forces.*
Figure 2-2

**Example 1:** For the crocodiles mentioned before, if the vectors (both 100 N) both point north, then the sum is a force of 200 N pointing north (Figure 2-3, where the sum is shown dashed).

If one vector points north and the other south, then the first tail coincides with the last tip and the sum is zero (Figure 2-3).

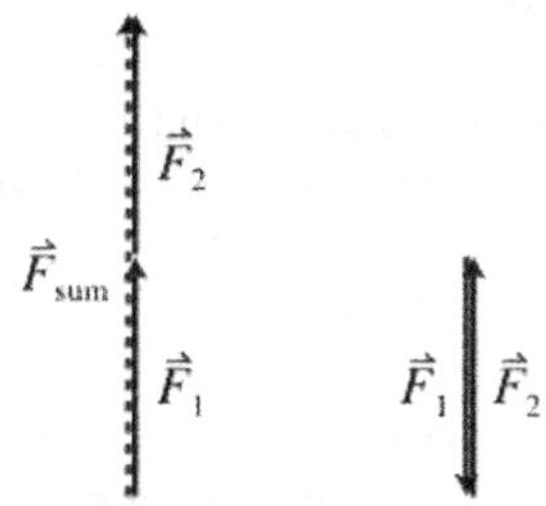


*Forces are vectors and they add according to the tip-to-tail method.*

Figure 2-3

**Example 2:** A crocodile has three forces acting on him: a 100-N force north, a 100-N force east, and a 100-N force southwest. What is the direction of the net force (that is, total force)?

**Solution:** We DRAW A DIAGRAM (Figure 2-4). The sum is a vector pointing northeast, about 40 N.

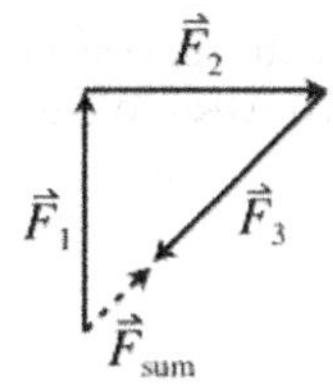


Figure 2-4

Note that, when you add vectors, the magnitude of the sum is equal to, at most, the sum of the individual magnitudes (and that only if they are pointing in the same direction). For instance, if three vectors of 100 N are acting on a crocodile, the sum can be anything from 0 N to 300 N, but no greater.

For ACT problems, vector addition need not get more sophisticated than this. It is useful to keep in mind the Pythagorean theorem and elementary trigonometry.

**Example 3:** A force of 4 N is acting to the north on a rock and a force of 3 N is acting to the east.

a. What is the magnitude of the total force?
b. What is $\cos\phi$, if $\phi$ is the deviation from north of the direction of the total force?

**Solution:** We DRAW A DIAGRAM (Figure 2-5). There is a right triangle, so we can write

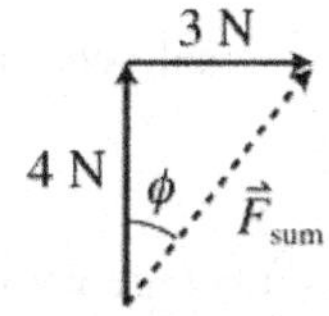


Figure 2-5

$$F_{sum}^2 = (3\text{N})^2 + (4\text{N})^2 = 25\text{N}^2$$

$$F_{sum} = 5 \text{ N}$$

Also we write

$$\cos\phi = 4\text{N}/5\text{N} = 0.8$$

If your trigonometry is rusty, now is a good time to relearn the definitions of sine, cosine, and tangent.

A vector is denoted by a half-arrow on top of a letter, $\vec{F}$ , for example.

## E. Position, Displacement, and Time

To specify position, we must give three coordinates $x$, $y$, and $z$, generally measured in [meters = m]. The symbol $s$ stands for the coordinates $(x, y, z)$. If an object moves from one position to another, the vector giving the change in position is the *displacement vector*, $\Delta s = s_2 - s_1$. The magnitude of the displacement vector is called the *displacement*.

Time is a fundamental quantity in classical physics, denoted $t$, measured in [seconds = s]. Often we will speak of a *time interval* $\Delta t = t_2 - t_1$, that is, the time between a beginning time $t_1$ and ending time $t_2$. An *instant* is a single moment of time.

## F. Velocity, Speed, and All That

We can think of the *velocity vector* in terms of a speedometer reading with units [meters/second = m/s] and a direction. The magnitude of the velocity vector (that is, just the speedometer reading) is called *speed*. The word "velocity" is sometimes used to refer to the vector and sometimes to the magnitude. When in doubt, you should assume it refers to the vector.

If an object is traveling such that its velocity vector is constant, we say it is in *uniform motion*. An example is a car going a constant 30 m/s (freeway speed) west. We can write the following equations for uniform motion in one dimension:

$$v = \frac{\Delta x}{\Delta t} \qquad (1a)$$

$$v = \frac{x_2 - x_1}{\Delta t} \qquad (1b)$$

When you see a formula in this text, instead of speeding by it, slow down and look at it. Ask yourself, "What is this equation telling me?" Equation (1a) is just another form of "distance equals rate times time" for an object in uniform motion. Since $v$ is constant, this tells you, for instance, that a car will travel twice as far if it travels for twice the time. This makes sense.

Equation (1b) is like the first, only $\Delta x$ is replaced by its definition $x_2 - x_1$. Do you see why it is this and not $x_1 - x_2$ or $x_2 + x_1$?

But in some problems the velocity does change, and we must pay attention to several velocities, that is,

$v_1$ initial velocity,

$v_2$ final velocity,

and

$v_{avg}$ average velocity.

The average velocity is defined as

$$v_{avg} = \frac{\Delta s}{\Delta t} \qquad (2)$$

This is different from equation (1). Equation (2) is the definition of an average velocity over a time interval when velocity is changing, whereas equation (1) defines a constant velocity and only holds for time intervals when the motion is uniform.

**Example:** A car goes west at 10 m/s for 6 s, then it goes north at 10 m/s for 5 s, and then it goes west again at 4 m/s for 15 s. What are $v_1$, $v_2$, and $v_{avg}$?

**Solution:** Well, we have $v_1$ = 10 m/s and $v_2$ = 4 m/s. For the average velocity we need to DRAW A DIAGRAM (Figure 2-6). The Pythagorean theorem gives us $\Delta s$ = 130 m. Thus

$$v_{avg} = 130\text{m}/26\text{s} = 5 \text{ m/s}$$

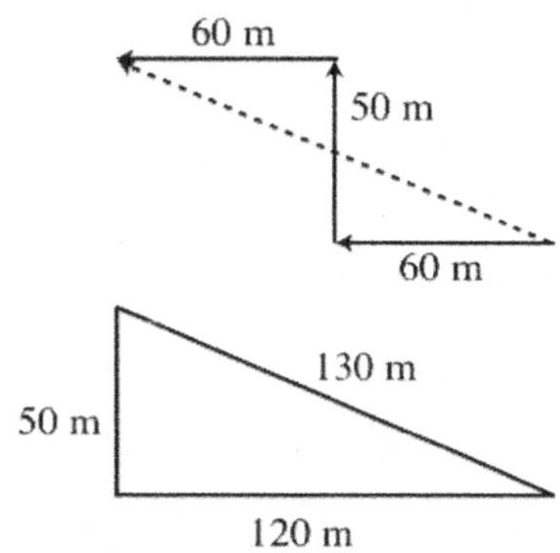


Figure 2-6

## G. Acceleration

When an object's velocity vector is changing, the object is *accelerating*. Examples include a car speeding up ("accelerating" in common parlance), slowing down or braking ("decelerating", but physicists prefer to say "negatively accelerating"), and turning. In three dimensions, we define acceleration by

$$\vec{a} = \frac{\Delta\vec{v}}{\Delta t} \qquad (3a)$$

$$= \frac{\vec{v}_2 - \vec{v}_1}{\Delta t}. \qquad (3b)$$

The numerator for equation (3a) gives the change in the velocity vector, so there is an acceleration if either the magnitude or the direction of the velocity vector change. In one dimension the definition of acceleration is

$$a = \frac{\Delta v}{\Delta t} \qquad (4a)$$

$$= \frac{v_2 - v_1}{\Delta t} \qquad (4b)$$

The units for acceleration are [(m/s)/s = m/s · 1/s = m/s$^2$].

**Example 1a:** Take north to be positive. A car is traveling south and speeding up. What is the sign of the acceleration?

**Solution:** Since the velocity vector points south and the car is speeding up, the acceleration vector must point south. With this sign convention, acceleration is negative.

**Example 1b:** Take north to be positive. A car traveling south speeds up from 10 m/s to 15 m/s in 10 s. What is its acceleration?

**Solution:** We write

$$a = \frac{-15\frac{m}{s} - (-10)\frac{m}{s}}{10s} = -0.5\frac{m}{s^2}$$

This confirms our thinking in Example 1a.

**Example 2a:** Take north to be positive. A car is traveling north and slowing for a red light. What is the sign of the acceleration?

**Solution:** The velocity vector points north. Since this vector is shrinking, the acceleration vector must point south. Thus the acceleration is negative.

**Example 2b:** What is the acceleration for the car in Example 2a slowing from 10 m/s to 8 m/s in 1 s?
**Solution:** We write

$$a = \frac{8\frac{m}{s} - 10\frac{m}{s}}{1s} = -2\frac{m}{s^2}$$

**Example 3:** An Oldsmobile takes a certain amount of time to accelerate from 0 to 60 mph. A Porsche takes less time by a factor of 3 to accelerate from 0 to 60 mph. How does the Porsche acceleration compare with that of the Oldsmobile?

**Solution:** We look at equation (4)

$$a = \frac{\Delta v}{\Delta t}$$

Since $\Delta v$ is constant, if $\Delta t$ is smaller by a factor of 3, then $a$ is larger by a factor of 3.

## H. Graphs

Now we have three quantities, position, velocity, and acceleration, all related to each other algebraically. Often it is helpful to visualize these quantities graphically. The following principles apply

1. Given a graph of $x$ versus $t$, the instantaneous slope at time $t$ is the velocity $v$ at time $t$.
2. Given a graph of $v$ versus $t$, the instantaneous slope at time $t$ is the acceleration $a$ at time $t$.
3. Given a graph of $a$ versus $t$, the area under the curve during interval $\Delta t$ gives the change in velocity $v$ during that interval.
4. Given a graph of $v$ versus $t$, the area under the curve during interval $\Delta t$ gives the change in position $x$ during that interval.

**Example 1:** The graph of $a$ versus $t$ for a car which undergoes constant acceleration is shown in Figure 2-7. Sketch the graph of $v$ versus $t$. Assume $v = 0$ m/s at $t = 0$ s.

**Solution:** The area under the curve between 0 and $\Delta t$ is shown in a "forward-slash" hatch. This area is $\Delta v$, that is, the change in velocity during $\Delta t$. The reason for principle 3 above becomes clear if we recall the formula for the area of the rectangle representing the hatched region:

$$\text{area} = \text{height x length}$$
$$\Delta v = a\Delta t$$

This is how we defined acceleration in equation (4). During the second interval $\Delta t$, the area is a $\Delta t$ again. Thus the change in velocity is the same, as shown in Figure 2-8. For the next intervals of time, the quantity $\Delta v$ is constant.

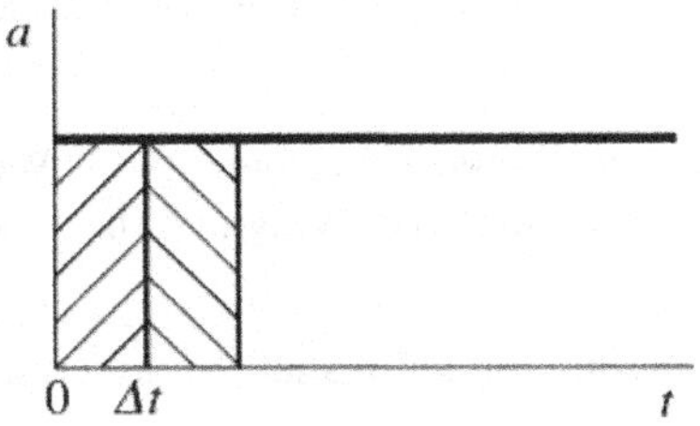

Figure 2-7

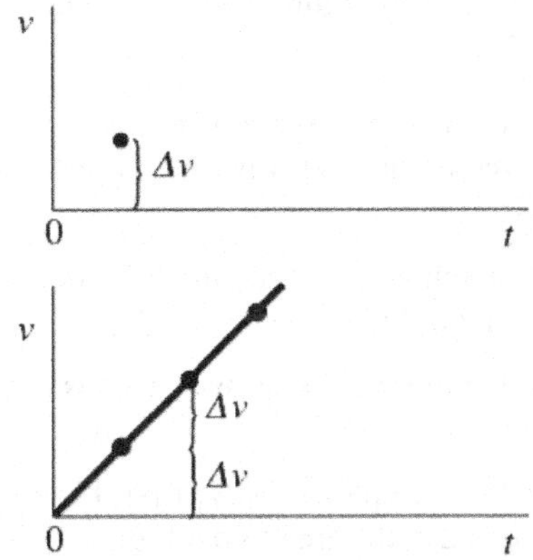

Figure 2-8

Note that Figures 2-7 and 2-8 give (almost) the same information in different forms. Figure 2-7 has the information that the acceleration is positive and constant, so the car is speeding up (if it is going forward). Figure 2-8 has the information that the velocity is increasing at a constant rate. This is the same thing.

Before you read the next example, consider an object thrown straight up. When it reaches the top of its path, what is the direction of its velocity? What is the direction of its acceleration?

**Example 2:** An apple is tossed straight up in the air. The graph of $y$ versus $t$ is shown in Figure 2-9. Sketch the graphs of $v$ versus $t$ and $a$ versus $t$.

**Solution:** To obtain an instantaneous slope, we can use an imaginary electron microscope to look at a small portion of the graph. A small section of Figure 2-9 has been enlarged using such a microscope. This portion looks almost straight, so we could calculate its slope if we had some numbers. We can at least read that the slope is positive and very large, hence the first point in Figure 2-10.

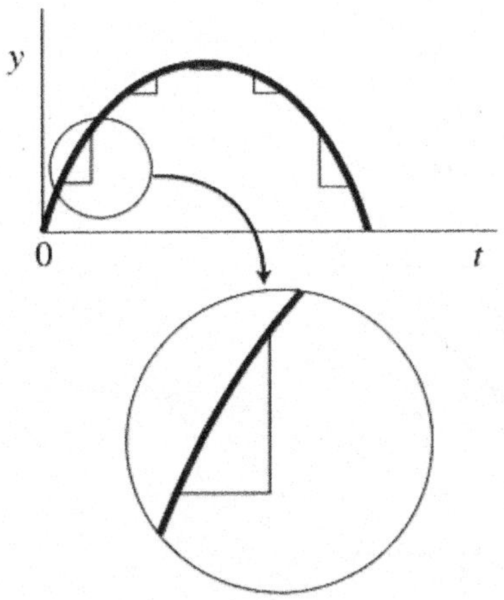

Figure 2-9

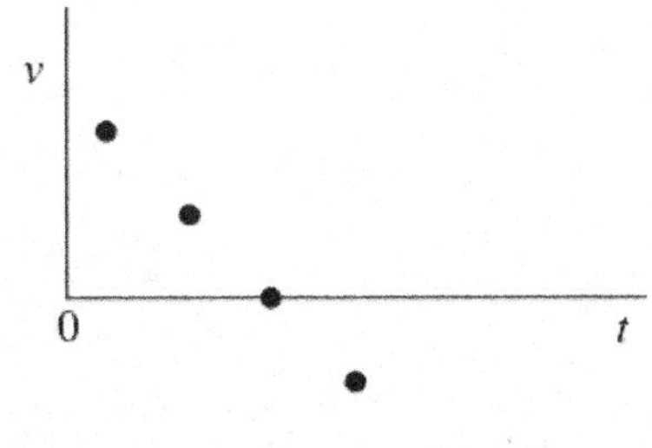

Figure 2-10

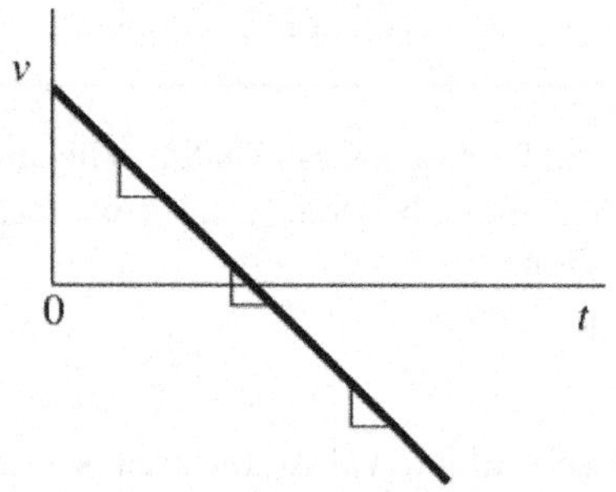


Figure 2-11

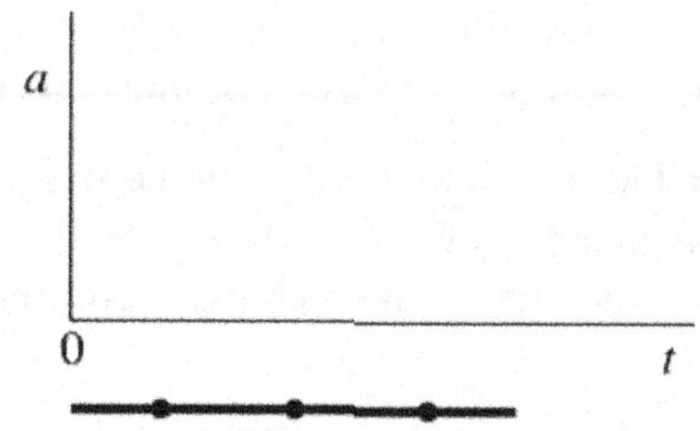


Figure 2-12

The second point on Figure 2-10 still has a positive slope, but smaller. The third point has a zero slope (see uppermost point in Figure 2-9). The fourth point has negative slope, and the fifth point has a slope more negative still.

It will not come as a surprise if we draw a straight line through these points, as in Figure 2-11. We take the slope at three points, but it is easy to see that the slope is constant and negative. We graph the acceleration in Figure 2-12.

Does this match your expectation? Particularly at the top of flight, did you know that the direction of the acceleration would be down?

**Example 3:** Figure 2-13 shows $v$ versus $t$ for a car. Sketch the graphs for $x$ versus $t$ and for $a$ versus $t$. (Say $x = 0$ at $t = 0$.)

**Solution:** Let's graph $x$ versus $t$ first. From $t = 0$ to 1 s, there is no area under the curve, so $x$ stays constant. From $t = 1$ to 2 s, the area under the curve is 0.5 m. (Recall the area of a triangle is $A = 1/2$ base x height.)

From $t = 2$ to 3 s the area under the curve is 1.5 m, so that the $x$ value jumps to 2 m (see the first graph of Figure 2-14). Between $t = 3$ and 4 s, the area is 2 m and $x$ jumps to 4 m, and so on. Figure 2-14 shows the result.

For the graph of $a$ versus $t$, the slope of $v$ versus $t$ for any point between $t = 0$ and 1 s is zero. The slope jumps to 1 m/s$^2$ for the interval from 1 to 3 s and drops back down to zero for times after 3 s. (See Figure 2-15.)

Think about all three graphs for a while and note how they give the same information in different forms.

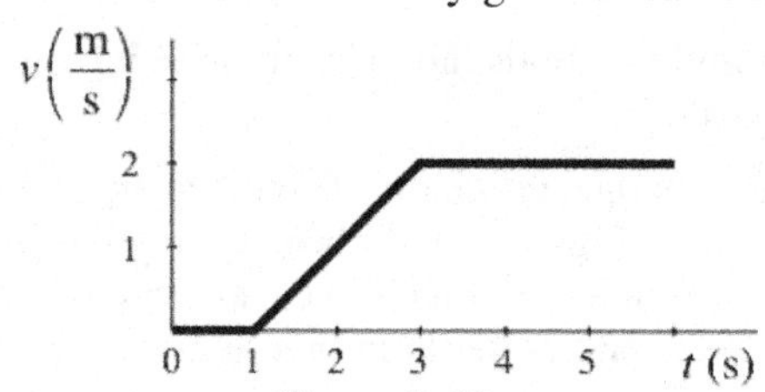


Figure 2-13

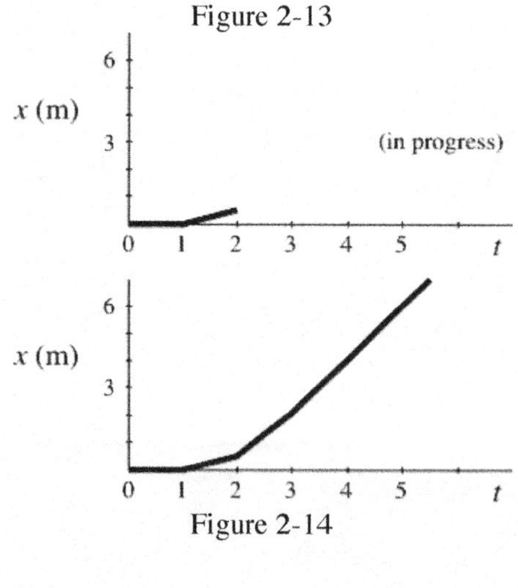


Figure 2-14

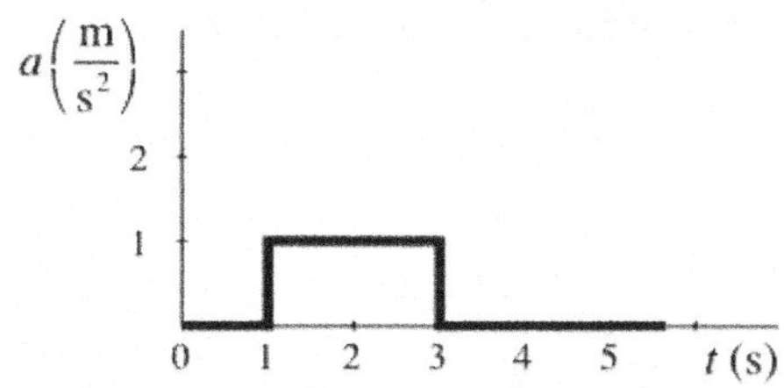


Figure 2-15

## I. Uniform Acceleration

If an object has a constant acceleration vector, we say it undergoes *uniform acceleration*. Most ACT problems involving acceleration will involve uniform acceleration. For uniform acceleration, we have the following:

$$v_{avg} = \frac{1}{2}(v_1 + v_2) \qquad (5)$$

that is, the average velocity over a period of time is the average of the beginning and ending velocities. This may seem like a natural definition of average velocity, but the definition of $v_{avg}$ is given by equation (2), and equation (5) holds only for uniform acceleration.

See Figures 2-7 and 2-8 for an example. The velocity $v_1$ is small, $v_2$ is large, and $v_{avg}$ is exactly between them.

If we start with the definition of average velocity, we can write

$$\Delta x = v_{avg}\Delta t$$

$$\Delta x = \frac{1}{2}(v_1 + v_2)\Delta t$$

This is a useful equation if you do not have and do not need the acceleration (see equation [7] below). Furthermore, if we substitute $v_2 = v_1 + a\Delta t$ (from equation [4]), then we obtain

$$\Delta x = \frac{1}{2}\left(v_1 + (v_1 + a\Delta t)\right)\Delta t$$

$$\boxed{\Delta x = v_1\Delta t + \frac{1}{2}a\Delta t^2 \qquad (6)}$$

This is the first equation which may seem a bit arcane. You should memorize it anyway. Working through the algebra will help you memorize it.

**Example:** A car is accelerating uniformly from rest. If it goes a distance $d$ in the first second, how far will it go in the first four seconds?

**Solution:** We want an equation involving the quantities mentioned in the problem, $a$, $v_1 = 0$, $\Delta x$, and $\Delta t$, so equation (6) is it. With $v_1 = 0$, we obtain

$$\Delta x = \frac{1}{2}a\Delta t^2$$

If $\Delta t$ increases by a factor of 4, the $\Delta x$ increases, and it increases by a factor of $4^2 = 16$.

## J. Kinematic Equations for Constant Acceleration

For uniform acceleration there are four equations you should know:

| | |
|---|---|
| $\Delta x = \frac{1}{2}(v_1 + v_2)\Delta t$ | (7) |
| $v_2 - v_1 = a\Delta t$ | (8) |
| $\Delta x = v_1\Delta t + \frac{1}{2}a\Delta t^2$ | (9) |
| $v_2^2 = v_1^2 + 2a\Delta x$ | (10) |

The first equation we have seen before, the modified "distance equals rate times time" when velocity is changing. It should be easy to remember. The second equation is just the definition of acceleration. The third equation was in the last section. The last equation is the only one which is new, obtained by eliminating $\Delta t$ from equations (7) and (8). It is useful for problems in which the time interval is neither specified nor desired.

**Example 1:** A cat drops from a ledge 2 m above the ground. If he accelerates 10 m/s$^2$ downward due to gravity, how much time does it take him to drop?

**Solution:** Let's choose "up" to be positive and DRAW A DIAGRAM (Figure 2-16). We write the quantities we know:

$$\Delta y = -2 \text{ m}$$

$$v_0 = 0 \text{ m/s}$$

$$a = -10 \text{ m/s2}$$

$$\Delta t = ?$$

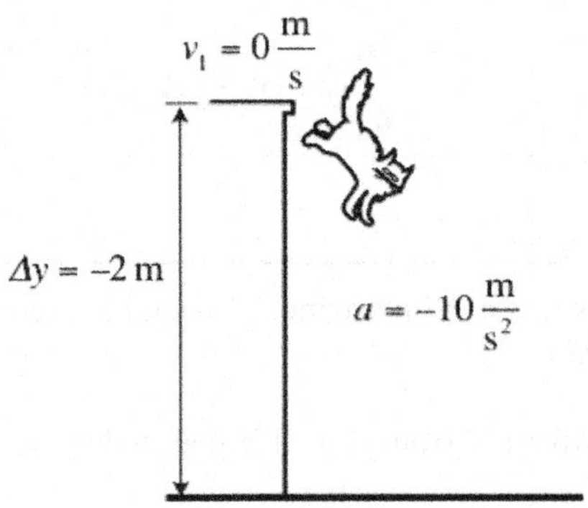


Figure 2-16

We look for an equation which involves these quantities and no others. Equation (9) fits, so that

$$-2\text{m} = 0\text{m} + \frac{1}{2}\left(-10\frac{m}{s^2}\right)\Delta t^2$$

$$\Delta t^2 = 0.4 \cdot \frac{\text{s}^2}{\text{m}}$$

$$\Delta t = 0.63\text{s}$$

**Example 2:** A man drops to his death from the sixth floor of a building (20 m). As he is falling, his acceleration is a constant 10 m/s$^2$ downward. What is his impact velocity? (He was a bad man, and if he had not died many other nice people would have.)

**Solution:** First we DRAW A DIAGRAM (Figure 2-17). The impact velocity is the man's velocity just before he hits the ground $v_2$. Thus our information summary is

$$v_1 = 0 \text{ m/s}$$

$$\Delta y = -20 \text{ m}$$

$$a = -10 \text{ m/s}^2$$

$$v_2 = ?$$

The formula which contains this information and nothing else is (10), so that

$$v_2^2 = v_1^2 + 2a\Delta y$$

$$v_2^2 = \left(0\frac{\text{m}}{\text{s}}\right)^2 + 2\left(-10\frac{\text{m}}{\text{s}^2}\right)(-20 \text{ m})$$

$$v_2^2 = 400\frac{\text{m}^2}{\text{s}^2}$$

$$v_2 = -20\frac{\text{m}}{\text{s}}$$

His impact velocity is 20 m/s.

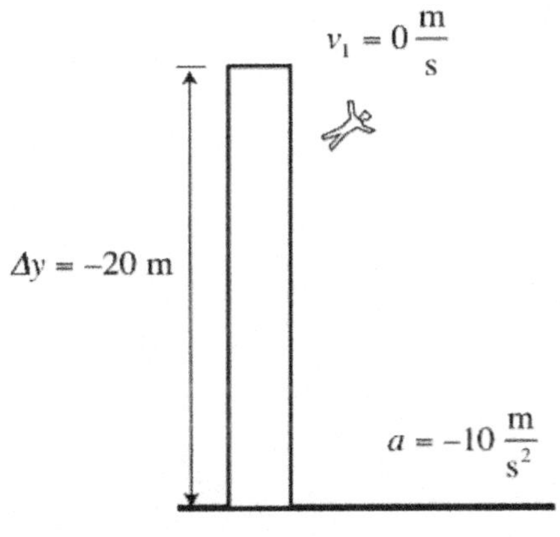


Figure 2-17

In this chapter we looked at the quantities which describe motion, that is, displacement, velocity, and acceleration, and the quantities which affect motion, that is, force and mass. Displacement is a change in location. Velocity is a measure of the change in location per unit time, while acceleration is a measure of the change in velocity per unit time. Displacement, velocity, acceleration, and force are all vectors, that is, they have direction as well as magnitude. Most of the mechanics problems on the ACT involve one dimension and uniform acceleration. In this case, we can derive four equations, shown in Section J.

In addition, you should know the equations for the definition of velocity for uniform motion and of average velocity.

# Physics Review 2 Problems

1. The gravitational field of a planet or spherical astronomical body depends on its mass and on its radius. When the Moon is compared to Earth, its smaller mass makes for a smaller gravitational field, while its smaller radius favors a larger one. The net effect is that the gravitational field of the Moon at its surface is one sixth that of the Earth. A 10,000-kg mobile unit is transported to the Moon. What is its mass on the Moon?
   A. 1/36 (10,000) kg
   B. 1/6 (10,000) kg
   C. 10,000 kg
   D. 60,000 kg

*Use the following information for questions 2 and 3:*

A car is driving north at 2 m/s, and a fly in the car is flying west at 0.3 m/s (relative to the car).

2. Which of the following best shows the appropriate diagram for the fly's velocity relative to the ground (thick arrow)?

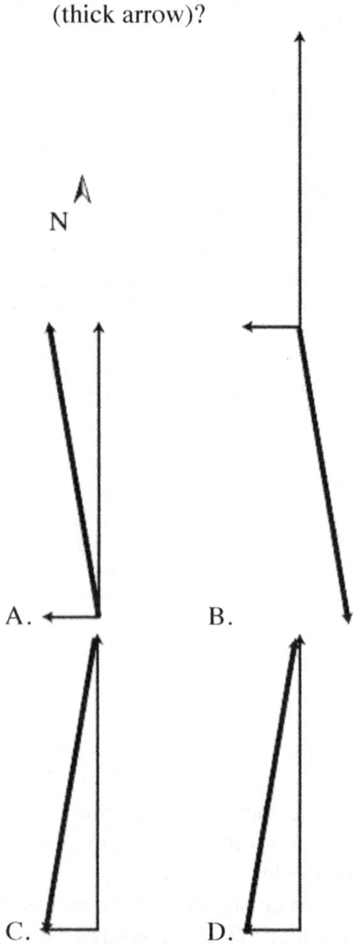


3. What is the speed of the fly?
   A. 1.7 m/s
   B. 2.02 m/s
   C. 2.3 m/s
   D. 4.09 m/s

4. The following diagram represents three vectors in a plane:

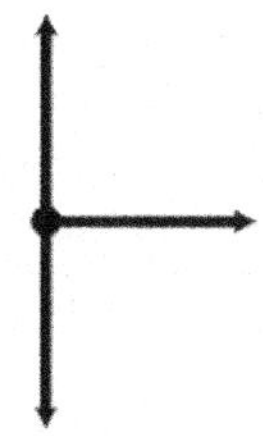

What arrow best represents the direction of the sum?
A. →
B. ←
C. ↑
D. ↓

5. The following diagram represents three vectors in a plane:

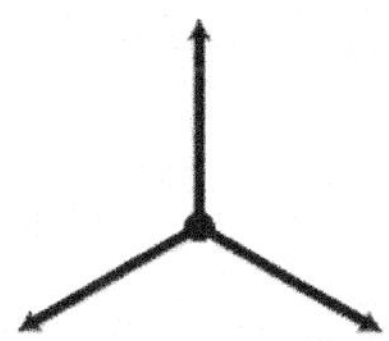

What arrow best represents the direction of the sum?
A. ↓
B. ↑
C. The sum is zero.
D. The diagram is invalid.

6. Two men pull on ropes connected to a large refrigerator with forces 3000 N and 4000 N. If there are no other (unbalanced) forces, which of the following is NOT a possibility for the magnitude of the net (total) force?
A. 500 N
B. 1000 N
C. 3500 N
D. 7000 N

7. If, in question 6, the two men are pulling at right angles to each other, what is the magnitude of the net force?
A. 1000 N
B. 5000 N
C. 7000 N
D. 8000 N

*Use the following information for questions 8–11:*

A woman is going to a friend's house to discuss opening a business. At 11:00 (exactly) she starts from rest and accelerates at a constant 2.5 $m/s^2$ for 9 s to get to her cruising speed. She then drives for 15 minutes at constant speed before she hits city traffic. She comes to a stop at her friend's house, which is 27 km away (straight-line distance), at 12:15 (exactly). Consider the interval from 11:00 to 12:15.

8. What is her initial velocity?
   A. 0 m/s
   B. 2.1 m/s
   C. 5.6 m/s
   D. 22.5 m/s

9. What is her final velocity?
   A. 0 m/s
   B. 2.1 m/s
   C. 5.6 m/s
   D. 22.5 m/s

10. What is her average velocity?
    A. 0 m/s
    B. 6 m/s
    C. 12 m/s
    D. 22.5 m/s

11. What is her velocity 9 s after 11:00?
    A. 0.28 m/s
    B. 6 m/s
    C. 11.25 m/s
    D. 22.5 m/s

*Use the following information for questions 12–14:*

A car accelerates uniformly in one dimension from 5 m/s to 30 m/s in 10 s.

12. What is the car's acceleration?
    A. 1.75 $m/s^2$
    B. 2.5 $m/s^2$
    C. 3.5 $m/s^2$
    D. 15 $m/s^2$

13. What is the car's average velocity for this time interval?
    A. 2.5 m/s
    B. 3.5 m/s
    C. 17.5 m/s
    D. This cannot be determined from the information given.

14. How far does the car travel during this time?
    A. 1.75 m
    B. 17.5 m
    C. 175 m
    D. This cannot be determined from the information given.

15. A car travels a certain distance at a constant velocity $v$ for a time $t$. If the car were to travel three times as fast, covering the same distance, then the time of travel would be
    A. decreased by a factor of 9.
    B. decreased by a factor of 3.
    C. increased by a factor of 3.
    D. increased by a factor of 9.

16. A sparrow cruising at 1.5 m/s begins to accelerate at a constant 0.3 m/s$^2$ for 3 s. What is its change in velocity?
    A. 0.9 m/s
    B. 1.5 m/s
    C. 1.95 m/s
    D. 2.4 m/s

*Use the following information for questions 17 and 18:*

A dropped ball falls from a height of 10 m to the ground. Its acceleration is a constant 9.8 m/s$^2$ downward.

17. How long does it take to fall?
    A. 1.02 s
    B. 1.43 s
    C. 2.04 s
    D. This cannot be determined from the information given.

18. What is its velocity just before hitting the ground?
    A. 0.98 m/s
    B. 14 m/s
    C. 98 m/s
    D. This cannot be determined from the information given.

19. A car is going 20 m/s in traffic. When the traffic breaks, the driver steps on the accelerator pedal, accelerating at a constant 1.2 m/s$^2$ for 5 s. How far does he travel during these 5 s?
    A. 30 m
    B. 115 m
    C. 130 m
    D. 160 m

20. A car is going up a slight slope decelerating at 0.1 m/s$^2$. It comes to a stop after going for 5 s. What was its initial velocity?
    A. 0.02 m/s
    B. 0.25 m/s
    C. 0.5 m/s
    D. 1.0 s

*Use the following information for questions 21 and 22:*

A ball is initially rolling up a slight incline at 0.2 m/s. It decelerates uniformly at 0.05 m/s$^2$.

21. At what time does the ball come to a stop?
    A. 2 s
    B. 4 s
    C. 8 s
    D. 16 s
22. What is the ball's net displacement after 6 s?
    A. 0.3 m
    B. 0.4 m
    C. 0.5 m
    D. 1.7 m

*Use the following information for questions 23 and 24:*

A car is going backwards at 5 m/s. After 10 s of uniform acceleration, the car is going forward at 10 m/s.

23. What is the acceleration?
    A. 0.5 m/s$^2$
    B. 0.75 m/s$^2$
    C. 1.5 m/s$^2$
    D. 5 m/s$^2$

24. What is the net distance traveled?
    A. 25 m
    B. 41.7 m
    C. 45 m
    D. 50 m

25. A bicycle traveling at speed $v$ covers a distance $\Delta x$ during a time interval $\Delta t$. If a car travels at speed $3v$, how much time does it take the car to go the same distance?
    A. $\Delta t + 3$
    B. $3\Delta t$
    C. $\Delta t/3$
    D. $\Delta t - 3$

26. A car is traveling 25 m/s when it passes kilometer-marker 3000. The car accelerates at 0.02 m/s$^2$ for the next 500 s. What kilometer marker will the car pass at that time?
    A. 3015 km
    B. 3030 km
    C. 12000 km
    D. 18000 km

27. A squirrel is running along a wire with constant acceleration. If it has an initial velocity 0.4 m/s and final velocity 1.8 m/s after 4 s, how far does it run in that time?
    A. 1.6 m
    B. 4.4 m
    C. 5.6 m
    D. 8.8 m

28. If a bicycle starts accelerating uniformly from rest (at $t = 0$), it attains a certain velocity $v$ after a time $t$. How fast would it be going after a time $3t$ (that is, a time $3t$ after the start $t = 0$)?
    A. $v + 9$
    B. $3v$
    C. $6v$
    D. $9v$

For questions 29 and 30, consider the following figure representing the velocity of a car along a street.

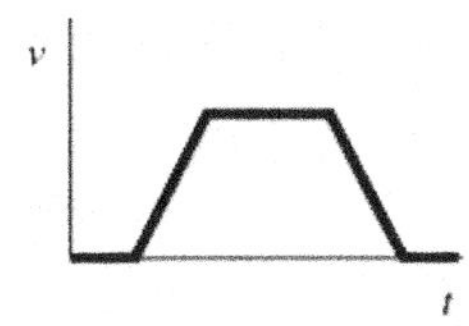


Consider also the following graphs:

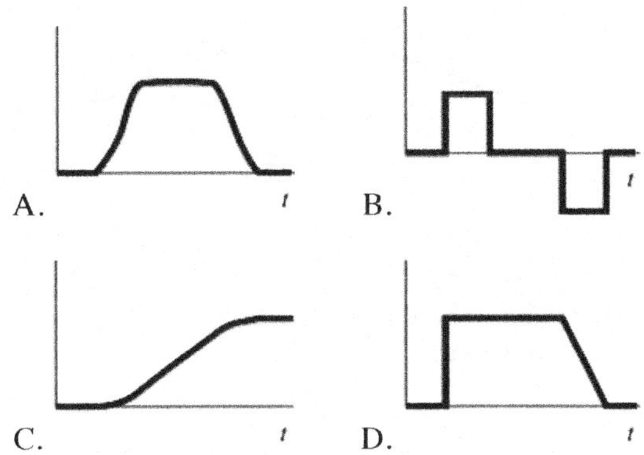


29. Which best represents the graph of displacement versus time?
    A. A
    B. B
    C. C
    D. D

30. Which best represents the graph of acceleration versus time?
    A. A
    B. B
    C. C
    D. D

For questions 31 and 32, consider the following figure representing the displacement of an object in one dimension.

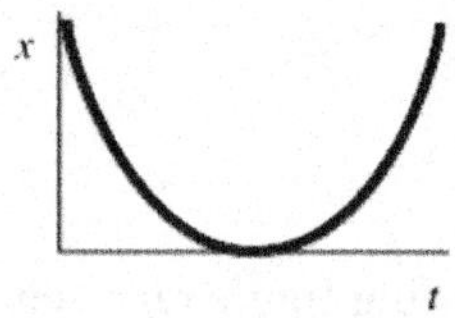


31. Which best represents the graph of acceleration versus time?

A. 

B. 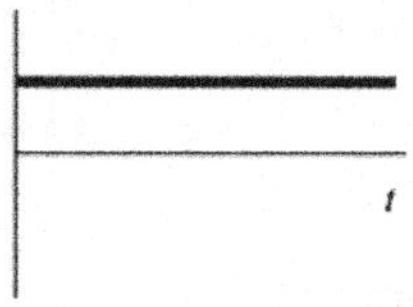


C. 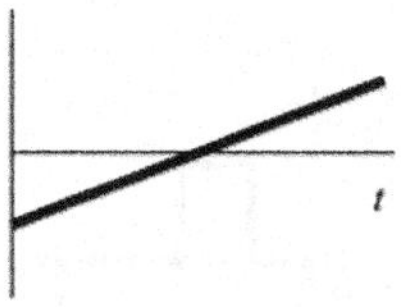


D. 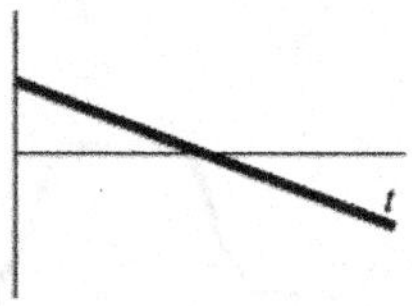


32. What can be concluded about the net displacement?
    A. It is zero.
    B. It is positive except for one point, where it is zero.
    C. It is negative, then zero, then positive.
    D. It is always positive?

For questions 33–35, consider the following figure representing the velocity of an object in one dimension.

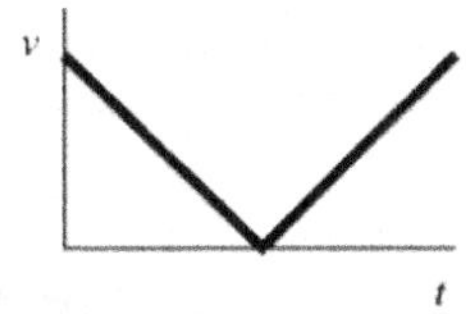


Consider also the following graphs:

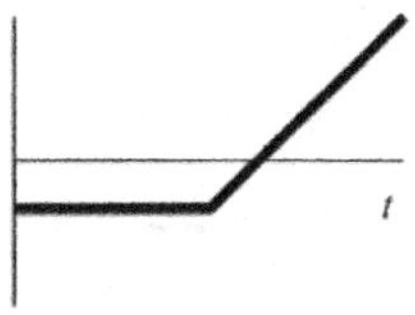


A.

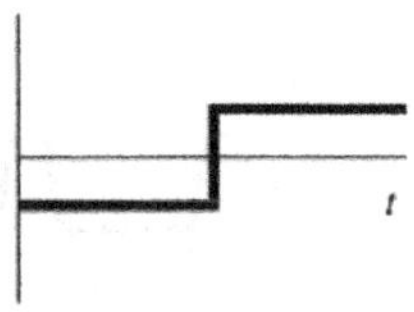


B.

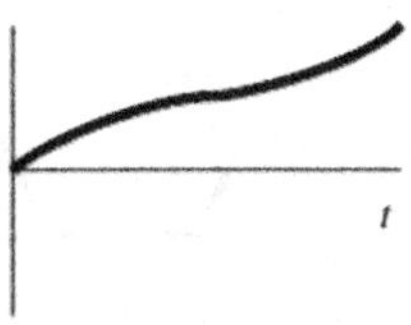


C.

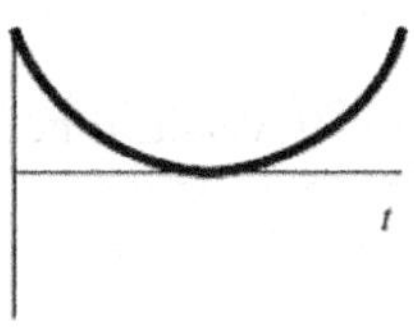


D.

33. Which best represents the graph of displacement versus time?
   A. A
   B. B
   C. C
   D. D

34. Which best represents the graph of acceleration versus time?
   A. A
   B. B
   C. C
   D. D

35. What can be said about the net velocity change $\Delta v$?
   A. It is positive.
   B. It is zero.
   C. It is negative.
   D. It is positive, except for one point.

*Use the following information for questions 36 and 37:*

A car backs up at constant velocity, then slows to a stop. After it is stopped for a while, it accelerates and then goes forward at constant velocity. Consider also the following graphs:

A.
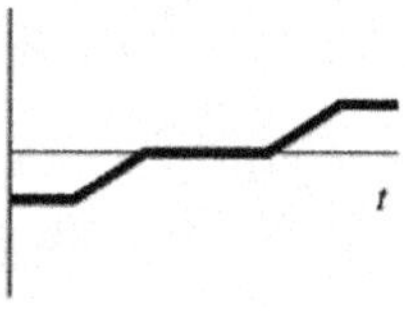


B.
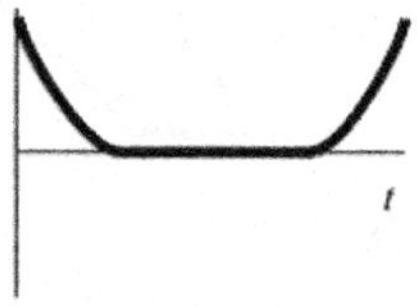


C.
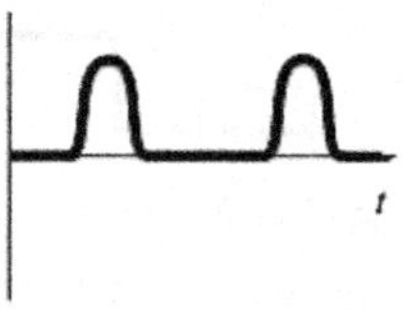


D.
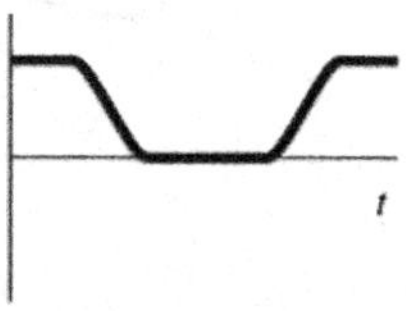


36. Which best represents the graph of velocity versus time?
    A. A
    B. B
    C. C
    D. D

37. Which best represents the graph of acceleration versus time?
    A. A
    B. B
    C. C
    D. D

## Passage 1

A man is driving out of his driveway by backing up. He realizes he has forgotten his lunch, so he pulls back into the driveway. Car experts agree that the best way to do this is to press on the brake until the car comes to a complete stop, shift from reverse into first gear, then accelerate forward.

The driver, however, shifts into first gear while the car is rolling backward and pushes on the accelerator until he is going forward. This causes some wear on the transmission. The following chart shows some data about his progress. (Negative velocity = backwards.)

| $t$ (s) | $x$ (m) | $v$ (m/s) |
|---|---|---|
| 0.0 | 1.35 | –1.8 |
| 0.5 | 0.60 | –1.2 |
| 1.0 | 0.15 | –0.6 |
| 1.5 | 0.00 | 0.0 |
| 2.0 | 0.15 | 0.6 |
| 2.5 | 0.60 | 1.2 |
| 3.0 | 1.35 | 1.8 |

38. What is the value of his initial velocity?
    A. –1.8 m/s
    B. 0.0 m/s
    C. 1.2 m/s
    D. 1.8 m/s

39. What is the value of his average velocity?
    A. –1.8 m/s
    B. 0.0 m/s
    C. 1.2 m/s
    D. 1.8 m/s

40. Which of the following is evidence that the acceleration is uniform?
    A. The displacement $x$ is always nonnegative.
    B. The velocity is always increasing.
    C. The velocity becomes zero at $t$ = 1.5 s.
    D. Equal intervals of time correspond to equal intervals of velocity.

41. What is the magnitude of the acceleration from $t$ = 1.0 s to 2.0 s?
    A. 0.0 m/s$^2$
    B. 0.6 m/s$^2$
    C. 1.0 m/s$^2$
    D. 1.2 m/s$^2$

42. What is the direction of the acceleration vector?
    A. forward
    B. backward
    C. up
    D. down

43. Which best represents the graph of acceleration versus time?

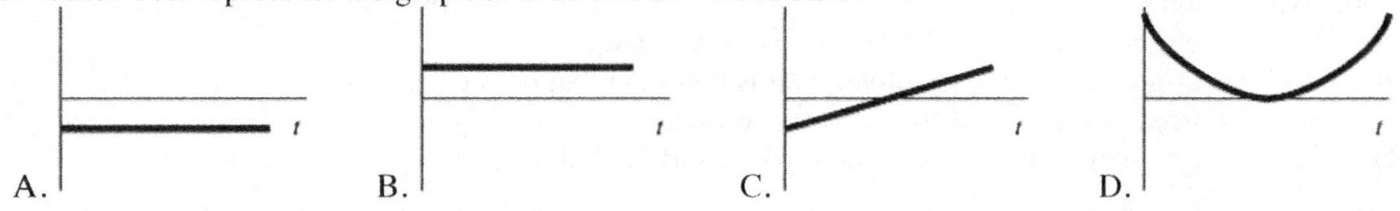

## Passage 2

A physics student leans out of the fortieth story of the physics building and drops two balls of the same size at the same time. One is 0.8 kg and made of iron, and the other is 1.2 kg and made of lead. Not only do the two balls hit the ground at the same time, the heights of the two balls match all the way down.

This somewhat counterintuitive result is an example of a general principle: If air resistance is negligible, then an object in free fall at the surface of the Earth has a downward acceleration of $g = 9.8\ \text{m/s}^2$. Free fall means that only the force of gravity is acting on an object.

In the following questions, consider a ball dropped from the fortieth story of a building, and consider "down" to be in the positive direction. Consider air resistance negligible unless noted otherwise.

44. How far does the object fall in the time interval from $t = 0$ to 4 s?
    A. 39.2 m
    B. 78.4 m
    C. 156.8 m
    D. 313.6 m

45. Which expression gives the change in velocity between $t_1 = 3$ s and $t_2 = 4$ s?
    A. $g(t_2 - t_1)/2$
    B. $g(t_2 - t_1)$
    C. $g(t_2 + t_1)/2$
    D. $g(t_2 + t_1)$

46. Which graph best represents velocity versus time?

A. 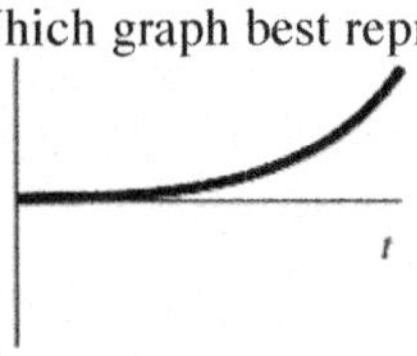


B. 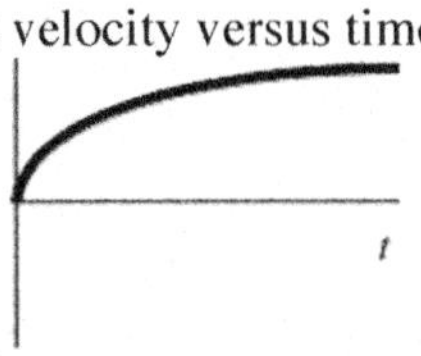


C. 

D. 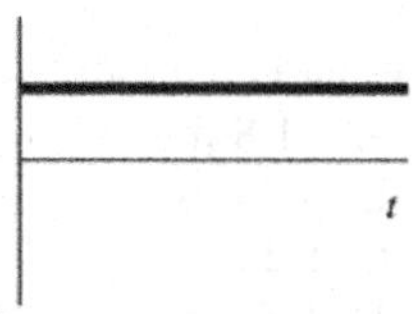


47. How does the change in velocity from $t = 1$ to 2 s compare with the change in velocity from $t = 3$ to 4 s?
    A. It is less.
    B. It is the same.
    C. It is greater.
    D. It depends on the object.

48. How does the change in height from $t = 1$ to 2 s compare with the change in height from $t = 3$ to 4 s?
    A. It is less.
    B. It is the same.
    C. It is greater.
    D. It depends on the object.

49. If an object falls a distance $\Delta x$ during the first $t$ seconds, how far does it fall during the first $3t$ seconds?
    A. $\Delta x + 3$
    B. $3\Delta x$
    C. $\Delta x + 9$
    D. $9\Delta x$

50. A Styrofoam ball of the same size as the lead ball takes a longer time to reach the ground. Which is a good explanation for this?
    A. The force of gravity does not act on the Styrofoam ball.
    B. The force of gravity on the Styrofoam ball is less than that on the lead ball.
    C. Air resistance is a significant force in this problem.
    D. There is a gravitational force between the ball and the building.

# Solutions

1. C.
The mass of the mobile unit does not change just because we transport it to a different place.

2. A.
We add the fly's velocity to the car's velocity in order to obtain its total velocity relative to the ground. We move the fly velocity vector so that its tail coincides with the tip of the car velocity vector. The resulting total is just west of north (see diagram). In the choices, only choice A shows the correct total vector. The choice does not show the fly vector already moved. You must do that.

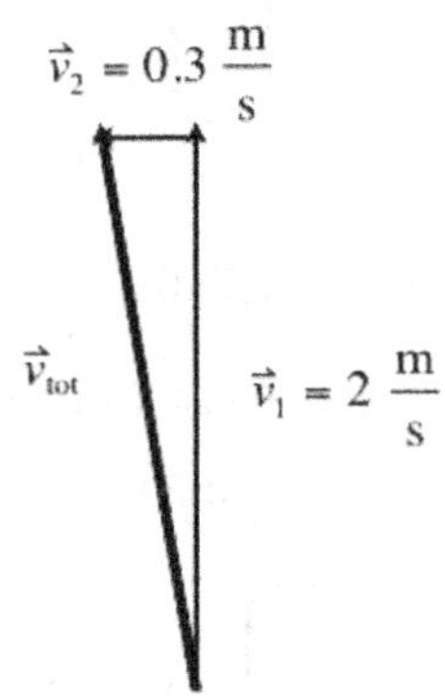


3. B.
From the diagram of the previous solution, we can see that the answer must be slightly larger than 2 m/s and certainly not so large as 2.3 m/s, so B is correct. To do the numbers, we use the Pythagorean theorem:

$$\begin{aligned}(v_{tot})^2 &= \left(2\frac{m}{s}\right)^2 + \left(0.3\frac{m}{s}\right)^2 \\ &= 4.09\frac{m^2}{s^2} \\ v_{tot} &= 2.02\frac{m}{s}\end{aligned}$$

4. A.
First we move the horizontal vector, so its tail is on the tip of the top vector. Then we move the bottom vector, so its tail is on the tip of the horizontal vector. The resulting sum is the arrow from the first tail to the last tip (see diagram).

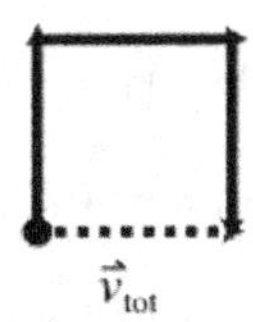


5. C.
Again we sequentially place the tail of one vector on the tip of the previous one. The resulting sum is zero (see diagram). We could see this in the diagram anyway, since the arrows seem to cancel out each other.

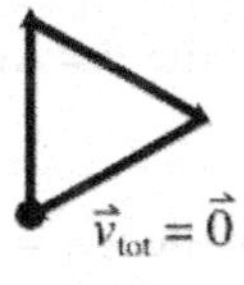

6. A.
They exert the largest net force when they pull the same direction (see figure), giving the total 7000 N. They exert the smallest net force when they are directly opposed to each other, giving 1000 N. Therefore it is not possible for the net force to be 500 N. If this is unclear, try drawing a few vector diagrams to get a total of 500 N.

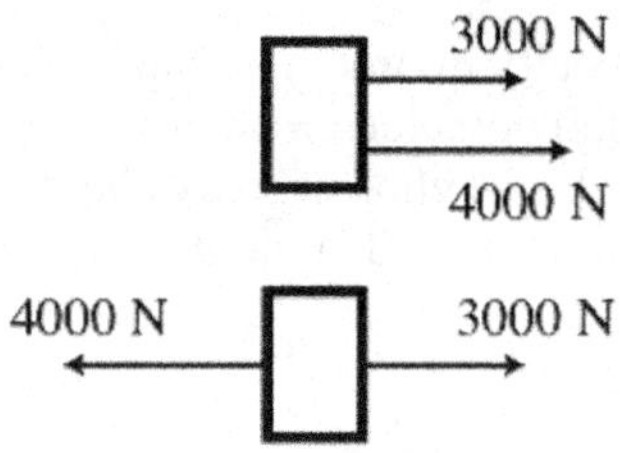


7. B.
If they pull at right angles, then we need to apply the Pythagorean theorem (see figure):

$$(F_{tot})^2 = (3000\text{ N})^2 + (4000\text{ N})^2$$

$$F_{tot} = 5000\text{ N}$$

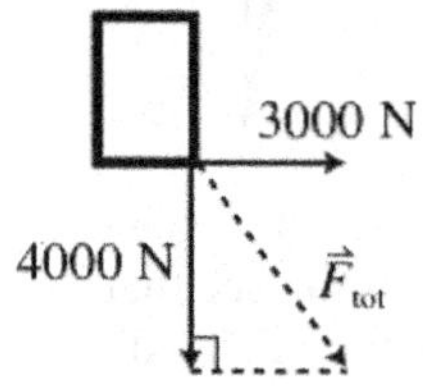


8. A.
Since she starts from rest, her initial velocity is zero.

9. A.
Again, she ends at rest.

10. B.
The net displacement is $\Delta s = 27$ km = 27,000 m. The total time is $\Delta t = 75$ min = 4500 s. Thus the average velocity is $v_{avg} = \Delta s/\Delta t = 6$ m/s.

11. D.
During these 9 s, she was accelerating uniformly, so we can write the equation:

$$v_2 - v_1 = a\Delta t$$

$$v_2 = v_1 + a\Delta t$$

$$= 0\text{ m/s} + (2.5\text{ m/s}^2)(9\text{s})$$

$$= 22.5\text{ m/s}$$

12. B.
Since the car is accelerating uniformly, we can write $a = \Delta v/\Delta t = (30\text{ m/s} - 5\text{ m/s})/10\text{ s} = 2.5\text{ m/s}^2$.

13. C.
Since the car is accelerating uniformly, we can write $v_{avg} = 1/2\,(v_1 + v_2) = 17.5$ m/s.

14. C.
We can write the equation

$$\Delta x = 1/2\,(v_1 + v_2)\Delta t$$

$$= 1/2\,(5\text{ m/s} + 30\text{ m/s})(10\text{s})$$

$$= 175\text{ m}$$

15. B.
We want a formula which relates distance, velocity, and time, so we write

$$v = \frac{\Delta x}{\Delta t}$$

Now $\Delta x$ is the same, while we want $v$ to increase by a factor of 3. To do this we need $\Delta t$ to decrease by a factor of 3. You could probably do this without writing down the formula. Other problems will not be so simple.

16. A.
We want to know change of velocity $\Delta v$, while we know acceleration $a = 0.3\text{ m/s}^2$, the time interval $\Delta t = 3$ s, and the initial velocity $v_1 = 1.5$ m/s. We need the definition of acceleration:

$$a = \Delta v/\Delta t$$

$$\Delta v = a\Delta t$$

$$= (0.3\text{ m/s}^2)(3\text{ s})$$

$$= 0.9\text{ m/s}^2$$

So we did not need the value of $v_1$ at all. The key to many problems is making an inventory of what we know and what we want.

17. B.
We know $\Delta y = -10$ m, $a = -10\text{ m/s}^2$ (approximately), $v_1 = 0$ m/s (because it is dropped). We want $\Delta t$. Thus we have

$$\Delta x = v_1\Delta t + \frac{1}{2}a\Delta t^2$$

$$-10\text{m} = \frac{1}{2}\left(-10\frac{\text{m}}{\text{s}^2}\right)\Delta t^2$$

$$\Delta t = \sqrt{2}\text{s} = 1.4\text{s}$$

18. B.
If we want $v_2$, we use the equation:

$$v_2^2 = v_1^2 + 2a\Delta x$$

$$v_2^2 = \left(0\frac{\text{m}}{\text{s}}\right)^2 + 2\left(-10\frac{\text{m}}{\text{s}^2}\right)(-10\text{m})$$

$$v_2 = 10\sqrt{2}\frac{\text{m}}{\text{s}}$$

$$= 14\frac{\text{m}}{\text{s}}$$

19. B.
We know $v_1 = 20$ m/s, $a = 1.2$ m/s$^2$, and $\Delta t = 5$ s. Thus $\Delta x = v_1\Delta t + 1/2\, a(\Delta t)^2 = 115$ m.

20. C.
We know $a = -0.1$ m/s$^2$, $v_2 = 0$ m/s (because it comes to a stop), and $\Delta t = 5$ s. We use the following equation:

$$v_2 - v_1 = a\Delta t$$

$$v_1 = v_2 - a\Delta t$$

$$v_1 = (0 \text{ m/s}) - (-0.1 \text{ m/s}^2)(5\text{s})$$

$$= 0.5 \text{ m/s}$$

21. B.
We know $v_1 = 0.2$ m/s and $a = -0.05$ m/s$^2$. If we want to know how long it takes to stop, we add the datum $v_2 = 0$ m/s. Thus we have

$$v_2 - v_1 = a\Delta t$$

$$0\frac{\text{m}}{\text{s}} - 0.2\frac{\text{m}}{\text{s}} = -0.05\frac{\text{m}}{\text{s}^2}\Delta t$$

$$\Delta t = \frac{-0.2\frac{\text{m}}{\text{s}}}{-0.05\frac{\text{m}}{\text{s}^2}}$$

$$= 4 \text{ s}$$

22. A.
We still use $v_1 = 0.2$ m/s and $a = -0.05$ m/s$^2$, but now we have $\Delta t = 6$ s. The net displacement is obtained from

$$\Delta s = v_1\Delta t + 1/2\, a\Delta t^2$$

$$= (0.2 \text{ m/s})(6 \text{ s}) + 1/2\,(-0.05 \text{ m/s}^2)(6 \text{ s})^2$$

$$= 0.3 \text{ m}$$

It is important to pay attention to signs in this problem, translating "up" into positive and "decelerating" into negative. Also note that the ball travels further than 0.3 m, going 0.4 m up the slope before heading back. The net displacement is the difference between final position and initial position.

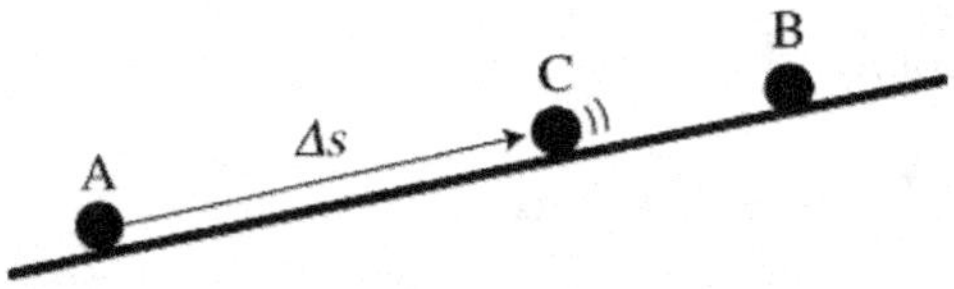


23. C.
Here we have $v_1 = -5$ m/s, $\Delta t = 10$ s, and $v_2 = 10$ m/s. We can obtain acceleration from its definition: $a = (v_2 - v_1)/\Delta t = 1.5$ m/s$^2$.

24. A.
We obtain the net displacement from $\Delta x = 1/2(v_1 + v_2)\,\Delta t = 25$ m.

25. C.
We want an equation which relates $v$, $\Delta x$, and $\Delta t$. This is $v = \Delta x/\Delta t$. If $v$ increases by a factor of 3, then $\Delta t$ must decrease by a factor of 3, so the answer is C.

26. A.
Here we know that $v_1 = 25$ m/s, $x_1 = 3000$ km, $a = 0.02$ m/s$^2$, and $\Delta t = 500$ s. We want $x_2$. We know that $\Delta x = x_2 - x_1$, so we can calculate

$$\begin{aligned}\Delta x &= v_1\Delta t + \frac{1}{2}a\Delta t^2\\ &=(25\frac{\text{m}}{\text{s}})(500\text{s}) + \frac{1}{2}\left(0.02\frac{\text{m}}{\text{s}^2}\right)(500\text{s})^2\\ &= 15000\text{m}\\ &= 15\text{km}\end{aligned}$$

Thus $x_2 = 3015$ km. Remember to be careful with the units.

27. B.
We calculate $\Delta x = 1/2\ (v_1 + v_2)\Delta t = 4.4$ m.

28. B.
We want an equation involving velocity and time and possibly acceleration. Let's look at the equation

$$a = \frac{\Delta v}{\Delta t}$$

Here $a$ is constant and $\Delta t$ increases by a factor of 3. Thus $\Delta v$ increases by a factor of 3 as well, giving B as an answer. Another way to see this is to rewrite the equation:

$$\Delta v = a\Delta t$$

so an increase in $\Delta t$ yields an increase in $v$.

29. C.
The figure in the problem shows a car which starts from rest, speeds up to a cruising speed which it maintains, then slows to a stop. Let's keep this in mind.

The figure above shows $v$ versus $t$, and we want to pick the best graph for $x$ versus $t$. From point A to point B, there is no area under the graph, so the displacement is zero, and all the choices show this. From B to C, there is an increasing amount of area under the curve, eliminating choices B and D which show a jump in $x$ versus $t$. From C to D, there is area under the curve, so $\Delta x$ is positive for every interval $\Delta t$. This means $x$ is increasing, which is shown only in choice C. After E, the velocity goes to zero, and $\Delta x$ goes to zero. This means that no more increments get added to x. *But this does not mean* that $x$ returns to the $x$-axis (as in choice A).

30. B.
Between A and B, the instantaneous slope is zero (see figure). Between points B and C, the instantaneous slope jumps to a constant value, eliminating A and C as choices. Between C and D, the instantaneous slope is zero, so the acceleration jumps back down to zero, so B is correct. Between points D and E, the acceleration is negative.

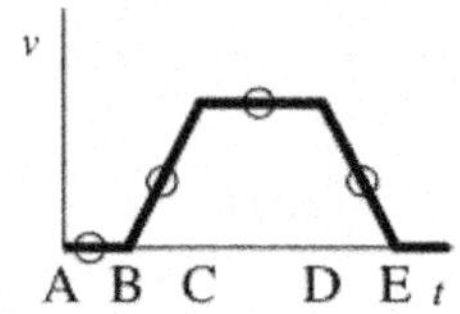


31. B.
The following figure is $x$ versus $t$. We can obtain $v$ versus $t$ by looking at the instantaneous slope at various times. This gives the points shown in the second figure.

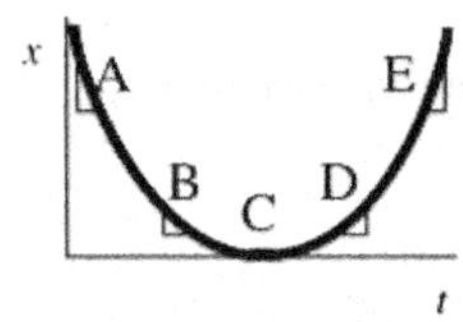


To obtain the acceleration, we need to take the instantaneous slope at various points in the second figure, but clearly the slope is always positive (and probably constant).

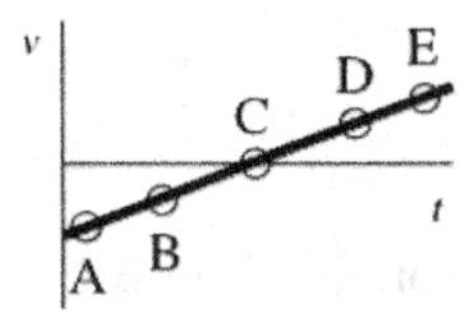


32. A.
The net displacement is $\Delta x = x_2 - x_1$, the difference between final and initial position. But in this case $x_1$ and $x_2$ are the same, so the net displacement is zero.

33. C.
In words, we can say that the velocity starts large and decreases to a stop. Then it increases again. Just this description alone might lead us to recognize C as the answer.

The figure shows how successive areas under the curve $v$ versus $t$ result in successive increases in the curve $x$ versus $t$. Note that the fact that $v$ is always positive translates into the fact that $x$ is always increasing.

We can also work this problem by taking the choices and working backward, taking instantaneous slopes, in order to find which would produce the given $v$ versus $t$.

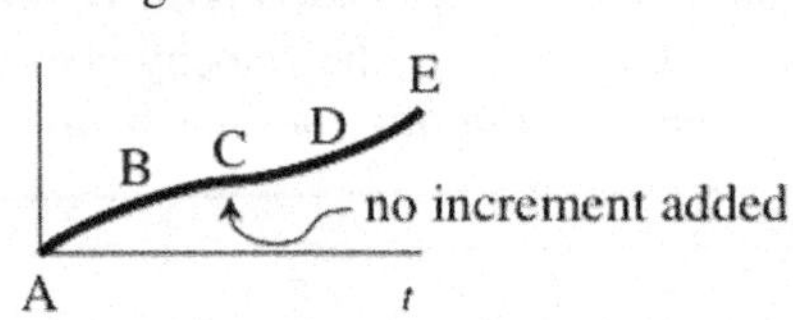

34. B.
Taking instantaneous slopes of $v$ versus $t$ gives an acceleration which is constant and negative until the cusp. Then it is constant and positive. This is shown in choice B.

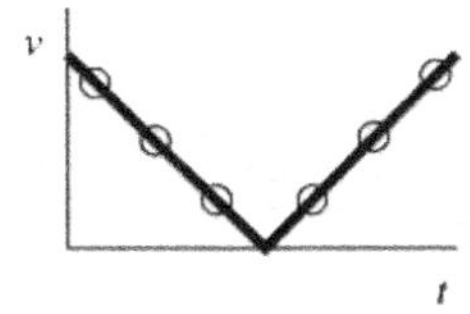

35. B.
The net change in velocity $\Delta v = v_2 - v_1$ is simply zero, since the velocity starts and ends with the same value.

36. A.
The car is backing up at the beginning of the problem, so the velocity is negative. The only choice to show this is A. The flat portion in the center of choice A is when the car is stopped. The later flat portion shows the constant velocity while going forward.

37. C.
The car accelerates exactly twice: while it slows to a stop going backward (positive acceleration) and while it speeds from a stop going forward (positive acceleration). When the car is cruising at constant velocity, the acceleration is zero, so C is the correct choice. By the way, choice B shows a sketch of $x$ versus $t$.

## Passage 1

38. A.
The initial velocity is just that at the beginning of the experiment.

39. B.
The average velocity is $v_{avg} = \Delta x/\Delta t$. The net displacement is $\Delta x = 1.35\text{ m} - 1.35\text{ m} = 0\text{ m}$.

40. D.
This question consists entirely of words, but let us write an equation anyway. Uniform acceleration means $a$ is constant, and $a = \Delta v/\Delta t$. Thus $\Delta v$ and $\Delta t$ are in a constant ratio. If one of the choices expresses this fact, then that would be the solution. (If not, we will have to think some more, perhaps find another equation.) D is the correct answer.

41. D.
We apply the definition of acceleration:

$$a = \frac{v_2 - v_1}{\Delta t} = \frac{0.6\frac{\text{m}}{\text{s}} - \left(-0.6\frac{\text{m}}{\text{s}}\right)}{1\text{s}} = 1.2\frac{\text{m}}{\text{s}^2}$$

42. A.
Since acceleration is positive, the vector points in the forward direction (according to the sign convention of the passage).

43. B.
For any 0.5-s interval in the chart, the acceleration ($\Delta v/\Delta t$) is a constant 1.2 m/s$^2$, even for the intervals near $t = 1.5$ s, where the velocity is zero. Thus B is the best answer.

## Passage 2

44. B.
We apply the formula $\Delta x = v_1\Delta t + 1/2\ a(\Delta t)^2 = (0 \text{ m/s})(4 \text{ s}) + 1/2\ (10 \text{ m/s}^2)(4 \text{ s})^2 = 80 \text{ m}$.

45. B.
The equation which involves change in velocity $\Delta v$ and time is the definition of acceleration:

$$g = \Delta v/\Delta t$$

$$\Delta v = g\Delta t$$

$$\Delta v = g(t_2 - t_1)$$

46. C.
Since velocity starts at zero, we can eliminate D. The instantaneous slope for the graph of $v$ versus $t$ must be a constant $g = 9.8 \text{ m/s}^2$. The graph to which this applies is C.

47. B.
Let's be clear about this by writing an equation. The equation involving $\Delta v$ is the definition of acceleration (see solution to problem 45). Thus $\Delta v = g\Delta t$. The acceleration $g$ is constant, and $\Delta t$ is the same for the two situations (both have $\Delta t = 1$ s). Thus $\Delta v$ is the same. The velocity increases at a constant rate throughout the fall.

48. A.
We can calculate the height at the four clock readings:

$$\Delta x\ (t = 1\text{s}) = v_1\Delta t + 1/2\ a\Delta t^2$$

$$= (0 \text{ m/s})(1 \text{ s}) + 1/2\ (10 \text{ m/s}^2)(1 \text{ s})^2$$

$$= 5 \text{ m}$$

$$\Delta x\ (t = 2 \text{ s}) = 20 \text{ m}$$

$$\Delta x\ (t = 3 \text{ s}) = 45 \text{ m}$$

$$\Delta x\ (t = 4 \text{ s}) = 80 \text{ m}$$

Thus from 1 to 2 s, the object falls 15 m, while from 3 to 4 s, the object falls 35 m. This confirms our intuition that the distance fallen is greater for the later time interval. Notice the difference between this problem and the previous one.

49. D.
We use the equation involving $\Delta x$, $\Delta t$, and the acceleration (since we know its value):

$$\Delta x = v_1\Delta t + 1/2\ g\Delta t^2$$

We have $v_1 = 0$ m/s, so we write

$$\Delta x = 1/2\ g\Delta t^2$$

If the time interval $\Delta t$ increases by a factor of 3, what happens to $\Delta x$? Apparently it increases by a factor of 9.

50. C.
Choice A is nonsense. Choice B is a true statement, but it cannot be an adequate explanation for the fact, since the lead and iron balls fall at the same rate, and the force of gravity is presumably different on those two balls as well. Choice C is a good candidate, since the passage mentions air resistance as a caveat. Choice D is irrelevant.

# Physics Review 3
# Laws of Motion

## A. First Law of Motion

The following is *not* a law of physics:

> If an object is moving, there must be a force on it.

Most people think the above statement is a law of nature. Some very intelligent thinkers thought it was a law of nature, including Aristotle (ancient Greek, no intellectual lightweight) and more recently Descartes (famous philosopher). Because it is common sense (right?) that if nothing pushes on an object, it eventually slows to a stop.

But sometimes closer scrutiny conflicts with common sense, and when that happens we have to change our thinking, to retune our intuition, so that what once seemed wrong now seems right. That can be difficult, but that's physics.

> **First Law of Motion**
> If the forces on an object are balanced, then the object moves with constant velocity (constant speed in a straight line). Conversely, if an object has constant velocity, then the forces on it are balanced. Some people use the term *inertia* to describe this property of matter.

Galileo discovered this law, although it's generally called Newton's first law of motion.

What does it mean for the forces to be balanced? Before we answer that question, let's look at a few cases. In the following figures (Figures 3-1a–f) we denote the motion of an object by "motion marks", so that

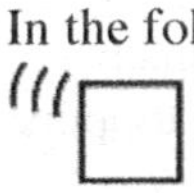 means the object is moving to the right.

**Case a.** There are no forces. In this case think of a rock in deep space moving along. The velocity vector is constant, meaning the rock continues traveling at constant speed to the right indefinitely.

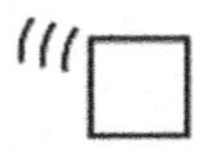

Figure 3-1a

**Case b.** There are two opposed forces, equal in magnitude, perpendicular to the motion. In this case think of a marble rolling along a smooth level floor (no friction). Gravity pulls down, but the floor pushes up. The velocity vector is constant.

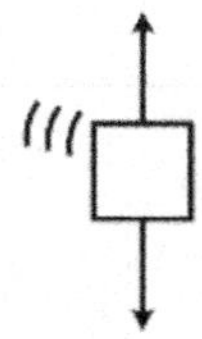

Figure 3-1b

**Case c.** This scenario is a nonexample, in which the right force is larger than the left force (hence unbalanced). The object will speed up.

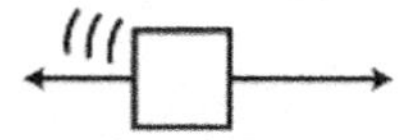

Figure 3-1c

**Case d.** This scenario is also a nonexample with the left force larger than the right force. The object will slow down.

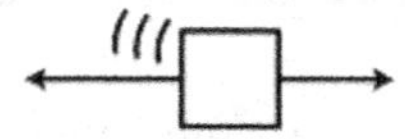

Figure 3-1d

**Case e.** Two opposed forces, left and right, are equal in magnitude. The case is between cases c and d. The object has constant velocity, that is, it will keep its speed indefinitely. Think about this one for a while. This stumps many people.

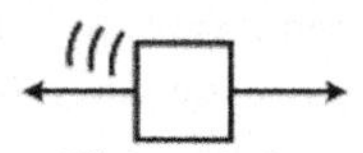

Figure 3-1e

**Case f.** The forces in all three directions are balanced. The object's velocity vector is constant.

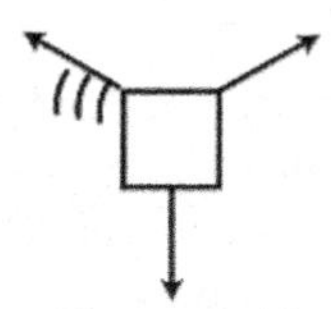

Figure 3-1f

The forces are balanced if the vector sum of all the forces on the object is zero. We define $F_{net}$ by

$$F_{net} = F_1 + F_2 + \ldots$$

where $F_1, F_2, \ldots,$ are all the forces acting on an object. The vector $F_{net}$ is the total force on the object.

**Example:** A woman kicks a soccer ball, and it rolls for a while at constant speed, then another woman stops it. Draw a diagram showing all the forces on the ball at the three times: kicking, rolling, and stopping.

**Solution:** Part a: The ball is kicked (see Figure 3-2). The vector $F_{grav}$ is the force of gravity, and $F_{kick}$ is the force of the foot on the ball. The symbol $N$ stands for "normal", a physics word meaning perpendicular to the ground. It is force the ground exerts on the ball.

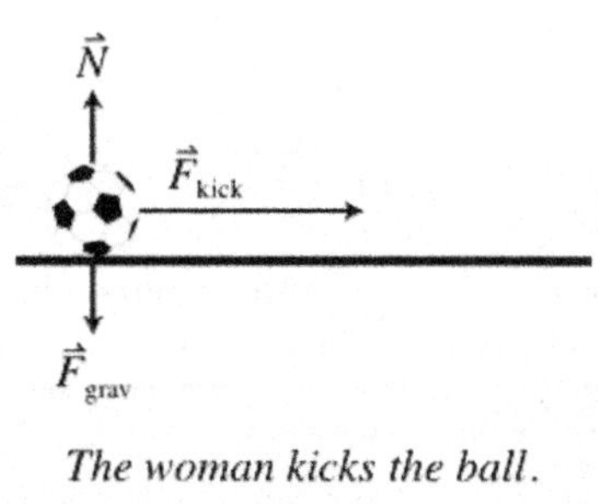


*The woman kicks the ball.*

Figure 3-2

Part b: The ball rolls along. It has only two forces acting on it (Figure 3-3). The ball does not remember (or care) what started it rolling. According to the first law, the balanced forces guarantee it will keep rolling indefinitely at constant speed.

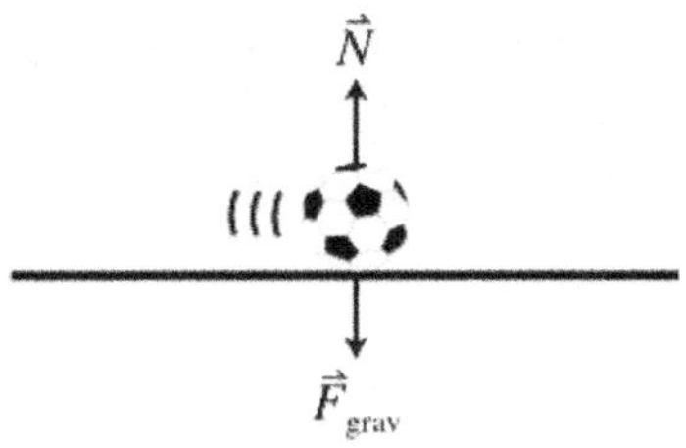


*The ball rolls without friction.*

Figure 3-3

Part c: The ball is stopped. Now there is a force of a foot on the ball as well (Figure 3-4).

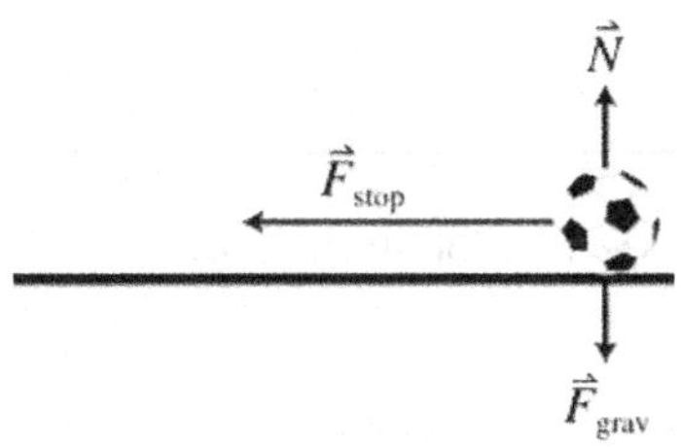


*The woman stops the ball.*

Figure 3-4

Over the next several chapters there will be many problems to test your intuition on the first law.

## B. Second Law of Motion

So what happens if the forces on an object are not balanced? If the net force on an object is nonzero, then the velocity vector changes. There must be acceleration. In fact the larger the force, the larger the acceleration. (See Figure 3-5.) On the other hand, if we apply the same push to both a small car and a large car (Figure 3-6), the small car will have the larger acceleration.

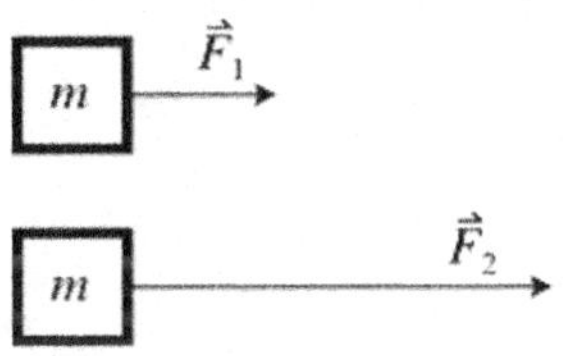


*A large force causes more acceleration than a small force.*

Figure 3-5

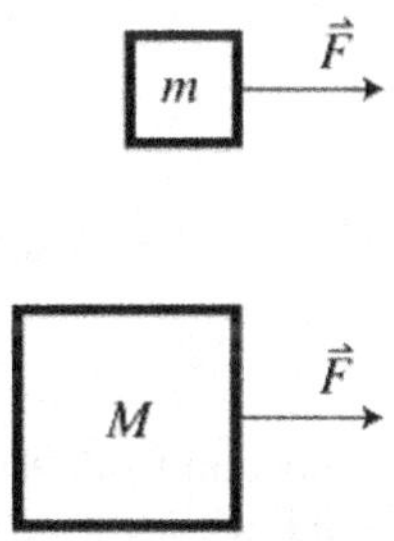


*A force on a small object causes more acceleration than the same force on a large object.*

Figure 3-6

We can write (in one dimension)

$$a = \frac{F_{net}}{m} \quad (1)$$

We have not proven this equation, but the discussion in the previous paragraph should make it seem reasonable to you.

In three dimensions, we write what is often called Newton's second law:

**Second Law of Motion**
If an object has forces $\vec{F}_1, \vec{F}_2, \cdots$, acting on it, then the total force on the object is the vector sum

$$\vec{F}_{net} = \vec{F}_1 + \vec{F}_2 + \cdots,$$

and the acceleration $\vec{a}$ of the object is given by

$$\vec{F}_{net} = m\vec{a}. \quad (2)$$

Most often, however, we will break up equation (2) into components:

$$(F_{net})_x = ma_x \quad (3a)$$
$$(F_{net})_y = ma_y \quad (3b)$$

Equation (3a), for example, states that the sum of all the horizontal forces is mass times the horizontal acceleration. We will discuss breaking vectors into vertical and horizontal components in Section 4.D.

Finally, we are able to make the connection between the units for force [N] and for mass [kg], introduced in Chapter 1. The Newton is defined by

$$N = \frac{kgm}{s^2}$$

**Example 1:** Bruce pushes a car (500 kg) on level ground starting from rest with a force 100 N. How long does it take to get the car rolling 1 m/s? (Assume no friction.)

**Solution:** We have the information $m = 500$ kg, $v_1 = 0$ m/s, $F = 100$ N, and $v_2 = 1$ m/s, and we want $\Delta t$. We can find acceleration from equation (1), so we obtain

$$a = 100\text{N}/500\text{kg}$$
$$= 0.2 \text{ N/kg}$$

Then we can find $\Delta t$ from

$$a = \frac{(v_2 - v_1)}{\Delta t}$$

$$\Delta t = 5 \text{ s}.$$

**Example 2:** A rocket provides a constant force to wagon A that rolls without friction. It starts from rest and after time t attains velocity v. A similar rocket providing the same force is attached to wagon B, which has five times the mass of A (Figure 3-7). How much time does it take wagon B to go from rest to velocity v? (Assume no friction.)

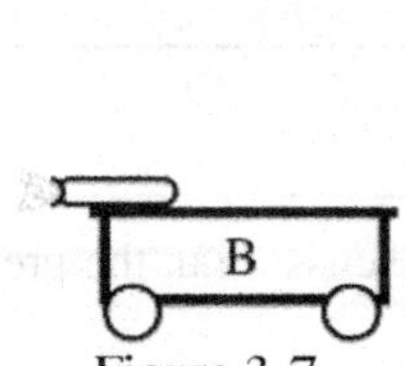


Figure 3-7

**Solution:** This one looks difficult, but if we write down the relevant equations, it is not so hard. We need to connect force and velocity, so we write,

$$F = ma$$

$$a = \frac{v_2 - v_1}{\Delta t} = \frac{v_2}{\Delta t}$$

We set $v_1$ to zero because the wagons start from rest. Substitution gives

$$F = m\frac{v_2}{\Delta t}$$

Since the problem asks about the change in time, we can solve for $\Delta t$ to obtain

$$\Delta t = m\frac{v_2}{F}$$

Now, $F$ and $v_2$ stay the same, but m is five times larger for wagon B, so $\Delta t$ is five times larger. The answer is $5t$.

## C. Third Law of Motion

The third law of motion is not so much a law about motion as it is a rule of thumb about pairs of forces. It is usually stated thus:

To every action there is an equal and opposite reaction.

This is certainly poetic, but what does it mean? More clear (and less poetic) is the following:

**The Third Law of Motion**
If object 1 exerts a force $\vec{F}_{12}$ on object 2,
then object 2 exerts a force $\vec{F}_{21}$ on object 1
which is equal in magnitude, opposite in direction, and of the same type (gravity, friction, etc.):
$\vec{F}_{12} = -\vec{F}_{21}$.

**Example 1:** The Sun and the Earth exert a force of gravity on each other. Draw a force diagram.

**Solution:** See Figure 3-8.

Figure 3-8

$F_{ES}$ = gravitational force of the Earth on the Sun,

$F_{SE}$ = gravitational force of the Sun on the Earth.

**Example 2:** Two spacecraft push off from each other. Draw a force diagram.

**Solution:** See Figure 3-9.

Figure 3-9

$F_{12}$ = contact force of craft 1 on craft 2,

$F_{21}$ = contact force of craft 2 on craft 1.

**Example 3:** A basketball player jumps up. While he is in the air, he pushes the basketball horizontally. Draw all the forces *while* he is pushing the ball. Ignore the tiny gravitational force between the player and the ball.

**Solution:** See Figure 3-10.

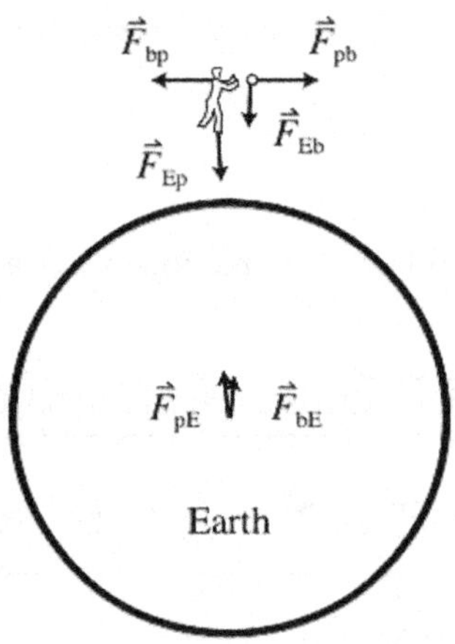


Figure 3-10

$F_{pb}$ = contact force of player on ball,
$F_{bp}$ = contact force of ball on player,
$F_{Ep}$ = gravitational force of Earth on player,
$F_{pE}$ = gravitational force of player on Earth,
$F_{Eb}$ = gravitational force of Earth on ball,
$F_{bE}$ = gravitational force of ball on Earth.

Notice that the magnitude of the force of the player on the ball is the same as that of the ball on the player. But the player moves hardly at all, while the ball springs toward another player. Why is the basketball affected more than the player? (Hint: Look at equation 1.)

## D. Force Diagrams

Already in this chapter we have seen a number of force diagrams. In this section we discuss some rules for drawing force diagrams. There are two types of force diagrams:

1. a diagram in which all the objects appear and the forces come in third-law pairs, and
2. a diagram featuring one object and all its forces (or maybe several objects, but not all the objects in the situation).

In most problems we will want the second type, but it is important to know how to draw both, and knowing diagrams of the first type will help with the second type.

To draw the first type of diagram, we ask four questions:

1. What gravitational forces are important?
2. What things are touching? (These give contact forces.)
3. Does the problem mention any specific forces?
4. Do the net forces in the diagram conform to expectation?

For each force we draw an arrow whose tail lies on the object *on which the force acts*. This may seem unnatural at first, but it makes things easier in the end. For some examples, look at the diagrams we drew in Section C.

**Example 1:** A girl jumps horizontally from a boat in the water. Draw all the forces on the boat, the girl, and the Earth. Ignore the tiny gravitational force between the girl and the boat, and ignore the drag force of the water on the boat.

**Solution:** First we add the gravitational forces in pairs (Figure 3-11).

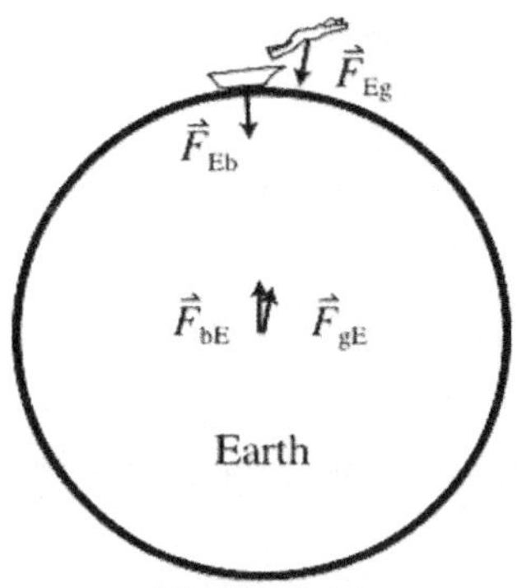


Figure 3-11

Then we add the forces due to the boat and girl touching ($F_{bg}$ and $F_{gb}$) and the contact force between the Earth and the bat ($N_{eb}$ and $N_{bE}$). See Figure 3-12.

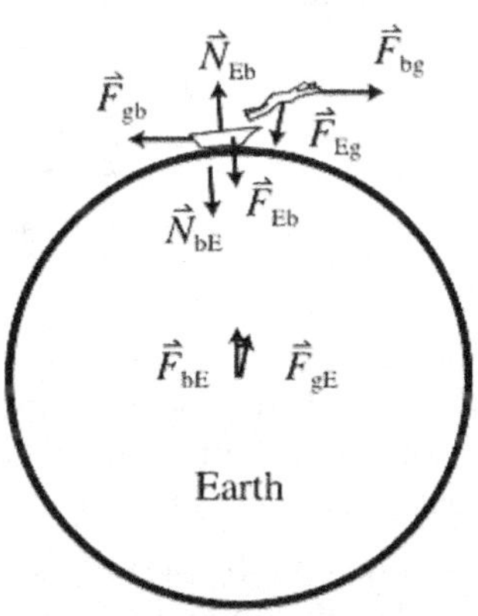


Figure 3-12

The net forces on the boat indicate that it accelerates backwards, which seems right. The net force on the girl indicates she would accelerate forward and down, which seems right. It is difficult to tell what is going on with the Earth.

**Example 2:** A vase sits on a table, which sits on the Earth. List all pairs of forces. Do this example on your own before you look at the solution.

**Solution:** See Figure 3-13.

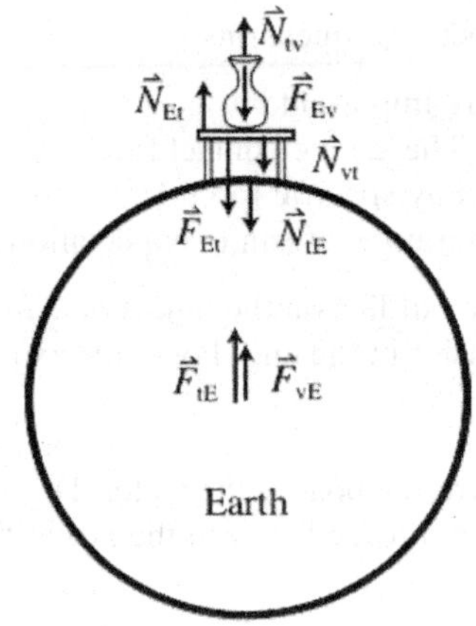


Figure 3-13

$F_{Ev}$ = gravitational force of Earth on the vase,
$F_{vE}$ = gravitational force of vase on the Earth,
$F_{Et}$ = gravitational force of Earth on the table,
$F_{tE}$ = gravitational force of table on the Earth,
$N_{tv}$ = contact force of table on vase,
$N_{vt}$ = contact force of vase on table,
$N_{Et}$ = contact force of Earth on table,
$N_{tE}$ = contact force of table on Earth.

Drawing a diagram of the second type is easier, but you have to be careful not to leave out any forces nor to add any ghost forces.

**Example 3:** A roller skate is rolling frictionlessly on level ground to the left. Draw all the forces on the roller skate. STOP! Try doing this problem before looking at the solution.

**Solution:** Gravity is pulling down (question 1). And the ground is touching the skate, pushing up (question 2). There is no friction, nor any other forces, so the force diagram is Figure 3-14.

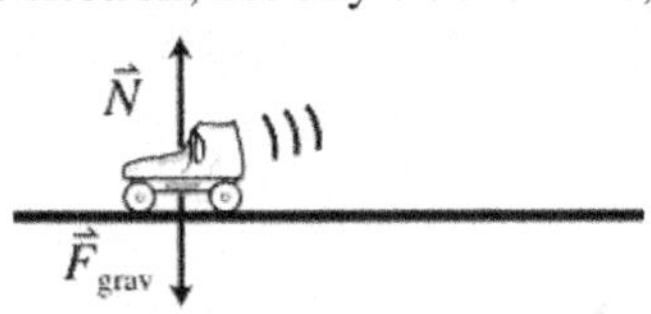


Figure 3-14

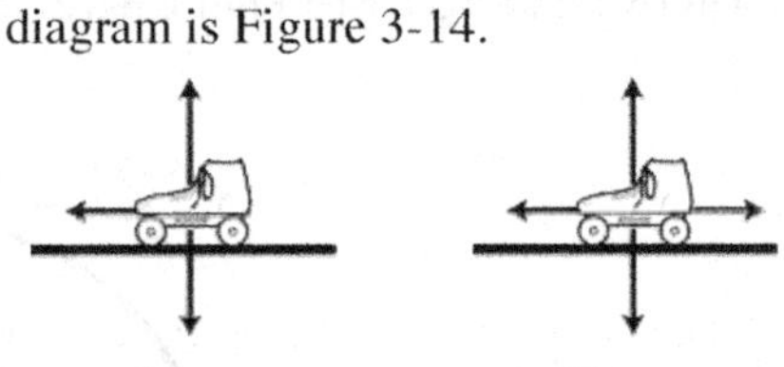


Figure 3-15

Did your diagram look like Figure 3-15 A or B? If so, you have not yet tuned your intuition about the first law of motion. Just because the skate is going to the left does not mean there is a force to the left. Only if the skate were speeding up to the left would we be forced to conclude that there was a force to the left.

In this chapter we studied Newton's laws of motion. In a sense, the first law of motion is the most subtle. If an object is moving at a constant velocity, then the forces on the object add to zero, and if the vector sum of the force vectors for an object is zero, then the object moves at constant velocity. Constant velocity means constant speed in a straight path. No force is required to keep an object moving.

The second law of motion concerns objects whose force vectors' sum is not zero: The acceleration of such an object is in the same direction as the total force, proportional to its magnitude and inversely proportional to the object's mass. That is $a = F_{net}/m$. Do you see why we use equations?

The third law states that forces come in pairs: If object 1 pushes object 2, then object 2 pushes object 1 in the opposite direction.

Pay especial attention to Section D on force diagrams. In solving problems, we are always interested in the forces on an object at a given instant in time. These include gravity, usually, and forces due to things touching the object at that moment. No other forces need to be included. In particular, do not include a force in a direction just because the object in moving in that direction.

# Physics Review 3 Problems

## Section A

1. Consider a paratrooper who has jumped from an airplane. After an initial accelerating plunge, he begins to fall at a constant speed in a straight vertical plunge (at *terminal velocity*). During the latter portion of his fall, are the forces on the paratrooper balanced?
   A. No, since a force balance exists only if an object is not moving.
   B. No, since gravity is not balanced by anything.
   C. No, since gravity is greater than the drag force.
   D. Yes, since he is moving at a constant velocity.

2. There is one force acting on an object. What can we definitely conclude from this?
   A. The object is speeding up or slowing down.
   B. The object is going at a constant speed, not necessarily in a straight line.
   C. The object is going at a constant speed in a straight line.
   D. None of the above may be definitely concluded.

3. A car's engine has died, and the car is slowing down as it coasts. What may we conclude about the forces acting on the car?
   A. There are no forces acting on the car.
   B. There are forces acting on the car, but the net force is zero.
   C. The net force acting on the car is not zero.
   D. None of the above may be definitely concluded.

4. An object is moving with uniform motion, that is, at constant speed in a straight line. There are two forces acting on the object. What can we definitely conclude from this?
   A. The net force is in the same direction that the object is moving.
   B. Both forces are in a direction perpendicular to the object's motion.
   C. The two forces have equal magnitudes but point in opposite directions.
   D. None of the above may be definitely concluded.

## Section B

5. In the following diagram, the magnitude of force $F_A$ is 400 N, and that of $F_B$ is 300 N. What is the magnitude of the net force in the three cases?

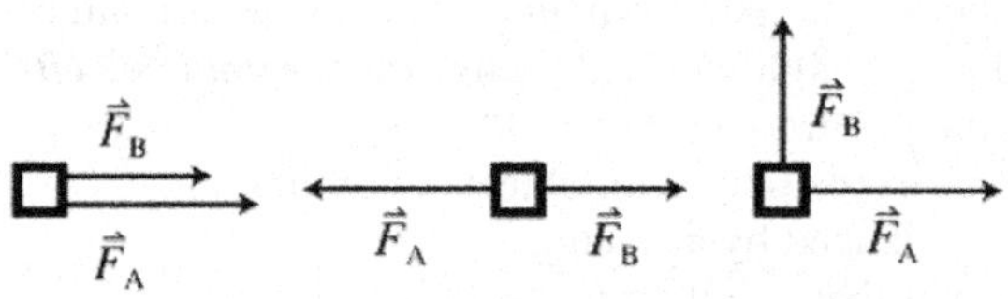


A. In case 1, the net force is 100 N; case 2, 700 N; and case 3, 500 N.
B. In case 1, the net force is 700 N; case 2, 100 N; and case 3, 500 N.
C. In case 1, the net force is 350 N; case 2, 50 N; and case 3, 450 N.
D. In case 1, the net force is 350 N; case 2, 50 N; and case 3, 550 N.

*Use the following information for questions 6 and 7:*

A man is pulling his son in a toy wagon. The son and the wagon are 60 kg. For 3 s the man exerts a force which has the effect of uniformly accelerating the wagon from 1.5 m/s to 3.5 m/s.

6. What is the acceleration of the wagon with the son?
A. 0.67 m/s$^2$
B. 0.84 m/s$^2$
C. 1.66 m/s$^2$
D. 20 m/s$^2$

7. What is the net force on the wagon and son?
A. 40 N
B. 50 N
C. 120 N
D. 1200 N

*Use the following information for questions 8–11:*

A tiger (100 kg) sees a wildebeest and accelerates uniformly from rest to 20 m/s in 12 s.

8. What is the acceleration of the tiger?
A. 0.60 m/s$^2$
B. 0.83 m/s$^2$
C. 1.67 m/s$^2$
D. 240 m/s$^2$

9. How much distance does the tiger cover in those 12 s?
A. 120 m
B. 240 m
C. 480 m
D. 960 m

10. What is the magnitude of the net force on the tiger?
A. 60 N
B. 83 N
C. 167 N
D. 24000 N

11. What are all the forces acting on the tiger?
    A. Gravity, down.
    B. Gravity, down; and the normal force, up.
    C. Gravity, down; the normal force, up, and a horizontal force of the ground pushing the tiger.
    D. None of the above is correct.

12. Three men push on a station wagon with a net force $F$, producing an acceleration. (Assume there is no friction.) If they push with the same net force on a compact car (with half the mass), the acceleration of the compact car is
    A. four times the acceleration of the station wagon.
    B. twice the acceleration of the station wagon.
    C. half the acceleration of the station wagon.
    D. one quarter the acceleration of the station wagon.

13. Object A is acted upon by a net force $F_A$ to produce an acceleration. If object B has three times the mass of A and is acted on by three times the force as A, the acceleration of B is
    A. 9 times that of A.
    B. 3 times that of A.
    C. the same as that of A.
    D. one third that of A.

14. At time $t = 0$ s, two dung beetles are pushing a small ball (0.5 g). One pushes east with a force 0.0015 N; the other pushes west with a force 0.0010 N. At $t = 0$ s, what is the acceleration of the ball?
    A. 1 m/s$^2$
    B. 2.5 m/s$^2$
    C. 3 m/s$^2$
    D. 5 m/s$^2$

15. A rocket ship (500 kg) is firing two jets at once. The two jets are at right angles, with one firing to yield a force of 5000 N and the other to yield a force of 12000 N. What is the magnitude of the acceleration of the rocket ship?
    A. 24 m/s$^2$
    B. 26 m/s$^2$
    C. 34 m/s$^2$
    D. 36 m/s$^2$

16. A girl shoves a 4-kg toy cart across the level floor with a speed of 15 m/s (so it is going 15 m/s when it leaves her hand). It slides to a rest in 5 s. Assuming a constant force slowing the cart, what is the magnitude of the force?
    A. 0.75 N
    B. 1.33 N
    C. 12 N
    D. 18.75 N

17. A piece of steel of mass 0.8 kg hangs from a string over the edge of a table. The string passes over a pulley and is connected in such a way as to maintain a tension force of 6 N (see figure).

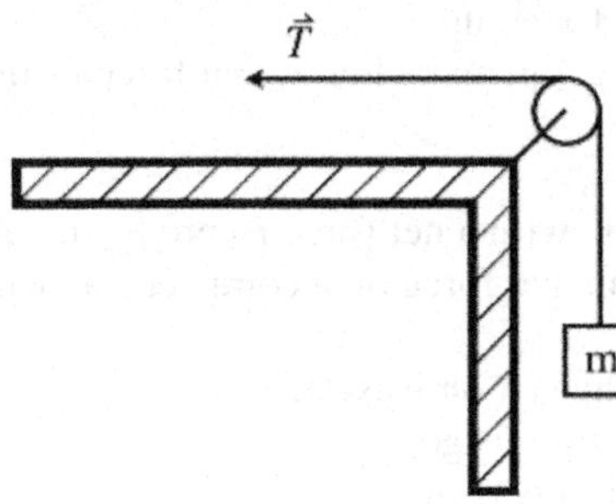


The force due to gravity on the steel is given by $F_{grav} = mg$, where $g = 10$ m/s$^2$ is the acceleration due to gravity. We allow the piece of steel to fall from rest for 5 s. What is the final acceleration of the piece of steel?
A. 2 m/s$^2$
B. 2.5 m/s$^2$
C. 10 m/s$^2$
D. 12.5 m/s$^2$

18. A woman is riding in an elevator which is going up at constant speed. The force of the floor against her feet
A. is less than the force of gravity on her.
B. is the same as the force of gravity on her.
C. is greater than the force of gravity on her.
D. has no relationship with the force of gravity on her that can be determined by the given information.

19. On September 12, 1966, astronauts conducted an experiment using the second law of motion. A Gemini spacecraft (measured to be 3400 kg) connected with an orbiting rocket case. The thrusters were fired to provide a force of 890 N for 7.0 s. The change in velocity of the spacecraft and rocket case was found to be 0.93 m/s. What was the mass of the rocket case?
A. 120 kg
B. 3300 kg
C. 10,100 kg
D. 15,400 kg

20. Car A starts from rest and accelerates uniformly for a period of time $t$ to travel a distance $d$. Car B, which has four times the mass of car A, starts from rest and accelerates uniformly as well. If the magnitudes of the forces accelerating A and B are the same, how long does is take B to travel the same distance $d$?
A. $16t$
B. $4t$
C. $2t$
D. $t/4$

21. A car trailer is connected to a car, and both are traveling forward at velocity 3 m/s at time $t = 5$ s. The force the car exerts on the trailer is 105 N, the force of friction on the trailer is 30 N, and the force of air resistance on the trailer is 70 N. The force of gravity on the trailer is 8000 N, and the road exerts an upward force of 8000 N. What conclusion may be drawn about the trailer?
A. It is speeding up.
B. It is staying the same speed.
C. It is slowing down.
D. It is speeding up or staying the same speed.

## Section C

22. A car is accelerating from rest at an intersection after the light has turned green. What is the force which accelerates the car?
    A. The horizontal force of the wheels on the road.
    B. The vertical force of the wheels on the road.
    C. The horizontal force of the road on the wheels.
    D. The normal force of the road on the car.

*Use the following information for questions 23–26:*

An antique stove is sitting on the ground. (See figure) Assume for this problem that the Earth is not rotating.

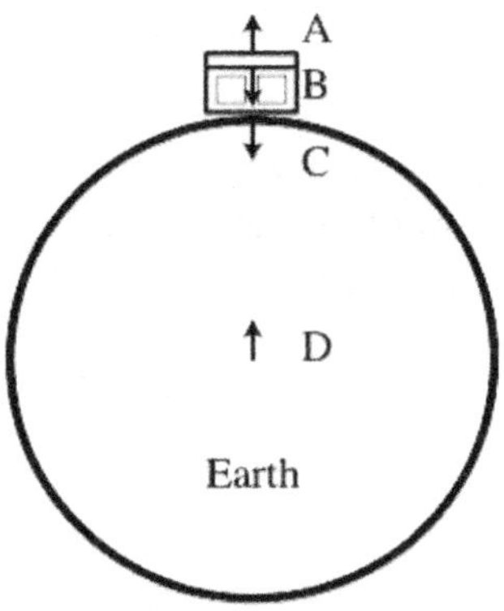


23. Which arrow represents the gravitational force of the Earth on the stove?
    A. A
    B. B
    C. C
    D. D

24. Which arrow represents the force paired with the gravitational force of the Earth on the stove, according to the third law of motion?
    A. A
    B. B
    C. C
    D. D

25. Why is the force vector A equal in magnitude to the force vector B?
    A. The first law of motion states that an object which is motionless has balanced forces.
    B. The first law of motion states that an object in motion will remain in motion only if acted upon by an *unbalanced* force.
    C. The second law of motion states that force and acceleration are proportional.
    D. The third law of motion states that forces come in equal and opposite pairs.

26. Why is the force vector A equal in magnitude to the force vector C?
    A. The first law of motion states that objects which are motionless have balanced forces.
    B. The first law of motion states that an object in motion will remain in motion unless acted upon by an unbalanced force.
    C. The second law of motion states that a force on an object and acceleration of the object are proportional.
    D. The third law of motion states that forces come in equal and opposite pairs.

27. A car's engine has died, and the car is slowing down as it coasts. What forces are acting on the car?
   A. Gravity, down.
   B. Gravity, down; and the road's force, up.
   C. Gravity, down; the road's force, up; and friction, backwards.
   D. Gravity, down; the road's force, up; friction, backwards; and a forward force.

28. An arrow is shot into the air. When the arrow is in the air, what forces are acting on the arrow?
   A. There are no forces.
   B. There is the force of gravity.
   C. There is the force of gravity and an upward normal force.
   D. There is the force of gravity and a forward force.

29. The planet Mars is traveling around the Sun. What forces are acting on Mars?
   A. There are no forces.
   B. There is the force of gravity.
   C. There is the force of gravity and a forward force.
   D. There is the force of gravity, a forward force, and an outward force.

# Solutions

1. D.
From the first law of motion, a force balance on an object implies it has constant velocity. From this we conclude that the force of gravity and the drag force due to the air are *exactly balanced.*

2. D.
Since there is only one force, there must be a net force on the object; there is no way for forces to be balanced. From the first law of motion, we conclude that the object is not undergoing uniform motion, so it is speeding up, slowing down, or changing direction. But none of the choices can be definitely concluded.

3. C.
Since the car is not undergoing uniform motion (it is slowing), it has a net force on it. In fact, the net force points in the opposite direction the car is going.

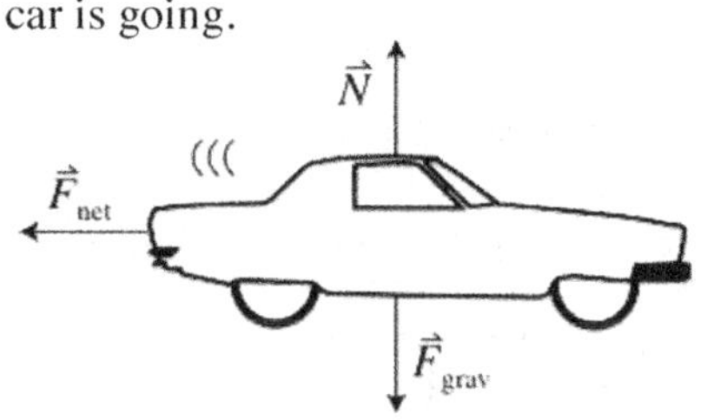


4. C.
Since the object is moving with uniform motion, the forces on the object must be balanced. Since there are exactly two forces, the only way for them to be balanced is that they be of equal magnitude in opposite directions, which is choice C. If you chose B, think of the example of the paratrooper falling in question 1, in which the downward force of gravity is balanced by the upward force of the air drag.

5. B.
In case 1, we move the tail of $F_B$ to the tip of $F_A$ (see figure). The sum, $F_{net}$, is drawn from the first tail to the last tip, giving a magnitude of 700 N. So choice B is correct. The same method for case 2 yields a magnitude 100 N. For case 3, we need to apply the Pythagorean theorem to obtain a magnitude 500 N.

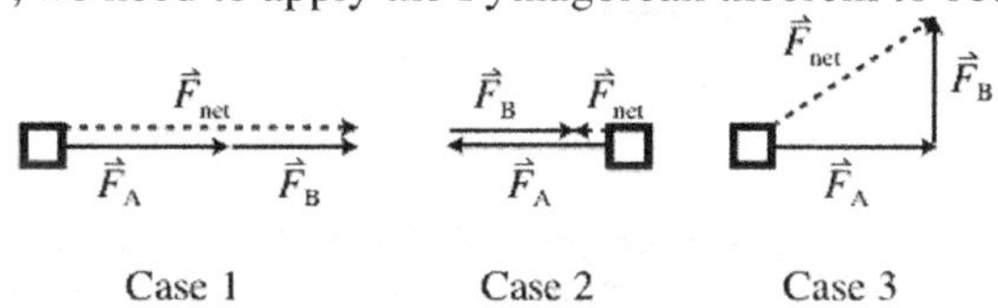


6. A.
(This was a review problem.) We calculate $a = (3.5\text{ m/s} - 1.5\text{ m/s})/3\text{ s} = 0.67\text{ m/s}^2$.

7. A.
We calculate $F_{net} = ma = (60\text{ kg})(0.67\text{ m/s}^2) = 40\text{ N}$.

8. C.
We calculate $a = (20\text{ m/s} - 0\text{ m/s})/(12\text{ s}) = 1.67\text{ m/s}^2$.

9. A.
We calculate (from Chapter 2) $\Delta x = 1/2\ (0\text{ m/s} + 20\text{ m/s})\ (12\text{ s}) = 120\text{ m}$.

10. C.
The magnitude of the net force is given by $F_{net} = ma = 167\text{ N}$.

11. C.
The vertical forces, which are balanced, are gravity (down) and the normal force of the ground (up). The one horizontal force accelerates the tiger, and it is due to the ground pushing forward. This may seem

strange, but it is a result of the third law of motion. The tiger pushes the ground backward. There is then an equal force of the ground on the tiger pushing forward, even though there is no "active" agent creating the force. But it has to be the ground pushing the tiger forward, since that is the only thing touching him. Certainly he does not push himself forward (what would that even mean?).

12. B.
We want an equation which connects force, acceleration, and mass. This is $F = ma$. Since we are looking for the change in acceleration, we can write this equation $a = F/m$. If the mass decreases by a factor of 2, then the acceleration increases by a factor of 2.

13. C.
As in question 12, we write $a = F/m$. If $F$ increases by a factor of 3 and $m$ increases by a factor of 3, then $a$ remains the same.

14. A.
We draw a diagram (see figure). We cannot calculate the acceleration using the methods of the previous chapter, but we can find a net force. In one dimension, we can call east positive, so $F_{net} = 0.0015$ N – 0.0010 N = 0.0005 N. Thus we can find acceleration $a = F_{net}/m = 0.0005$ N/0.0005 kg = 1 m/s$^2$.

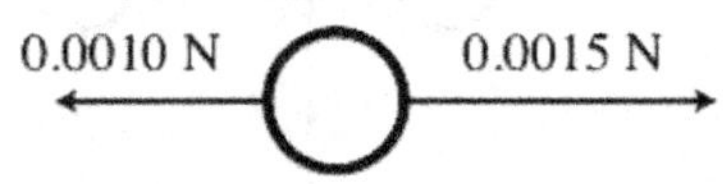


15. B.
We draw a diagram (see figure). We can obtain the net force by applying the Pythagorean theorem. The net force is 13000 N. From this we obtain the magnitude of the acceleration $a = F_{net}/m = 26$ m/s$^2$.

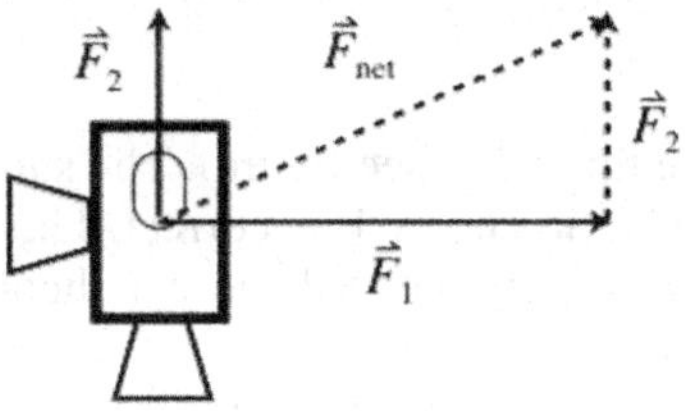


16. C.
We draw a diagram (see figure). We include the vertical forces of gravity and the normal force, which add to zero (that is, balance each other). There is nothing touching the truck except the ground (the girl has already let go), but there must be another force because the truck is changing velocity. This force is the drag force, pointing backward. (There is no forward force.) We want to know the magnitude of the drag force, since this is also the net force. We obtain the acceleration $a = (0$ m/s – 15 m/s)/5 s = –3 m/s$^2$. Thus, in magnitude, the net force is $F_{net} = ma = 12$ N.

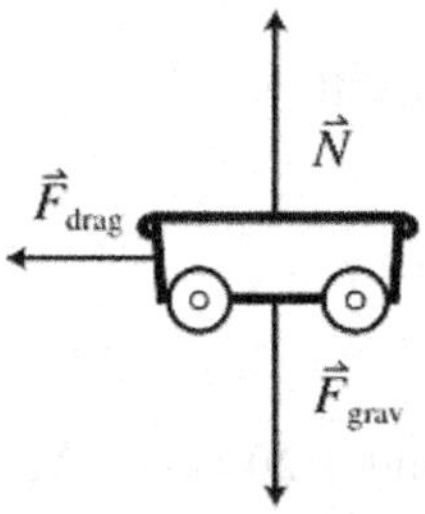


17. B.
We draw a diagram showing the forces on the mass (see figure). There is the force of gravity and the tension in the string, which pulls up. The net force (with down as positive) is $F_{net} = (0.8$ kg) (10 m/s$^2$) – 6 N = 2 N. The sign indicates the net force is down. Now we can calculate the acceleration $a = F_{net}/m = 2$ N/0.8 kg = 2.5 m/s$^2$.

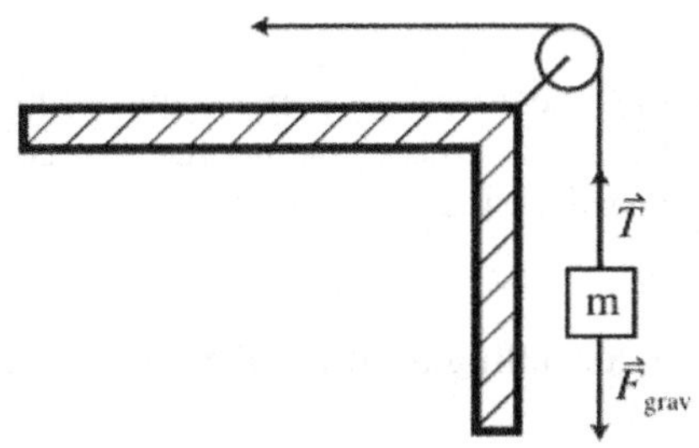


18. B.
We draw a diagram showing the forces on the woman. Since she is traveling at constant velocity, the net force on her is zero, and the forces are balanced. Thus the force of the floor against the woman's feet has the same magnitude as the force of gravity on her. This is just the first law of motion. If you chose C, then you need to study the first law of motion again.

19. B.
We can calculate an estimate of acceleration $a = \Delta v/\Delta t$ = (1 m/s)/(7 s) = 1/7 m/s$^2$. Thus the total mass is given by $m = F_{net}/a$ = 900/(1/7) kg = 6300 kg. The mass of the rocket case is then (6300 – 3300) kg = 3000 kg. This is close to B.

20. C.
We need an equation which connects time and mass (which differs from A to B). We also have information about $F$, $v_1$, and $d$. We can use $F = ma$ and $d = v_1\Delta t + 1/2\ a(\Delta t)^2$. Combining these, we obtain

$$d = v_1 + \frac{1}{2}a\Delta t^2$$

$$= \frac{1}{2}\left(\frac{F}{m}\right)\Delta t^2$$

$$\Delta t = \sqrt{\frac{2dm}{F}}$$

So if $m$ increases by a factor of 4, then time increases by a factor of

$$\sqrt{4} = 2$$

21. A.
First we draw a diagram (see figure). The vertical forces balance. The net horizontal force is $F_{net}$ = 105 N – 30 N – 70 N = 5 N. The net force is forward, so the car is speeding up.

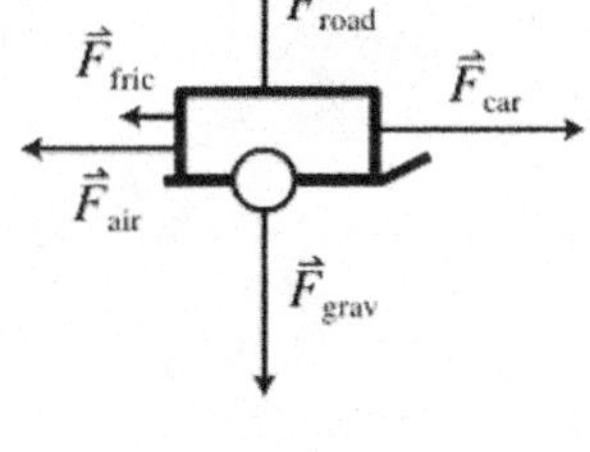

22. C.
The force which accelerates the car must be a force of some agent acting on the car, so that fact narrows the choices to C and D. The acceleration is horizontal, so the net force must be horizontal, therefore C is correct. The car exerts a force on the road, so that (third law) the road exerts a force on the car.

23. B.
The gravitational force acts on the stove, pulling it toward the Earth. The answer is B.

24. D.
The paired force is the gravitational force of the stove on the Earth, that is, arrow D.

25. A.
Forces A and B represent the normal force and gravitational force on the stove. If these were not balanced, the stove would not undergo uniform motion (first law), so A is correct. Choice B is not a true statement. Choice C is not even a correct statement of the second law of motion, which states that the acceleration is proportional to *net* force. Choice D concerns two forces which act on different objects.

26. D.
Forces A and C are paired. Force A is the upward contact force of ground on the table, and force C is the downward contact force of the table on the ground.

27. C.
There is certainly gravity, and there is an upward force from the road to balance it. There must be another force, because the car has an acceleration vector pointing backwards, hence the frictional force. The correct choice is C. There is no force pushing the car forward. It continues going forward for a while because that is what cars do, by the first law of motion. The force diagram is shown in the solution for problem 3.

28. B.
Gravity acts, of course. What else is touching the arrow? Nothing, so B is correct. (See figure.)

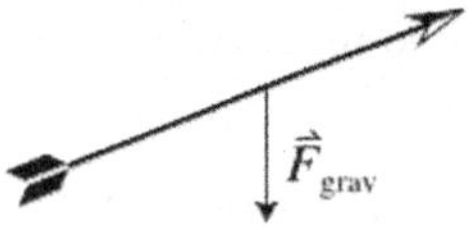


29. B.
The Sun's gravity acts on Mars. What else touches the planet? Nothing, so B is correct. The Sun's gravity changes the course of Mars, which would be a straight line away from the solar system, into the circular orbit. (See figure.)

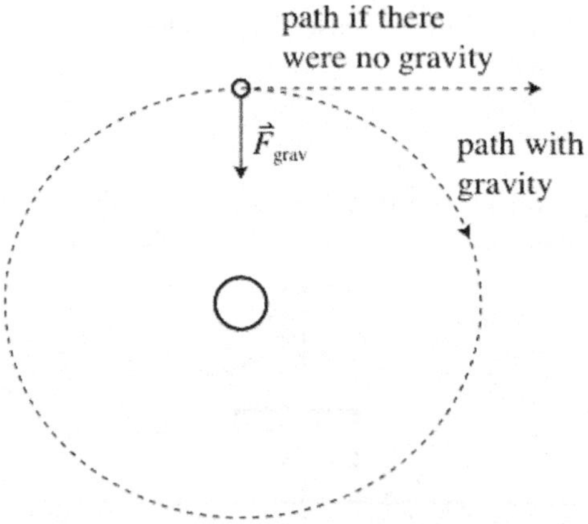

# Physics Review 4
# Gravitation

## A. The Law of Gravitation

When a person does a great deal of work in a scientific field, it often happens that that person eventually receives credit for almost everything done by anybody (see Matthew 13:12 in the Christian Bible). For instance, Newton is given credit, at least in popular accounts, for almost every interesting thing that happened in science during the Renaissance. In fact, then as now, science is the activity of a community, with many people contributing to the revolution in thinking. For example, the essentials of Newton's first law of motion were discovered by Galileo, and Robert Hooke surmised the essential parts of the law of gravitation.

Newton's genius lay in his ability to see a simple underlying law for very different phenomena and to synthesize diverse branches of science. An example of this is his realization that both the motion of the Moon and the motion of a falling apple could be explained by the same force, the force of gravity. In this chapter we will study the physics of the gravitational force.

Newton's law of gravity states that any two objects exert an attractive force on each other given by

$$F_{grav} = \frac{Gm_1m_2}{d^2} \qquad (1)$$

Here $F_{grav}$ is the magnitude of the gravitational force between two objects, $m_1$ and $m_2$ are the masses of the objects, $d$ is the distance between the centers of the objects, and $G$ is a universal constant $6.67 \times 10^{-11}$ m$^3$/kg s$^2$. Do not memorize $G$, but do remember the equation. We discussed it in Chapter 1.

**Example 1:** What is the force on a cow (200 kg) standing on the surface of the Earth? (Assume $M_{Earth} = 6.0 \times 10^{24}$ kg, $R_{Earth} = 6.4 \times 10^6$ m.)

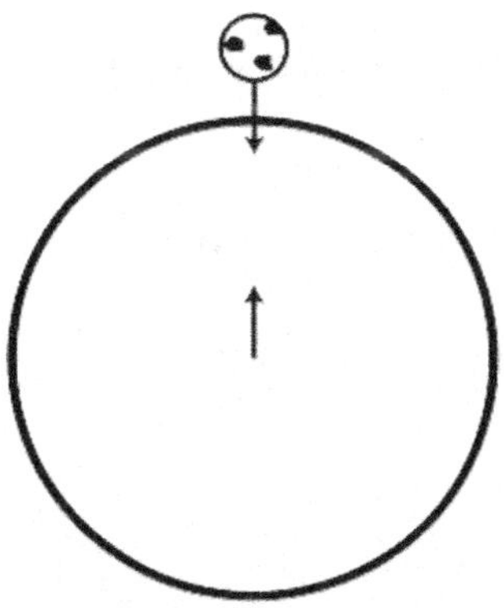

Figure 4-1

**Solution:** Let's assume we have a spherical cow (Figure 4-1). We calculate

$$F_{\text{grav}} = \frac{GM_{\text{Earth}} m_{\text{cow}}}{R_{\text{Earth}}^2}$$

$$= \frac{\left(6.67\times10^{-11}\frac{\text{m}^3}{\text{kgs}^2}\right)\left(6.0\times10^{24}\text{kg}\right)}{\left(6.4\times10^6\text{m}\right)^2} m_{\text{cow}}$$

$$= m_{\text{cow}}\left(9.8\frac{\text{m}}{\text{s}^2}\right)$$

$$= 1960 \text{ N}$$

**Example 2:** What is the motion of an apple (0.1 kg) which has let go of its tree?

**Solution:** First, we DRAW A DIAGRAM (Figure 4-2) showing all the forces on the apple while it is falling. There is only the force of gravity (nothing else is touching it), so we write

$$F_{\text{grav}} = \frac{GM_{\text{Earth}} m_{\text{apple}}}{R_{\text{Earth}}^2}$$

$$= \frac{\left(6.67\times10^{-11}\frac{\text{m}^3}{\text{kgs}^2}\right)\left(6.0\times10^{24}\text{kg}\right)\left(0.1\text{kg}\right)}{\left(6.4\times10^6\text{m}\right)^2}$$

$$= 0.98 \text{ N}$$

Figure 4-2

Since the total force is simply the gravitational force, we write

$$m_{\text{apple}} a = F_{net}$$

$$a = \frac{0.98\frac{\text{kgm}}{\text{s}^2}}{0.1\text{kg}} = 9.8\frac{\text{m}}{\text{s}^2}$$

So the apple accelerates downward at 9.8 m/s$^2$.

Equation (1) is easy to use in two types of problems:

1. obtaining the force between two planets (*d* is much larger than the radii of the planets), and
2. obtaining the force between a planet and a small object on its surface (*d* is essentially the radius of the planet).

The previous examples illustrated this second use.

## B. Surface of the Earth

Most of us will spend most of our lives on or near the surface of the Earth. What is the force of gravity of the Earth on an object of mass *m*? The force is

$$F_{\text{grav}} = \frac{GM_{\text{Earth}}m}{R_{\text{Earth}}^2} = \left(9.8\frac{\text{m}}{\text{s}^2}\right)m$$

where we use a calculation from the previous section. This number 9.8 m/s$^2$ comes up so often in introductory physics that we have given it a name: The *acceleration due to gravity* is

$$g = 9.8\frac{m}{s^2} \qquad (2)$$

In working problems, however, we *always* approximate this as $g$ = 10 m/s$^2$ (even if the problem says to use 9.8 m/s$^2$).

Please note: *Whenever* there is gravity and we are at the surface of the Earth, which is in most problems, we will use

$$F_{grav} = mg \qquad (3)$$

## C. Free Fall

An object is said to be in *free fall* when nothing is touching it, so that the only force on it is gravity. Such an object is called a *projectile*. The simplest problem in free fall involves dropping objects near the surface of the Earth. We want to know which falls faster, a heavy object or a light one? Things become complicated if the object is too light, like a leaf fluttering to the ground, so at first we will consider two objects for which air resistance is only a small consideration. Let us start by doing a pair of examples.

**Example 1:** How long does it take a small rock (0.02 kg) to fall from rest 2 meters to the ground?

**Solution:** First, we DRAW A DIAGRAM (Figure 4-3). There is only one force since nothing touches the rock and we are neglecting air resistance. Second, we find the net force

$$\begin{aligned} F_{\text{net}} &= F_{\text{grav}} = -mg \\ &= -(0.02 \text{ kg})(10 \text{ m/s}^2) \\ &= -0.2 \text{ N} \end{aligned}$$

where the negative sign reminds us that gravity points down.

Figure 4-3

Third, we obtain acceleration by writing

$$F_{\text{net}} = ma$$
$$a = F_{\text{net}}/m = -0.2 \text{ N}/0.02 \text{ kg} = -10 \text{ m/s}^2$$

(This result should look familiar.)

We also have

$$v_1 = 0$$
$$\Delta y = -2 \text{ m}$$

so that

$$\Delta y = v_1 \Delta t + 1/2\, a_y \Delta t^2$$
$$-2 \text{ m} = 1/2\,(-10 \text{m/s}^2)\Delta t^2$$
$$\Delta t = 0.63 \text{ s}$$

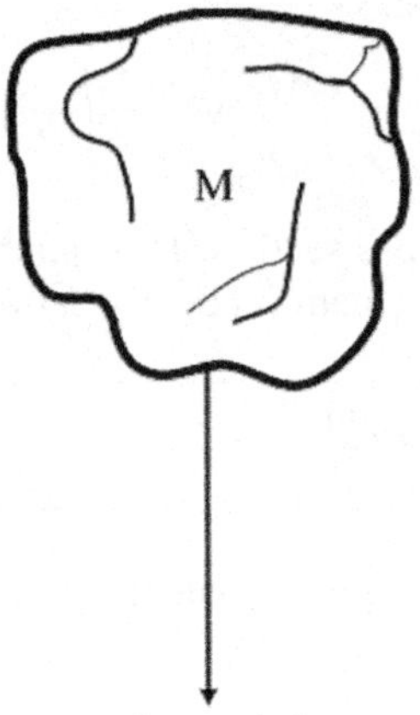


Figure 4-4

**Example 2:** How long does it take a medium-sized rock (0.2 kg) to fall 2 meters from rest?

**Solution:** The rock is larger, and so is the force of gravity (and the force arrow, see Figure 4-4). We write

$$F_{net} = F_{grav} = -Mg = -(0.2 \text{ kg})(10 \text{ m/s}^2) = -2 \text{ N}$$

The acceleration is

$$a = F_{net}/M = -2 \text{ N}/0.2 \text{ kg} = -10 \text{ m/s}^2$$

The rest of the problem is the same, so $\Delta t = 0.63$ s.

WHAT HAPPENED? The force of gravity was larger in Example 2, BUT the acceleration is inversely proportional to mass in the second law of motion. The net effect is that the acceleration is the same 9.8 m/s$^2$ for both rocks. Try this at home with a pen and a stapler or some such thing. It is difficult to gain an intuitive grasp of this situation, but think about it until you also understand *why* the two rocks have the same acceleration.

Let us revisit Example 2 in Section 2.H. There we saw an apple tossed straight up. It rose, came to a stop, and fell. We claimed that the acceleration was always negative (that is, down), even at the top point. We are now at a better position to understand why. Once the apple leaves the hand, there is only one force on the apple, the downward force of gravity. (See Figure 4-5.) The acceleration must be down as well. In fact, we now know the acceleration is a constant 9.8 m/s$^2$, down.

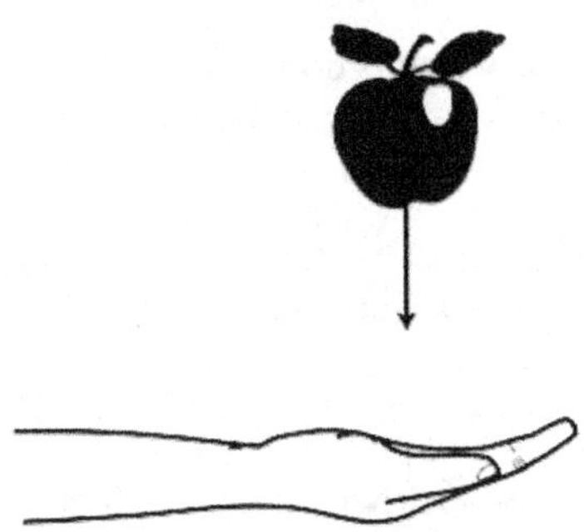

*A tossed apple at the top of its flight is accelerating down at* $9.8\frac{\text{m}}{\text{s}^2}$.

Figure 4-5

## D. Horizontal and Vertical Motion

This section has no new equations but it does present one new idea, so prepare your imagination. A pencil and paper may prove handy.

Imagine you are sitting at the shore of a bay, and you see a boat traveling along at constant speed in a straight line. A sailor at the top of a vertical mast drops a grapefruit. (See Figure 4-6.) We will pretend air resistance plays no role (mostly true). Where will the grapefruit land?

A. In front of the mast.
B. At the foot of the mast.
C. Behind the mast.

Choose your answer before you read any further.

*A boat moves with uniform motion.*
*A grapefruit falls from the mast.*
*Where does it land?*
Figure 4-6

Few people choose A. Not many people choose B. If you are like most people, you chose C, thinking that somehow the boat moves out from under the grapefruit. If that was your answer, then you need to do some rethinking. Here is what really happens. Figure 4-7 shows two ships at four successive times, one ship at rest and the other in uniform motion. Sailors at the tops of the masts drop grapefruits at the same time. Both grapefruits drop to the foot of the corresponding mast.

*The grapefruit on the moving ship retains its horizontal motion regardless of vertical motion, and it drops vertically regardless of its horizontal motion.*
Figure 4-7

What is going on in the previous example? Just after the grapefruit is released from the hand on the second ship, it still has its horizontal motion. If air resistance does not affect it, then it maintains its same horizontal motion from start to finish. The vertical motion, on the other hand, proceeds on schedule regardless of the horizontal motion. At $t = 0.2$ s, both grapefruits have moved *vertically* 0.05 m (the second has moved horizontally as well). As time goes on, the second grapefruit keeps up with the ship, and both grapefruits hit the deck at the same time.

If you do not believe the figure, try the experiment yourself. While walking at a constant speed, release an apple above your head (and a little forward). It will fall in front of your face and land at your feet. (See Figure 4-8.)

Figure 4-8

The picture of a boat in the bay and the resulting principle are due to Galileo:

> Horizontal and vertical motion are independent. That is, motion in the $x$- and $y$-directions can be considered independently.

Now let's leave the bay and travel to the edge of a cliff with a large plain at the bottom. We have two cannonballs, and at the same time, we shoot one horizontally off the cliff and drop the other. Which will hit the plain first?

Again, most people will say the dropped ball hits first. As long as air resistance plays at most a small role, they will hit at the same time. (See Figure 4-9.) For the second ball, the vertical motion of falling is not affected at all by its horizontal motion. It may help your intuition to realize that the shot cannonball does have a larger total velocity all the way down. Notice that Figure 4-9 looks just like Figure 4-7 with the boats removed. It is the same physical principle.

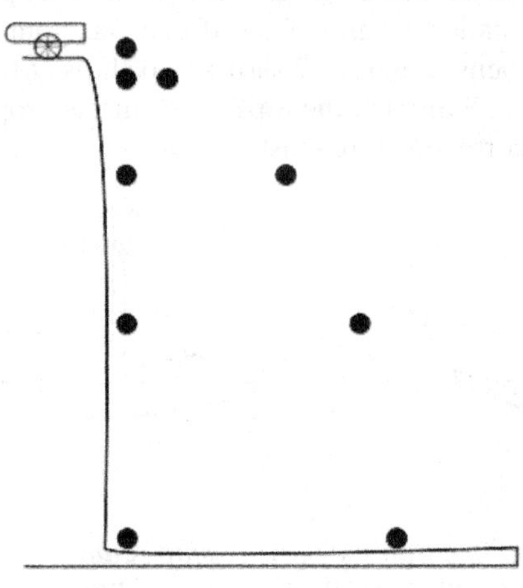

*A dropped cannonball falls at the same rate as one shot horizontally.*

Figure 4-9

Up until now we have been using $v_1$, $v_2$, and $a$ to denote velocities and acceleration in one dimension. For these problems we will need to keep track of the vertical and horizontal pieces separately, so we need the symbols $v_{1x}$, $v_{2x}$, $a_x$, $v_{1y}$, $v_{2y}$, and $a_y$.

The following box shows how this principle gets translated into equations:

If the net vertical force is $(F_{net})_y$, then we can determine the vertical motion using

$$(F_{net})_y = ma_y,$$

$$\Delta y = \frac{1}{2}(v_{1y} + v_{2y})\Delta t,$$

$$v_{2y} = v_{1y} + a_y \Delta t,$$

$$\Delta y = v_{1y}\Delta t + \frac{1}{2}a_y \Delta t^2,$$

$$v_{2y}^2 = v_{1y}^2 + 2a_y \Delta y.$$

Similarly, we can determine the horizontal motion using similar equations with $y$ replaced by $x$.

Objects in free fall experience only the force of gravity, so we can say more:

For an object in free fall at the surface of the Earth, we have

$$a_y = -9.8\frac{\text{m}}{\text{s}^2}, \quad (4a)$$

$$a_x = 0\frac{\text{m}}{\text{s}^2}, \quad (4b)$$

where "up" is positive and we use *the estimate* $10\frac{\text{m}}{\text{s}^2}$.

**Example 1:** A cliff stands 80 m above a flat plane. One cannonball is dropped, and another is fired horizontally at 120 m/s at the same time. How far from the first ball will the second ball land?

**Solution:** The first ball falls straight down, of course. Let's DRAW A DIAGRAM for the second ball while it is in flight (Figure 4-10). Note that the cannon exerts a force on the cannonball while it is in the cannon, but after the ball leaves the cannon, *the only force is gravity*.

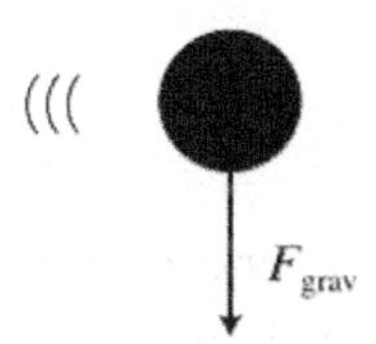


Figure 4-10

We record the information we have. We do not know the mass of the ball, but from Section B we know we do not need it. The acceleration vector is 10 m/s$^2$, down.

| Vertical | horizontal |
|---|---|
| $v_{1y} = 0$ | $v_{1x} = 120$ m/s |
| $\Delta y = -80$ m | $\Delta x = ?$ |
| $a_y = -10$ m/s$^2$ | $a_x = 0$ m/s$^2$ |
| $\Delta y = v_{1y}\Delta t + 1/2\, a_y \Delta t^2$ | |
| $-80 \text{ m} = 1/2\,(-10 \text{ m/s}^2)\Delta t^2$ | |
| $\Delta t = 4$ s | $\Delta t = 4$ s |
| | $\Delta x = v_{1x}\Delta t + 1/2\, a_x \Delta t^2$ |
| | $\Delta x = (120 \text{ m/s})(4 \text{ s})$ |
| | $\Delta x = 480$ m |

We solved the vertical problem first because we had more vertical information than horizontal information. The time $\Delta t = 4$ s was the connection between the horizontal and vertical parts. You should work through this example yourself without looking at the book.

The following example involves a *projectile*.

**Example 2:** A cannon is fired on level ground, so that the ball's initial velocity is 300 m/s and directed 30° up from the horizontal. How far from the cannon will the ball fall? (See figure 4-11.)

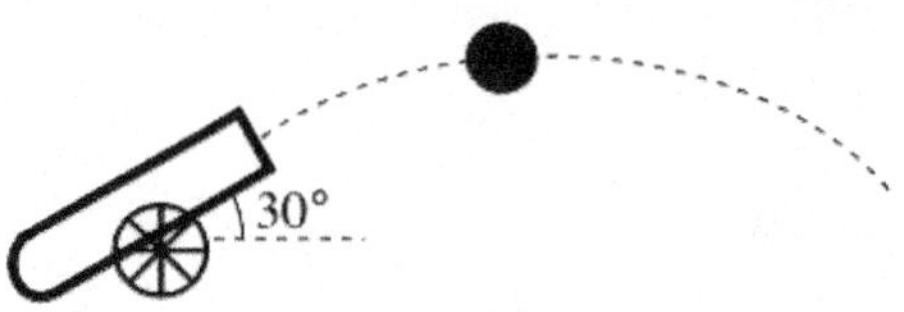


Figure 4-11

**Interruption:** We need to know the horizontal and vertical components of the initial velocity $v_1$. We need to find a horizontal vector $v_{1x}$ and a vertical vector $v_{1y}$ so that their sum is the original vector $v_1$ (see Figure 4-12). We can find the magnitudes of $v_{1x}$ and $v_{1y}$ using simple trigonometry. (You may need to review trigonometry at this point.)

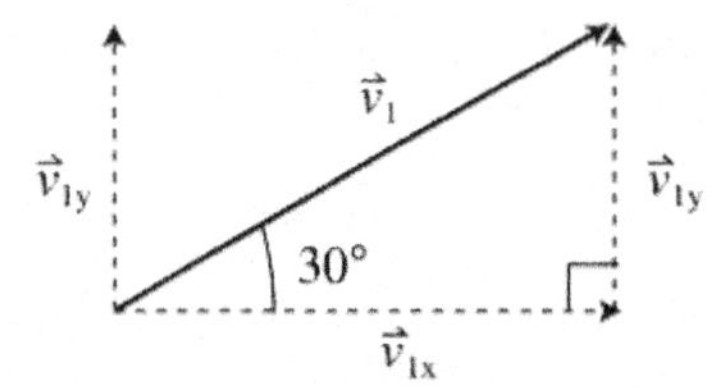


Figure 4-12

$$\frac{v_{1y}}{v_1} = \sin 30^\circ \qquad \frac{v_{1x}}{v_1} = \cos 30^\circ$$

$$v_{1y} = v_1 \sin 30^\circ \qquad v_{1x} = v_1 \cos 30^\circ$$

$$v_{1y} = 300\left(\frac{1}{2}\right)\frac{\text{m}}{\text{s}} \qquad v_{1x} = 300\left(\frac{\sqrt{3}}{2}\right)\frac{\text{m}}{\text{s}}$$

$$v_{1y} = 150\frac{\text{m}}{\text{s}} \qquad v_{1x} = 150\sqrt{3}\frac{\text{m}}{\text{s}}$$

**Solution**: The force diagram is the same as in Figure 4-10. The cannonball rises and then falls to the same height from which it started, so we have $\Delta y = 0$ m.

| vertical | horizontal |
|---|---|
| $\Delta y = 0$ m | $\Delta x = ?$ |
| $v_{1y} = 150$ m/s | $v_{1x} = 150\sqrt{3}\frac{\text{m}}{\text{s}}$ |
| $a_y = -10$ m/s$^2$ | $a_x = 0$ m/s$^2$ |
| $\Delta y = v_{1y}\Delta t + 1/2\, a_y \Delta t^2$ | |
| $0 = v_{1y}\Delta t + 1/2\, a_y \Delta t^2$ | |
| $0 = v_{1y} + 1/2\, a_y \Delta t$ | |

In the last line, we have divided by a factor of $\Delta t$.

$$0 = (150\text{ m/s}) + 1/2\,(-10\text{ m/s}^2)\,\Delta t$$
$$\Delta t = 30\text{ s}$$

$$\Delta t = 30\text{ s}$$
$$\Delta x = v_{1x}\Delta t + 1/2\,a_x\Delta t^2$$
$$\Delta x = \left(150\sqrt{3}\,\frac{\text{m}}{\text{s}}\right)(30\text{s})$$
$$\Delta x = 4500\sqrt{3}\text{m}$$
$$\Delta x = 7800\text{ m}$$

We have been talking about grapefruits and cannonballs so far. Objects with a more complicated shape obey the same rules, as long as we use the center of mass to talk about the position of the object. Figure 4-13 shows a baseball bat fired from a cannon. The center of mass moves in a parabola, just like the cannonball in the previous example, even though the bat is rotating. In fact this is a definition of the center of mass. If an object is set to freely rotating, the center of mass is the point which refuses to rotate. The gravitational force acts as if it were exerted only at the center of mass.

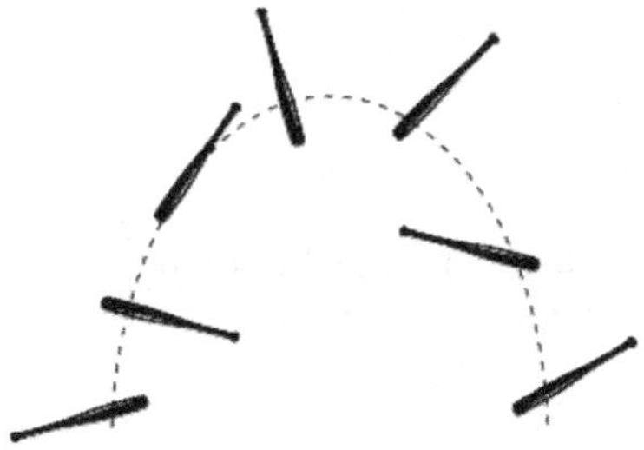

Figure 4-13

In this chapter we looked at the law of gravitation, whose grand form is

$$F_{\text{grav}} = Gm_1m_2/d^2$$

For most problems near the surface of the Earth we can use simply

$$F_{\text{grav}} = mg$$

where $g = GM_{\text{Earth}}/R_{\text{Earth}}{}^2 = 10\text{m/s}^2$

When any object near the Earth's surface has only gravity acting on it (freefall), it has a downward acceleration vector of magnitude 10 m/s$^2$. This curious result comes from the fact that the force of gravity is proportional to the mass, while the second law of motion states that the acceleration in inversely proportional to the mass.

We also explored the principle that horizontal and vertical motion are independent. This allows us to solve problems involving projectiles, that is, objects with only gravity acting on them. We will see more of this principle in the following chapter.

# Physics Review 4 Problems

## Sections A–C

*Use the following information for questions 1–3:*

In a binary star system, two stars revolve about their combined center of mass. They are held together by the force of gravity. For a certain system, the force of gravity between the stars (masses $M_1$ and $M_2$) is $F_0$.

1. If the mass of one of the stars could somehow be decreased by a factor of 2 at a given moment, how would this affect the force between them?
   A. It would decrease by a factor of 4.
   B. It would decrease by a factor of 2.
   C. It would stay the same.
   D. It would increase by a factor of 2.

2. If the distance between the stars were doubled, how would this affect the force between them?
   A. It would decrease by a factor of 4.
   B. It would decrease by a factor of 2.
   C. It would increase by a factor of 2.
   D. It would increase by a factor of 4.

3. In some binary systems, one star transfers material onto the other. If star 1 dumps half of its material onto star 2, which thus increases its mass by a factor of 5, how is the force affected if the distance between the stars' centers remains the same?
   A. The force increases by a factor of 5.
   B. The force increases by a factor of 2.5.
   C. The force remains the same.
   D. The force decreases by a factor of 2.5.

*Use the following information for questions 4–6:*

For a spherical planetary body, the gravitational field depends on its mass and radius. The strength of the gravitational field determines the acceleration due to gravity of freely falling objects at the planet's surface. At the surface of the Earth, for example, the acceleration due to gravity is $g_{\text{Earth}} = 9.8\text{ m/s}^2$. The Moon has a smaller mass and a smaller radius. The net result is that the acceleration due to gravity is $g_{\text{Moon}} = 1.6\text{ m/s}^2$. A metal block (12 kg on Earth) is taken to the Moon.

4. What is the mass of the block on the Moon?
   A. 2 kg
   B. 12 kg
   C. 24 kg
   D. 72 kg

5. The block is dropped from a height of 2 m to the surface of the Moon. How long does it take to drop to the surface?
   A. 1.6 s
   B. 2.4 s
   C. 6.3 s
   D. There is not enough information to answer this question.

6. What is the weight of the block on the Moon?
   A. 2 kg
   B. 12 kg
   C. 19 N
   D. 118 N

*Use the following information for questions 7 and 8:*

The colonization of Mars is a favorite topic among science fiction writers. It is a smaller planet than Earth and further from the Sun. Its radius is 0.5 times that of Earth, and its mass is 0.1 times that of Earth. It is 1.5 times further from the Sun than Earth. (Note: On Earth, the acceleration due to gravity is 10 $m/s^2$. Also, we have $G = 6.67 \times 10^{-11}$ N $m^2/kg^2$.)

7. If a person were standing on the surface of Mars and dropped a Martian rock, what is the approximate acceleration of the rock's fall?
   A. 0.3 $m/s^2$
   B. 1 $m/s^2$
   C. 2 $m/s^2$
   D. 4 $m/s^2$

8. How does the force of attraction between the Sun and Mars compare with that between the Sun and Earth?
   A. 0.04 times weaker.
   B. 0.07 times weaker.
   C. 0.10 times weaker.
   D. There is not enough information to answer this question.

*Use the following information for questions 9 and 10:*

The Earth and the Moon attract each other with the force of gravity. The Earth's radius is 3.67 times that of the Moon, and the Earth's mass is 81 times that of the Moon. The acceleration due to gravity on the surface of the Moon is one sixth the acceleration due to gravity on the Earth's surface.

9. How would the force of gravity between the Earth and the Moon be affected if the distance between the Earth and the Moon were decreased by a factor of 3?
   A. It would decrease by a factor of 9.
   B. It would decrease by a factor of 3.
   C. It would stay the same.
   D. It would increase by a factor of 9.

10. How does the Earth's gravitational pull on the Moon differ from the Moon's pull on the Earth?
    A. It is the same.
    B. It is 3.67 times larger.
    C. It is 6 times larger.
    D. It is 81 times larger.

*Use the following information for questions 11 and 12:*

A new planet is discovered whose mass is the same as that of Earth, although the acceleration of freely falling objects at the surface of this planet is three times larger than that corresponding to Earth. (That is, we have $g_{\text{Earth}} = 9.8\ \text{m/s}^2$ and $g_{\text{new}} = 29.4\ \text{m/s}^2$.)

11. What is the radius of this new planet?
    A. 9 times smaller than the radius of Earth.
    B. 3 times smaller than the radius of Earth.
    C. 1.7 times smaller than the radius of Earth.
    D. 3 times larger than the radius of Earth.

12. If an object were dropped from a height *h* on this planet, how much time would it take to reach the ground compared to the time it would take on Earth to drop the same distance *h*? (Assume no atmospheric resistance.)
    A. It would take the same amount of time.
    B. The time would be less by a factor of 1.7.
    C. The time would be less by a factor of 3.
    D. The time would be less by a factor of 9.

*Use the following information for questions 13–15:*

An astronaut is in a spaceship traveling at constant velocity toward the star Rigel and is far away from any other objects (planets and stars, etc.). There is racquet ball moving at 0.5 m/s on the spaceship. An astronaut blows air on the racquet ball, producing a small force of 0.08 N in the direction opposite the ball's velocity. The ball slows to a stop in 4 s.

13. What is the weight of the racquet ball?
    A. 0 N
    B. 0.016 N
    C. 0.4 N
    D. 6.4 N

14. What is the magnitude of the acceleration of the racquet ball?
    A. $0\ \text{m/s}^2$
    B. $0.125\ \text{m/s}^2$
    C. $0.2\ \text{m/s}^2$
    D. $0.4\ \text{m/s}^2$

15. What is the mass of the racquet ball?
    A. 0 kg
    B. 0.64 kg
    C. 1.6 kg
    D. 2.5 kg

## Section D

*Use the following information for questions 16–20:*

In a certain sports event, called the appliance toss, men and women test their strength by hurling an electric can opener horizontally over a cliff. (See figure.) The cliff is very high, standing over a plain, and judges at the bottom determine where the openers land. When Barbara heaves her opener at $t = 0$ s, the opener has horizontal velocity 1.5 m/s when it leaves her hand. After 2 s, it is still in the air (point B in the diagram). (Use $g = 10$ m/s$^2$.)

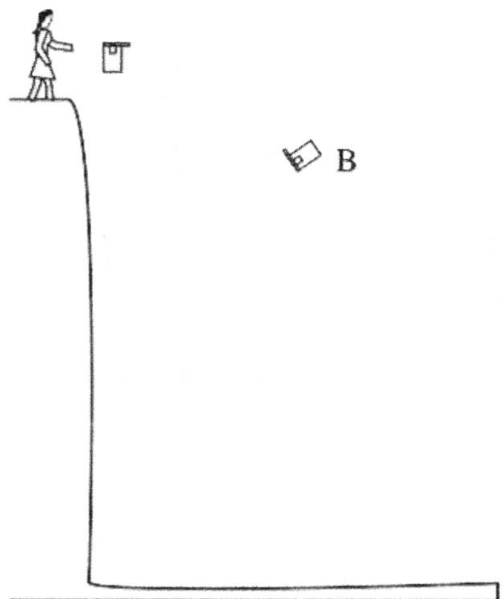


16. Which of the following best shows a force diagram for the can opener at point B?

A. 

B. 

C. 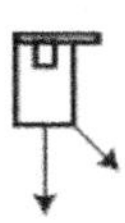

D. 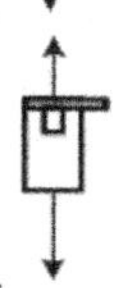

17. What vertical distance has the can opener fallen in the two seconds?
    A. 3 m
    B. 20 m
    C. 20.2 m
    D. 23 m

18. What horizontal distance has the can opener traversed in the two seconds?
    A. 3 m
    B. 20 m
    C. 20.2 m
    D. 23 m

19. At $t = 2$ s, what is the opener's vertical velocity?
    A. 0 m/s
    B. 1.5 m/s
    C. 20 m/s
    D. 21.5 m/s

20. At $t = 2$ s, what is the opener's horizontal velocity?
    A. 0 m/s
    B. 1.5 m/s
    C. 20 m/s
    D. 21.5 m/s

*Use the following information for questions 21–27:*

A ball (0.2 kg) rolls along a level table at 1.5 m/s and then rolls off the edge. The table is 1.25 m off the floor. Consider the time from the moment just after the ball leaves the table till the moment just before the ball touches the floor. See figure. (Use $g = 9.8\ m/s^2$, and ignore any air resistance.)

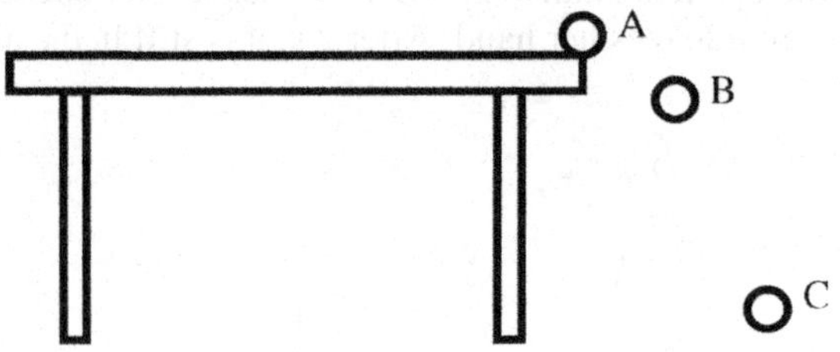


21. What is the ball's initial horizontal velocity?
 A. 0 m/s
 B. 1.5 m/s
 C. 5.0 m/s
 D. 5.2 m/s

22. What is the ball's initial vertical velocity?
 A. 0 m/s
 B. 1.5 m/s
 C. 5.0 m/s
 D. 5.2 m/s

23. When the ball is in midair (point B), what is the net force on the ball?
 A. 1.96 N
 B. 3.92 N
 C. 9.8 N
 D. 19.6 N

24. What is the vertical displacement of the ball during the fall?
 A. 0.76 m
 B. 1.47 m
 C. 1.25 m
 D. 2.25 m

25. How much time does the drop take?
 A. 0.26 s
 B. 0.51 s
 C. 0.83 s
 D. 1.01 s

26. What is the horizontal acceleration of the ball at point B?
 A. 0 $m/s^2$
 B. 0.2 $m/s^2$
 C. 1.96 $m/s^2$
 D. 9.8 $m/s^2$

27. What is the horizontal displacement of the ball during the fall?
 A. 0.76 m
 B. 1.47 m
 C. 1.25 m
 D. 2.25 m

*Use the following information for questions 28–31:*

Two girls are sitting on the edge of a building tossing coins over the edge. Alice is actually dropping her coins, each of which is 10 g. Barbara is tossing her coins horizontally at 0.3 m/s, and her coins are 40 g each. (See figure.) (Ignore air resistance.)

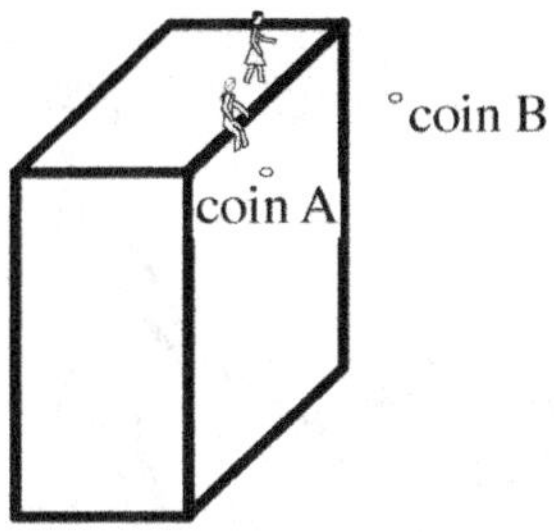


28. When the coins are in midair, how does the gravitational force on one of Alice's coins compare with the force on one of Barbara's?
A. It is one fourth as large.
B. It is the same.
C. It is four times as large.
D. It depends on the height at which the force is recorded.

29. When the coins are in midair, how does the acceleration of one of Alice's coins compare with the acceleration of one of Barbara's?
A. It is one fourth as large.
B. It is the same.
C. It is four times as large.
D. It depends on the height at which the acceleration is recorded.

30. How does the time to reach the ground for one of Alice's coins compare with the time of fall for Barbara's?
A. The time for Alice's coins is less.
B. The times are the same.
C. The time for Alice's coins is greater.
D. It depends on the height of the building.

31. Just before Alice's coin reaches the ground, it has speed $s_A$. Just before Barbara's coin reaches the ground, it has speed $s_B$. How does $s_A$ compare with $s_B$?
A. The speed $s_A$ is less than $s_B$.
B. The speed $s_A$ is the same as $s_B$.
C. The speed $s_A$ is greater than $s_B$.
D. It depends on the height of the building.

*Use the following information for questions 32–36:*

A woman (50 kg) is pulling a wagon behind her. In the wagon is her daughter by her first marriage; the daughter and the wagon are 60 kg. (See figure.) The woman pulls the handle with a tension 200 N, and the handle makes a 30° angle with the horizontal. There is a horizontal force of friction, and the wagon moves at a constant 2 m/s. (Use $g = 10\ m/s^2$.)

32. What is the horizontal component of the force of the wagon handle on the wagon body?
    A. 0 N
    B. (200 N) (sin 30°)
    C. (200 N) (cos 30°)
    D. 200 N

33. What is the vertical component of the force of the wagon handle on the wagon body?
    A. 0 N
    B. (200 N) (sin 30°)
    C. (200 N) (cos 30°)
    D. 200 N

34. What is the horizontal component of the gravitational force on the wagon and daughter?
    A. 0 N
    B. (600 N) (sin 30°)
    C. (600 N) (cos 30°)
    D. 600 N

35. What is the vertical component of the gravitational force on the wagon and daughter?
    A. 0 N
    B. (600 N) (sin 30°)
    C. (600 N) (cos 30°)
    D. 600 N

36. What is the magnitude of the net force?
    A. 0 N
    B. 173 N
    C. 600 N
    D. 800 N

*Use the following information for questions 37–39:*

A bale of hay (500 kg) is dropped from the second story of a barn (9 m) with no initial velocity. A rope is tied to the hay to control its fall. The rope extends up from the bale and maintains a tension of 4000 N. (Use $g = 10\ \text{m/s}^2$.)

37. What are the forces acting on the bale of hay?
   A. The tension in the rope only.
   B. Gravity only.
   C. The tension in the rope and gravity.
   D. The tension in the rope, gravity, and an additional downward force once the bale is moving.

38. What is the magnitude of the net force on the bale of hay?
   A. 0 N
   B. 1000 N
   C. 5000 N
   D. 9000 N

39. What is the acceleration of the bale of hay?
   A. $0\ \text{m/s}^2$
   B. $2\ \text{m/s}^2$
   C. $10\ \text{m/s}^2$
   D. $18\ \text{m/s}^2$

40. On the Moon there is very little atmosphere (several centimeters of thin gas). An astronaut drops a hammer and a feather at the same time from about shoulder height. What happens?
   A. The hammer lands first.
   B. The feather lands first.
   C. The hammer and the feather land at the same time.
   D. Neither lands but instead fly off the Moon's surface.

*Use the following information for questions 41 and 42:*

A large ball (2 kg) is rolling (at the surface of the Earth) on a large, flat plain, so large that we will idealize it as an infinite plain. It is rolling 0.3 m/s to the right at time $t = 0$ s. Take the acceleration of gravity to be $10\ \text{m/s}^2$. Assume there is no friction.

41. Which of the following represents the best force diagram at times after $t = 0$ s?

A. B. C. D.

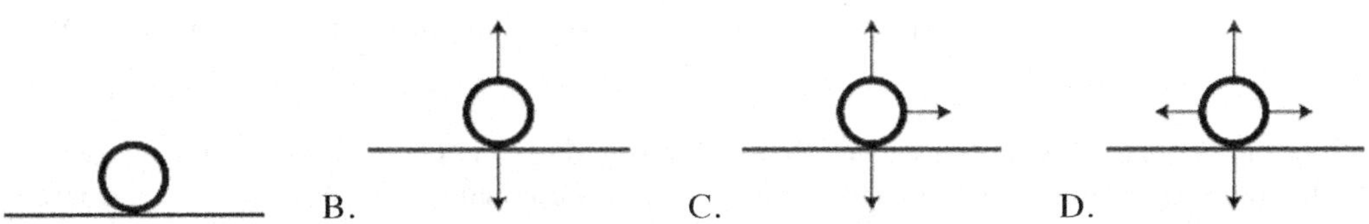

42. How long would it take the ball to come to a stop?
   A. 0.3 s.
   B. 3.33 s.
   C. 30 s.
   D. There is no answer.

# Solutions

1. B.
The fact that the stars revolve about their combined center of mass is irrelevant for calculating the force of gravity between them. That depends only on their masses and the distance between their centers. Since we have $F_{grav} = Gm_1m_2/d^2$, if one of the masses decreases by a factor of 2, then $F_{grav}$ decreases by a factor of 2.

2. A.
The factor of $d^2$ in the denominator indicates that if $d$ increases by a factor of 2, then $F_{grav}$ decreases by a factor of $2^2 = 4$.

3. B.
If $m_1$ decreases by a factor of 2, and $m_2$ increases by a factor of 5, then $F_{grav}$ changes by a factor of 5/2.

4. B.
I hope this one did not fool you. The mass of an object does not change just because you transport it somewhere.

5. A.
Objects in free fall on the Moon have a constant acceleration due to gravity 1.6 m/s$^2$ for the same reason that free-fall objects have $a = 9.8$ m/s$^2$ here on Earth. This comes from setting $F_{net} = ma$ equal to $F_{grav} = mg$, so that $a = g$. In this case, we have

$$a = -1.6 \text{ m/s}^2$$

$$v_1 = 0 \text{ m/s}$$

$$\Delta y = -2 \text{ m}$$

$$\Delta t = ?$$

$$\Delta y = v_1\Delta t + 1/2\, a\Delta t^2$$

$$-2 \text{ m} = 1/2\,(-1.6 \text{ m/s}^2)\Delta t^2$$

$$\Delta t^2 = 2.5 \text{ s}^2$$

We only need to work as far as $\Delta t^2$ in order to eliminate choices B, C, and D.

6. C.
The weight of the block on the Moon is an exact analogy to its weight on Earth: $F_{grav} = mg_{Moon} = 20$ N.

7. D.
On Earth (or on any planetary body), we get the acceleration due to gravity on the surface by assuming that gravity is the only force acting on an object. So we set $F_{net} = ma$ equal to $F_{grav} = Gm_{planet}m/r^2$ , where $r$ is the radius of the planet:

$$ma = \frac{Gm_{planet}m}{r^2}$$

$$a = \frac{Gm_{planet}}{r^2}$$

See Example 2 in Section A. Now if the radius of the planet were multiplied by 0.5, then the acceleration due to gravity would be *divided* by $(0.5)^2$, giving us 40 m/s$^2$. (We expect an increase, since the acceleration increases if radius decreases.) But the mass of Mars is 0.1 of Earth's mass, so a factor of 0.1 in the numerator brings the acceleration down to 4 m/s$^2$.

8. A.
This time the relevant equation is

$$F_{grav} = \frac{Gm_{sun}m_{planet}}{d^2}$$

where $d$ is the distance from the Sun to that planet. The mass of Mars is 0.1 times that of Earth, so that is a factor of 0.1. Another factor comes from the distance from the Sun to Mars, which is 1.5 times greater, so we *divide* by $(1.5)^2$, yielding about 0.04.

9. D.
The equation we need for this problem and the next is

$$F_{grav} = \frac{Gm_{Earth}m_{moon}}{d^2}$$

where $d$ is the distance between the Earth and the Moon. If $d$ decreases by a factor of 3, then $F_{grav}$ increases by a factor of $3^2 = 9$.

10. A.
The Earth's gravitational pull on the Moon is the same as the Moon's on the Earth. That's the third law of motion. (Why does the Moon in its orbit move so much more than the Earth, then?)

11. C.
In problem 7, we worked out the surface acceleration due to gravity of a planet. Since we are looking for radius, we can solve for it, so we have

$$r = \sqrt{\frac{Gm_{planet}}{a}}$$

If the new planet has the same mass as Earth, but a larger acceleration due to gravity, then the radius must be smaller. And if $a$ changes by a factor of 3, then $r$ changes by a factor of $\sqrt{3}$, so the answer is B. (We don't need to know $\sqrt{3}$ to figure out which choice is right.)

12. B.
We want to relate acceleration and time, so we want an equation that involves these quantities. (We have seen enough of these problems to realize $v_1 = 0$.) We can use

$$h = v_1\Delta t + 1/2\ g\Delta t^2$$
$$h = 1/2\ g\Delta t^2$$
$$\Delta t = \sqrt{\frac{2h}{g}}$$

If $g$ increases by a factor of 3, then $\Delta t$ decreases by a factor of $\sqrt{3}$.

13. A.
Since weight is the force of gravity on an object, and there are no massive planetary bodies around, the weight is zero.

14. B.
The acceleration is $a = (0\text{ m/s} - 0.5\text{ m/s})/(4\text{ s}) = -0.125\text{ m/s}^2$. (This was a review question.)

15. B.
Once we have the acceleration, we can calculate a mass $m = F/a = (0.08\text{ N})/(0.125\text{ m/s}^2) = 0.64\text{ kg}$.

16. A.
Certainly there is the force of gravity, down. There is nothing else touching the can opener, so gravity is the only force. Since there is no horizontal force, there is constant horizontal motion. But horizontal motion *does not* imply a horizontal force (see the first law of motion).

17. B.
Since we have a force diagram, we next need to inventory the information relevant to the vertical motion. We have $a_y = -10$ m/s$^2$ and $\Delta t = 2$ s. Because the opener is traveling *horizontally* when it leaves Barbara's hand, we have $v_{1y} = 0$ m/s. Since we want to know $\Delta y$, we use the equation $\Delta y = v_{1y}\Delta t + 1/2\, a_y (\Delta t)^2 = 20$ m.

18. A.
Since we are looking for $\Delta x$, we now inventory the horizontal information. We have $a_x = 0$ m/s$^2$ and $v_{1x} =$ 1.5 m/s. Thus we have $\Delta x = v_{1x}\Delta t + 1/2 a_x (\Delta t)^2 = 3$ m.

19. C.
The problem asks for $v_{2y}$, so we write $v_{2y} = v_{1y} + a_y\Delta t = -20$ m/s. Since only positive choices are listed, we choose the magnitude 20 m/s.

20. B.
We can calculate $v_{2x} = v_{1x} + a_x\Delta t = 1.5$ m/s. Of course, we don't really have to do a calculation, since we know that the horizontal velocity is constant as long as there are no horizontal forces on the opener.

21. B.
As the ball just leaves the table, it is going 1.5 m/s horizontally, so $v_{1x} = 1.5$ m/s.

22. A.
Since the initial velocity is horizontal, we have $v_{1y} = 0$ m/s.

23. A.
We draw a diagram including all the forces (see figure). Since nothing is touching the ball, gravity is the only force. Thus $F_{grav} = mg = (0.2 \text{ kg})\ (10 \text{ m/s}^2) = 2$ N.

24. C.
Since the ball simply drops the height of the table, we have $\Delta y = -1.25$ m (choosing "up" to be positive).

25. B.
For horizontal information, we have $a_x = 0$ m/s$^2$ and $v_{1x} = 1.5$ m/s, which is not enough to obtain $\Delta t$. For vertical information, we have

$$\Delta y = -1.25 \text{ m}$$
$$a_y = -10 \text{ m/s}^2$$
$$v_{1y} = 0 \text{ m/s}$$

We can calculate

$$\Delta y = v_{1y}\Delta t + 1/2\, a_y\Delta t^2$$
$$\Delta t^2 = 2\Delta y/a_y$$
$$\Delta t^2 = 0.25 \text{ s}^2$$

We do not really need to take the square root to figure out the answer. If choice A is right, then $(\Delta t)^2 =$ $(0.26 \text{ s})^2 \approx 0.12 \text{ s}^2$ (wrong). If choice B is right, then $(\Delta t)^2 = (0.5 \text{ s})^2 = 0.25 \text{ s}^2$. (Do not feel the need to work every arithmetic problem to its end.)

26. A.
There is no horizontal force on the ball.

27. A.
We want $\Delta x$. Since we now have $\Delta t$, we have enough information to calculate

$$\Delta x = v_{1x}\Delta t + 1/2\, a_x \Delta t^2$$

$$\Delta x = 0.76 \text{ m}$$

28. A.
Since gravitational force is given by $F_{grav} = mg$, and since Barbara's coins have four times the mass of Alice's coins, the force on them is four times as large.

29. B.
The acceleration of Alice's coins is a constant
9.8 m/s$^2$ down, as is the acceleration of Barbara's.

30. B.
The vertical acceleration $a_y$ is the same for both coins, and the initial vertical velocity $v_{1y}$ is the same for both coins. The vertical displacement for the fall is certainly the same, so in the equation $\Delta y = v_{1y}\Delta t + 1/2\, a_y (\Delta t)^2$, all the parameters are the same. This question combines the two ideas in this chapter, namely, for *freely falling bodies* near the surface of the Earth the vertical acceleration is 9.8 m/s$^2$ down, and for *all bodies* vertical motion is independent of horizontal motion.

31. A.
Both coins have the same vertical velocity. In addition, Barbara's coins retain their horizontal velocity.

32. C.
First, let's draw a diagram showing all the forces on the wagon body.

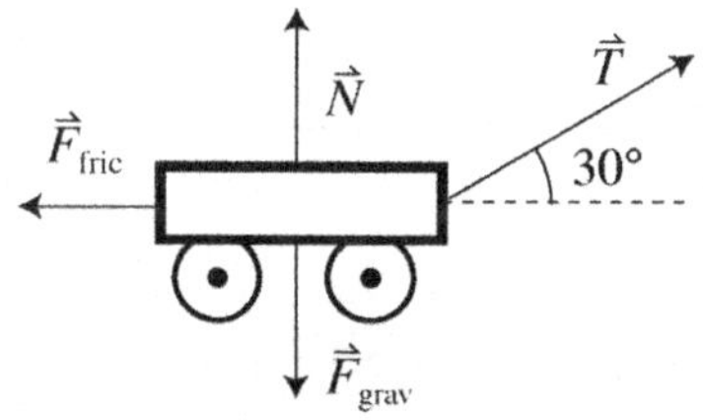


The handle and the ground are touching the wagon, so we include the tension force of the handle $T$ and the upward force of the ground $N$. The problem mentions friction, so we include that as well. There are four forces on the wagon. We can redraw the force of the handle as the sum of two components, as shown in the second figure.

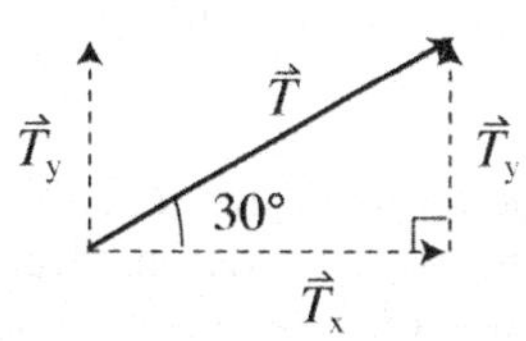


We calculate $T_x$ as follows:

$$T_x/T = \cos 30°$$

$$T_x = T\cos 30°$$

33. B.
We calculate $T_y$ as follows:

$$T_y/T = \sin 30°$$
$$T_y = T\sin 30°$$

34. A.
The gravitational force is vertical, so its horizontal component is zero.

35. D.
The vertical component of the gravitational force is the magnitude of the force, which is $F_{grav} = mg$.

36. A.
We don't have enough information to add up the force vectors, but we don't need to. The problem states the wagon is nonaccelerating (velocity is a constant 2 m/s), so $a = 0$ m/s$^2$ and $F_{net} = 0$ N.

37. C.
Gravity is acting on the bale, of course. The rope exerts a tension force. The force diagram is shown. Note that the bale is *not* in free fall.

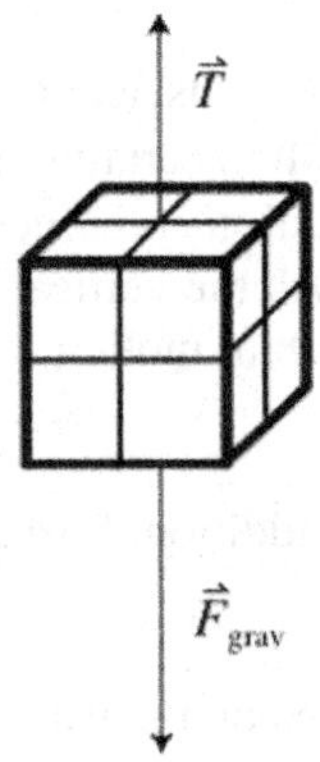


38. B.
If "up" is positive, then the net force is given by $F_{net} = 4000 \text{ N} - (500 \text{ kg})(10 \text{ m/s}^2) = -1000$ N. We have used a negative sign for gravity since it points down.

39. B.
We can calculate the acceleration $a = F_{net}/m = 1000 \text{ N}/500 \text{ kg} = 2 \text{ m/s}^2$.

40. C.
On Earth, the hammer lands first because of air resistance, but if there is no air, then they have the same acceleration (about 1.6 m/s$^2$). Concerning choice D, of course they do not fly off, but they move along with the Moon in the same way that the grapefruit moves along with the ship. See the example in the text. This is the same way that a passenger moves along with a swiftly moving train.

41. B.
There is the force of gravity, down, and the normal force, up. The only thing touching the ball is the floor, which can cause only a normal force and possibly a frictional force. Since the problem states there is no friction, choice B is correct. Concerning choice C, there is nothing pushing the ball to the right, since it simply retains its initial motion. Choice D adds in a leftward force, presumably of friction, which the problem says is not there.

42. D.
If, as the problem states, the ball is going to the right at 0.3 m/s at time $t = 0$ s, then without any horizontal forces, it will continue to go 0.3 m/s forever. The problem is idealized, but without any information about a horizontal force, there is no way to calculate a horizontal acceleration or a horizontal change in velocity.

# Physics Review 5
# Planes and Circles

## A. Horizontal and Vertical Motion, Again

In the last chapter we solved problems with gravity as the only force. Well, gravity is a fine force indeed, but we need to understand problems in which other forces are present. That is the goal of this chapter.

In the last chapter we discussed the independence of vertical and horizontal motion for objects in freefall, but it turns out the principle works when other forces are acting as well:

> **Independence of Vertical and Horizontal Motion**
>
> An object has forces $\vec{F}_1, \vec{F}_2, \ldots,$ acting on it. If $F_{1y}, F_{2y}, \ldots,$ are the vertical components of the forces, then we have
>
> $$F_{\text{net,y}} = F_{1y} + F_{2y} + \cdots, \tag{1}$$
>
> $$a_y = \frac{F_y}{m}. \tag{2}$$
>
> Similarly, if $F_{1x}, F_{2x}, \ldots,$ are the horizontal components of the forces, then we have
>
> $$F_{\text{net,x}} = F_{1x} + F_{2x} + \cdots, \tag{3}$$
>
> $$a_x = \frac{F_x}{m}. \tag{4}$$

This is a more useful form of $F_{net} = ma$, the equation that we discussed in Section 3.B. An example will help illustrate how the principle in the above box is used to solve problems.

In the following example, we find a toy wagon rolling on the ground. In general, when an object is on the ground or some other surface, that surface exerts one or two forces on it: always a normal force $N$ pointing perpendicular to the surface and sometimes a frictional force $F_{\text{fric}}$ pointing parallel to the surface.

**Example:** A boy pulls a red wagon (10 kg) with a constant force 20 N. The handle makes an angle 30° with the horizontal. Assume there is no friction.

a. If the wagon starts from rest, how fast is it going after 3 seconds?
b. What is the normal force acting on the wagon?

**Solution:** First we DRAW A DIAGRAM (Figure 5-1) showing all the forces on the wagon body. The handle and the ground are the only two things which touch it. So in addition to gravity, there are the tension due to the handle and the normal force. The tension $T$ points along the handle, that is, 30° from the horizontal.

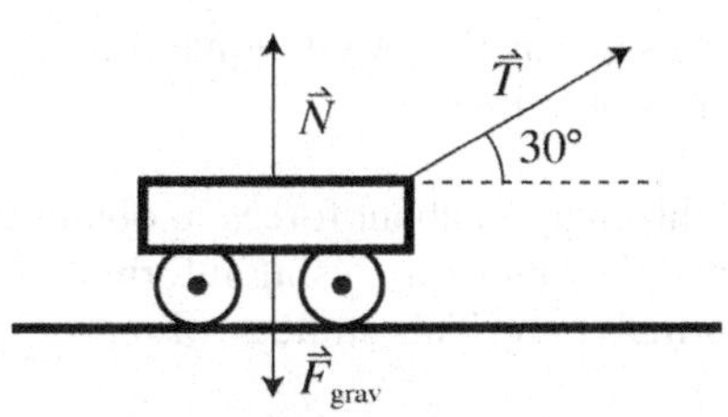


Figure 5-1

In Figure 5-2 we resolve the tension into components (recall trigonometry), so we have

$$\frac{T_y}{T} = \sin 30°$$
$$T_y = (20\text{N})\sin 30°$$
$$T_y = 10\text{N}$$

$$\frac{T_x}{T} = \cos 30°$$
$$T_x = (20\text{N})\cos 30°$$
$$T_x = 17\text{N}$$

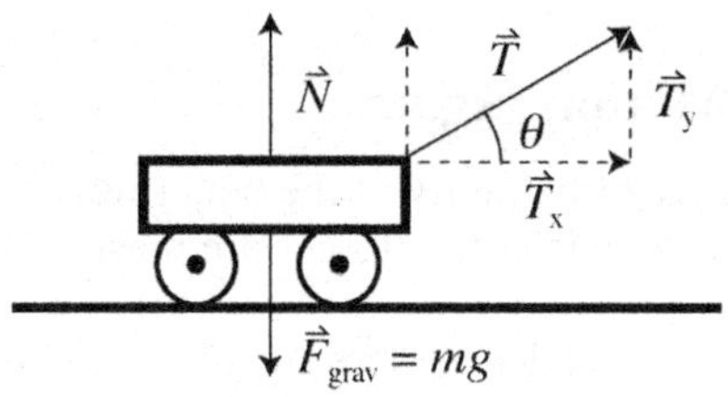


Figure 5-2

The normal force and gravity do not have horizontal components, so the only horizontal force is $T_x$. Thus we write

$(F_{net})_x = ma_x,$
$T_x = ma_x,$
$a_x, = T_x/m = 17\text{N}/10\text{kg} = 1.7 \text{ m/s}^2$

Using this and $v_{1x} = 0$ m/s and $\Delta t = 3$ s, we derive a horizontal velocity

$v_{2x} = v_{1x} + a_x\Delta t$
$= 0 \text{ m/s} + 1.7 \text{ m/s}^2(3\text{s}) = 4.1 \text{ m/s}$

which is the answer to part a.

If we consider the vertical components of force, then we can find $(F_{net})_y$ just by looking at the diagram, so we write

$(F_{net})_y = N + T_y - F_{grav}$

where we use the positive sign for forces which point up; negative for down. Now the second law of motion connects this with vertical acceleration, so that

$(F_{net})_y = ma_y$

But the cart is not moving vertically, so we know that $v_y$ is constant (and zero) and thus $a_y$ is zero. This means $(F_{net})_y$ is zero, so we have

$0 = N + T_y - F_{grav},$
$N = F_{grav} - T_y$
$= mg - T_y$
$= (10\text{kg})(10\text{m/s}^2) - 10\text{N}$
$= 90 \text{ N}.$

Notice that the normal force is not the same as the gravitational force, a mistake often made by students. Why is the normal force not the same as gravity?

Note that in solving part a, we used information about forces to obtain the horizontal acceleration $a_x$ and then the answer. In part b we reasoned the other way, using information about the vertical acceleration $a_y = 0$ to obtain information about the normal force. This strategy of reasoning in both directions will be useful in many problems.

## B. Inclined Planes and Force Components

The following method generally works for force problems in two dimensions:

1. DRAW A DIAGRAM.
2. Draw all the forces on the object(s) in question (see Section 3.D).
3. Decide the orientation of the axes ("horizontal" and "vertical").
4. Divide the forces into components if necessary.
5. Solve $(F_{net})_y = ma_y$ and $(F_{net})_x = ma_x$.

For the last step, note that we often have $a_y = 0$, leading to $(F_{net})_y = 0$. Also, be on the lookout for the words "constant velocity" or the equivalent, since that implies $a_x = 0$ and $a_y = 0$, a force balance.

The principle of the independence of the components of the $F = ma$ equation is valid even when the axes are tilted, as the next example will show.

**Example:** A toy car of mass 40 grams is released at the top of an incline of plastic track, inclined 30° from the horizontal. The car starts from rest and travels 4 m to the floor. Assuming there is no friction, how much time does it take the car to reach the floor?

**Solution:** First we DRAW A DIAGRAM (Figure 5-3). In addition to gravity, the track touches the car and exerts a normal force. Since the track is inclined, the normal force points not up but perpendicular to the surface. There is no frictional force.

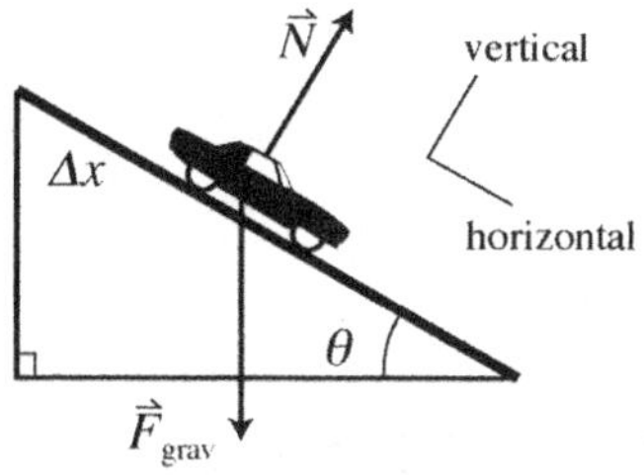


Figure 5-3

We will call "horizontal" the direction along the track and "vertical" the direction perpendicular to the track (Figure 5-3).

Next, we divide the gravitational force into components (Figure 5-4). Note that $F_x$ and $F_y$ are not new forces but pieces of $F_{grav}$. If we add $F_x$ and $F_y$ together like vectors (tip-to-tail), then we get $F_{grav}$. It may not be obvious that the two angles shown in Figure 5-4 should both be $\theta = 30°$. Note that both angles labeled $\theta$ are complementary to the angle between $F_x$ and $F_{grav}$. On the other hand, in physics it is generally true that two small angles which look congruent are congruent(!).

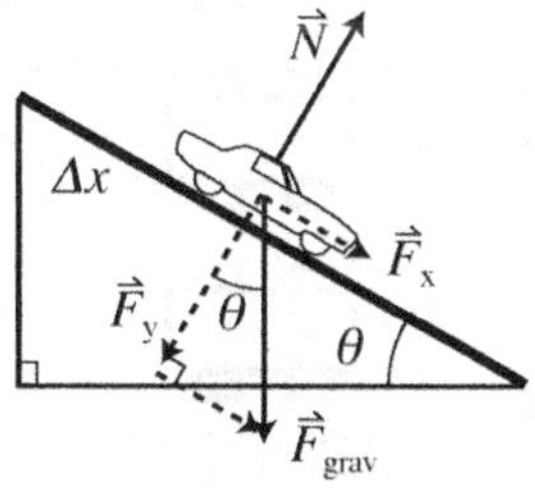


Figure 5-4

Now if we look at the triangle in Figure 5-4, we can write

$$\frac{F_x}{F_{grav}} = \sin\theta \qquad \frac{F_y}{F_{grav}} = \cos\theta$$

$$F_x = F_{grav}\sin\theta \qquad F_y = F_{grav}\cos\theta$$

$$F_x = mg\sin\theta \qquad F_y = mg\cos\theta$$

We care about the horizontal motion, so we write

$$(F_{net})_x = ma_x$$

The only horizontal force is $F_x$, so we write

$$F_x = ma_x$$

$$mg \sin \theta = ma_x$$

$$g \sin \theta = a_x$$

$$a_x = (10 \text{ m/s}^2)\sin 30° = 5 \text{ m/s}^2$$

We have $v_{1x} = 0$ m/s and $\Delta x = 4$ m, so we can find $\Delta t$ by writing

$$\Delta x = v_{1x}\Delta t + \frac{1}{2}a_x\Delta t^2$$

$$\Delta t = \sqrt{\frac{2\Delta x}{a_x}}$$

$$= \sqrt{\frac{2(4\text{m})}{5\frac{\text{m}}{\text{s}^2}}}$$

$$= 1.3 \text{ s}$$

## C. Circular Motion, Qualitative Description

Let's think a minute about a toy car moving along the floor and pretend the movement is frictionless. Figure 5-5 shows a top view, with the car moving to the right (see the motion marks behind the car). If we were to apply a rightward force $F_A$, by blowing the car with a portable hair dryer, for instance, this would clearly speed up the car. If we were to apply a leftward force $F_B$, this would slow it down. A force $F_C$ applied (for just a moment) perpendicular to the motion would neither speed it up nor slow it down, but it would cause the car to veer from its straight path.

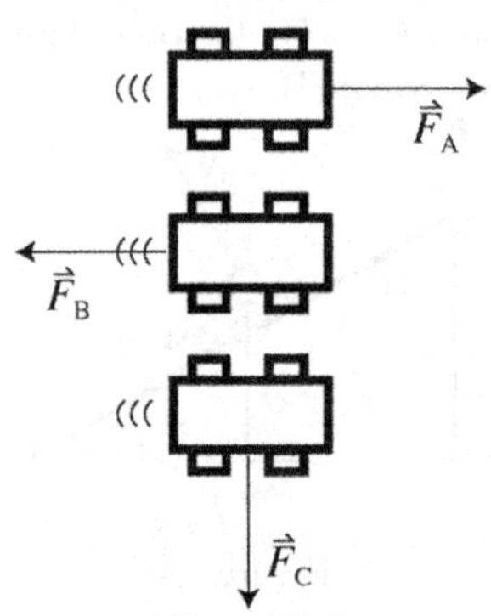

Figure 5-5

If we are using a hair dryer to exert $F_C$, then we can keep adjusting the direction of the force to keep it perpendicular to the motion of the car. The car will end up traveling in a circle (Figure 5-6). Given this discussion, the following box should seem reasonable.

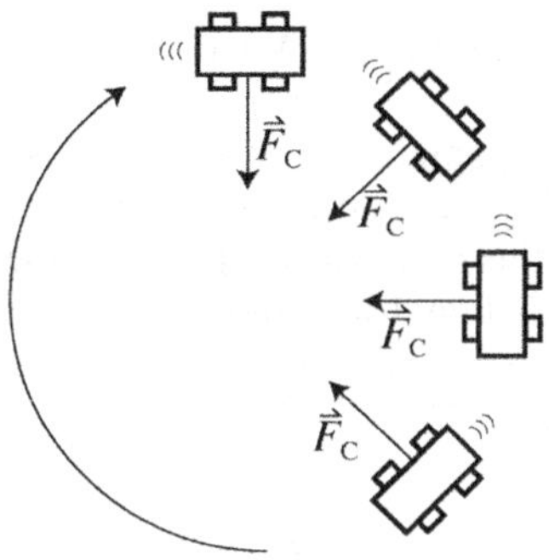


*An object moving in a circle with constant speed has an acceleration vector pointed towards the center of the circle.*

Figure 5-6

> If an object is moving at constant speed in a circle, then the net force on the object points toward the center of the circle, and the acceleration vector points towards the center as well.

In normal English, we do not say that the car is accelerating when it is turning. But in physics language, an object that is moving at constant speed in a circle is "accelerating toward the center of the circle", because the *velocity vector* is changing. We call this *centripetal acceleration*, which is Latin for "toward the center". (Parenthetically, "centrifugal" means "away from the center".) The force which provides the centripetal acceleration is the *centripetal force*.

**Example 1:** The Earth moves around the Sun. What force provides the centripetal force?

**Solution:** The gravitational force provides the centripetal force (Figure 5-7).

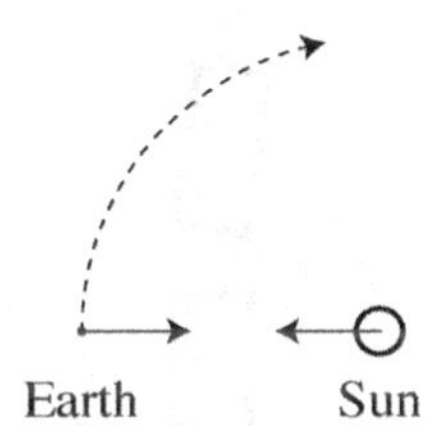


Figure 5-7

**Example 2:** A car goes around a curve to the left. What force provides the centripetal force?

**Solution:** This example is a bit more complicated. The driver turns her wheels to the left, so the wheels exert a force on the road to the right. By Newton's Third Law of Motion, the road exerts a force to the left on the wheels, turning the car left. In this case friction provides the centripetal force (Figure 5-8). (Think what would happen if there were no friction, for instance, if there were oil on the road.)

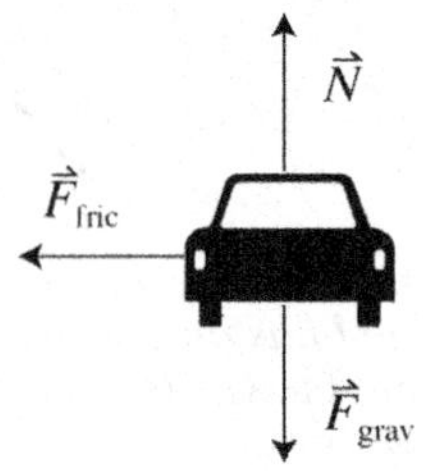


*The car, seen from the rear, is turning left, and the frictional force provides the centripetal acceleration.*

Figure 5-8

**Example 3:** A father is driving a car and turns to the left. There is a sack of groceries in the front passenger seat which crashes into the passenger door. The little brother Samson in the back seat asks the father why the groceries crashed into the door. The father says that was due to centrifugal force. The older sister Cadenza rolls her eyes at this, thinking about how much physics her brother will have to unlearn as he grows up. What is the correct explanation for the groceries' crashing into the door?

**Solution:** Figure 5-9 gives the real story. The groceries are going along a straight path, as we would expect according to the first law of motion. The car door is pulled into the path of the groceries. The father invents the word "centrifugal force" in order to hide his ignorance. Whenever you are tempted to explain something by centrifugal force, it is likely that you can explain it better with the ideas of first law of motion and a turning frame of reference (like the car).

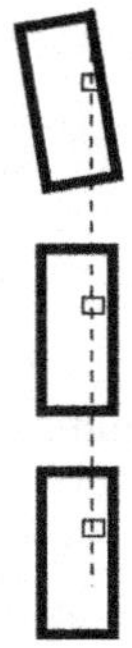

*The groceries in the passanger's seat crash into the car door because the door turns into their path (not because of "centrifugal force").*

Figure 5-9

## D. Circular Motion, Quantitative Description

An object moving in a circle has an acceleration which has, in general, two components: the centripetal acceleration and the tangential acceleration. The former is directed toward the center and is responsible for changing the direction of the object. The latter is responsible for changing the speed. The acceleration is given by

| | | |
|---|---|---|
| Tangential acceleration | $a_{tang} = \dfrac{\Delta\text{speed}}{\Delta t}$ , | (5) |
| Centripetal acceleration | $a_{cent} = \dfrac{v^2}{r}$ . | (6) |

**Example 1:** A bombardier beetle sits on a blade of a windmill which is going counterclockwise and slowing down. The blade is at the top of its cycle. Sketch the acceleration vector.

**Solution:** In Figure 5-10, if the beetle is going left and slowing down, then the tangential acceleration must be to the right. The centripetal acceleration vector is down, and the total acceleration $a$ is shown.

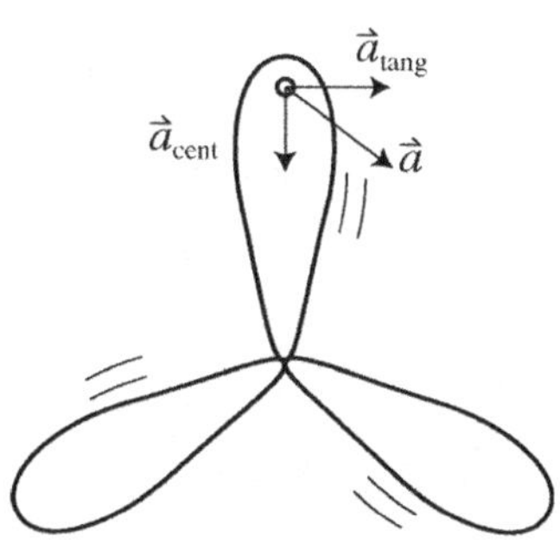


Figure 5-10

**Example 2:** Use the following information to find the mass of the sun:

$G = 6.67 \times 10^{-11}$ m$^3$/kg s$^2$
$R = 1.5 \times 10^{11}$ m (distance from the Sun to the Earth)

**Solution:** First, we DRAW A DIAGRAM showing all the forces (Figure 5-11). Some students are tempted to draw two forces on the Earth: a gravitational and a centripetal force. But the only force is gravity, and this provides the centripetal force in this problem. This last sentence provides the clue for solving the problem. We know expressions for gravitation and for centripetal acceleration, so that we have

$$F_{grav} = F_{cent}$$

$$\frac{GM_{sun}M_{Earth}}{R^2} = M_{Earth}\frac{v^2}{R} \qquad (7)$$

We use $M_{Earth}$ on the right-hand side of the equation, because it is the Earth's acceleration we are concerned about. The Sun's acceleration is much smaller because its mass is larger. (Recall that the force the Earth exerts on the Sun is the same as that which the Sun exerts on the Earth.) Note that $M_{Earth}$ cancels.

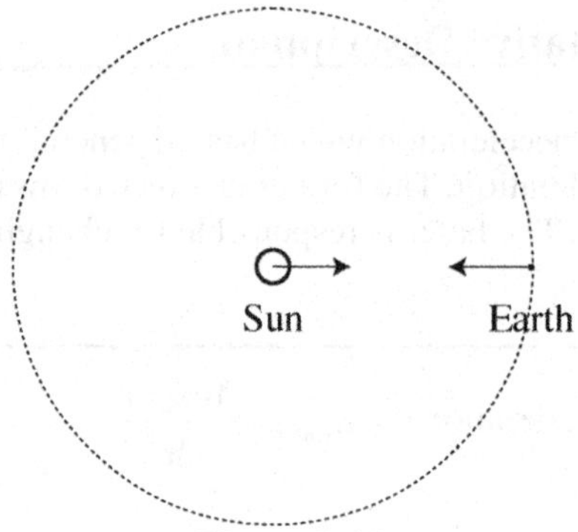


Figure 5-11

What expression shall we use for $v$? What is the velocity of the Earth? If the Earth travels a full circuit in a year, then velocity is simply distance per time, where the distance is the circumference of the circle. Thus we write

$$v = \frac{2\pi R}{T} \qquad (8)$$

where $T$ is 1 year. Do not simply memorize this, but take a minute to think about why this equation is true, so it will come immediately to mind in any similar situations. Substituting this into equation (7) and doing some algebra gives

$$M_{sun} = \frac{4\pi^2 R^3}{GT^2}$$

into which we can substitute the values given in the problem, along with

$$T = 365\text{days}\left(\frac{24 \text{ hours}}{1 \text{ day}}\right)\left(\frac{3600 \text{ s}}{1 \text{ hour}}\right) = 3\times10^7 \text{ s}$$

to yield

$$M_{sun} = 2 \text{ x } 10^{30} \text{ kg}$$

The importance of this example does not lie in the arithmetic. The important parts are the method of setting two expressions for the same force equal to each other and the use of equation (8).

**Example 3:** A fan spins at a frequency of 50 cycles per second, and its plastic blades are 0.4 meters long. What is the centripetal acceleration of a piece of plastic at the tip of one of the blades?

**Solution:** You should try to work this out before you read the solution.

If the fan spins at 50 cycles per second, it must undergo one cycle in one fiftieth of a second, so $T$ = 1/50 s = 0.02 s. The velocity is given by

$$v = \frac{2\pi r}{T} = \frac{2\pi(0.4\text{m})}{0.02\text{s}} = 126\frac{\text{m}}{\text{s}}$$

so the acceleration is given by

$$a = \frac{v^2}{r} = \frac{\left(126\frac{\text{m}}{\text{s}}\right)^2}{0.4\text{m}} = 4\times10^4 \frac{\text{m}}{\text{s}^2}$$

**Example 4:** A space warrior must fly his spacecraft at constant speed around a spherical space station "Bad Star". Bad Star is large but not large enough to have an atmosphere or gravity worth considering. It is important for the warrior to stay close to the surface and maintain constant speed. A conventional rocket provides the thrust to maintain course, so the plume appears in the opposite direction of the desired thrust. In which direction (Figure 5-12) would we see the plume?

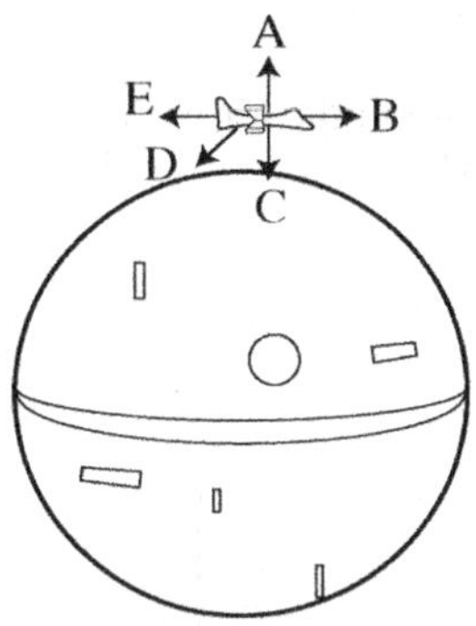


Figure 5-12

Stop! Think about his question and answer it before you look at the solution.

**Solution:** No one chooses B. Few people choose A. In fact, almost everyone chooses E, because they are thinking about a car driving on the surface of the Earth. But a car on the Earth's surface encounters the force of drag, both by friction and by air resistance, whereas the space warrior encounters neither. No force is required to maintain constant speed. If you chose E, it means you need to study the first law of motion again.

Okay, what is the correct way to think of this problem? The net force of the spacecraft is down, toward the center of Bad Star, because the craft is moving at constant speed in a circle. there is no gravity, no friction, no air drag, so the only force on the craft is due to the rockets. Therefore the rocket plume points in the direction of A.

Some students object that a rocket firing in the direction of A would push the craft into the Bad Star. This is the same as objecting that if the Earth pulls on the Moon, it ought to fall down. What holds up the Moon? The answer in both cases is that the centripetal force is large enough to keep the object (spacecraft or Moon) from moving away but not so large as to pull them into the ground. That is, in Figure 5-13, path 1 is the path the spacecraft takes if the warrior does not fire his rockets at all (no force). Path 2 is the path if he does not fire the rockets enough; and path 4, if he fires them too much. Path 3 is just right.

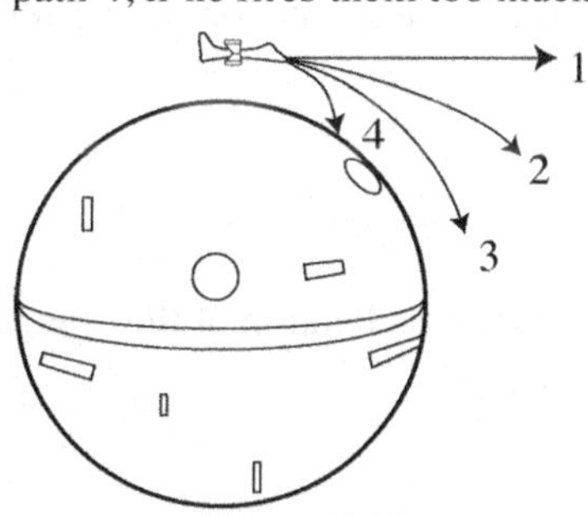


Figure 5-13

The most important concept in this chapter is the independence of horizontal and vertical motion. We saw this concept for the first time in Chapter 4. In this chapter we have used the concept to solve problems involving inclined planes and oblique forces, that is, forces which are neither horizontal nor vertical.

We also looked at circular motion. If an object is moving in a circle, its velocity vector is not constant, and the object must be accelerating. If the object is moving at constant speed, then the direction of the acceleration vector is toward the center, called centripetal acceleration, and its magnitude is $a_{cent} = v^2/r$.

# Physics Review 5 Problems

## Section A

*Use the following information for questions 1–5:*

A cannon shoots an orange (3 kg) straight up in the air with initial velocity 5 m/s (see figure). A horizontal wind exerts a force of 6 N on the orange while it is in the air. Use 10 m/s$^2$ for the acceleration due to gravity.

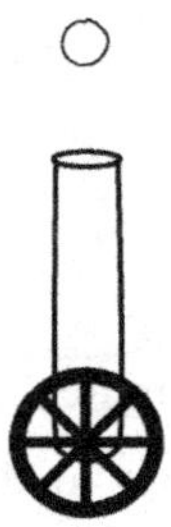

1. Which is the best force diagram while the orange is in the air?

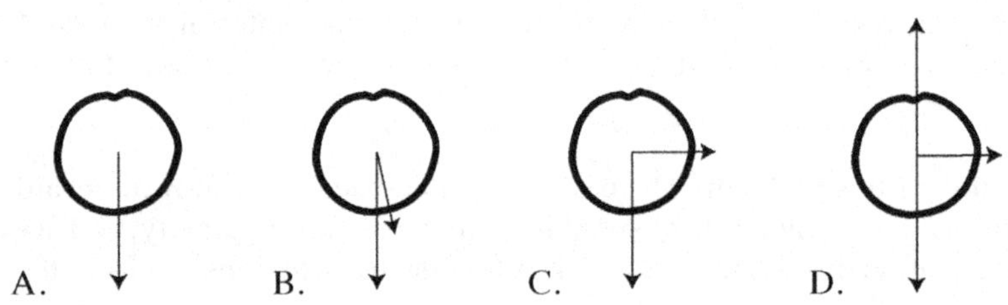


2. What is the vertical component of the net force on the orange while it is in the air?
   A. 0 N
   B. 2 N
   C. 30 N
   D. 32 N

3. How much time does it take the orange to reach the top of its path?
   A. 0.25 s
   B. 0.4 s
   C. 0.5 s
   D. 2.5 s

4. What is the horizontal component of the acceleration of the orange while it is in the air?
   A. 0 m/s$^2$
   B. 2 m/s$^2$
   C. 10 m/s$^2$
   D. 18 m/s$^2$

5. What is the horizontal velocity of the orange at the top of its path?
   A. 0 m/s
   B. 1 m/s
   C. 5 m/s
   D. 1.25 m/s

6. A shoe is being dragged to the right at constant velocity along a level floor by its string. The string is horizontal and bears a tension $T$ in magnitude. The magnitude of the frictional force is $F$, the magnitude of the gravitational force is $G$, and the magnitude of the normal force is $N$. What can be definitely be concluded?
   A. $T < F$
   B. $T = F$
   C. $T > F$
   D. $T + F = G + N$

*Use the following information for questions 7–11:*

A winch pulls a crate of oranges (300 kg) up an incline (30° with the horizontal) by maintaining a tension on a rope over a pulley. The crate is moving at a constant speed 0.2 m/s. There is no friction. (Using $g$ = 10 m/s$^2$.)

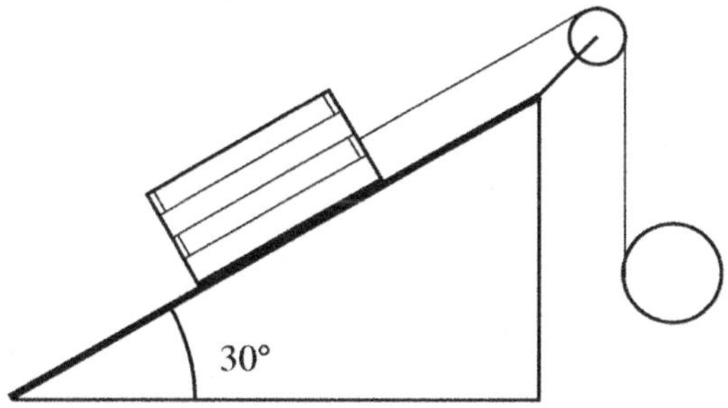


7. What is the component of the gravitational force perpendicular to the surface on which the crate sits?
   A. 0 N
   B. (3000 N) sin30°
   C. (3000 N) cos30°
   D. 3000 N

8. What is the component of the gravitational force parallel to the surface on which the crate sits?
   A. 0 N
   B. (3000 N) sin30°
   C. (3000 N) cos30°
   D. 3000 N

9. What is the normal force on the crate?
   A. 0 N
   B. (3000 N) sin30°
   C. (3000 N) cos30°
   D. 3000 N

10. What is the net force on the crate?
    A. 0 N
    B. (3000 N) sin30°
    C. (3000 N) cos30°
    D. 3000 N

11. What is the tension on the rope?
    A. 0 N
    B. (3000 N) sin30°
    C. (3000 N) cos30°
    D. 3000 N

*Use the following information for questions 12–17:*

A boy (60 kg) is pushing a sled (5 kg) with a stick, so that the stick makes a 30° angle with the vertical. He applies a force of 20 N, so that the force acts along the stick (i.e., there is no shear or friction force). There is negligible friction between the sled and the ground. At time $t = 0$ s, the sled is at rest.

12. Which of the following is the best force diagram for the sled after $t = 0$ s?

A. 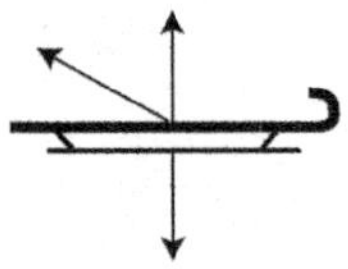

B. 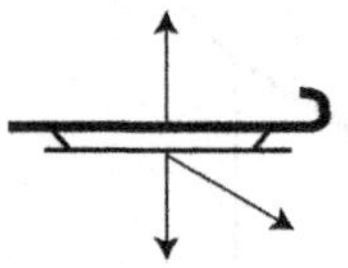

C. 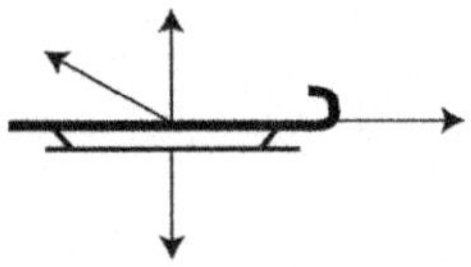

D. 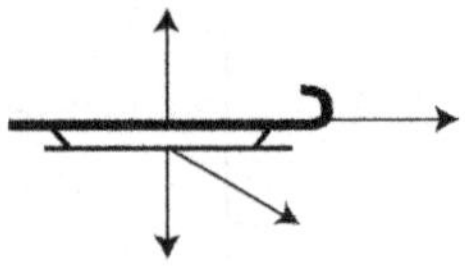

13. What is the horizontal component of the force due to the stick?
 A. 10 N
 B. 17 N
 C. 20 N
 D. 50 N

14. What is the vertical component of the force due to the stick?
 A. 10 N
 B. 17 N
 C. 20 N
 D. 50 N

15. What is the normal force of the ground on the sled?
 A. 17 N
 B. 33 N
 C. 50 N
 D. 67 N

16. What is the net force on the sled?
 A. 0 N
 B. 10 N
 C. 50 N
 D. 67 N

17. What is the acceleration of the sled?
 A. 0 $m/s^2$
 B. 0.5 $m/s^2$
 C. 2.0 $m/s^2$
 D. 9.8 $m/s^2$

## Section B

18. A woman is driving a car along a road when she realizes, almost too late, that she needs to make a left hand turn. She quickly turns the wheel, and the books which were in the passenger seat go crashing against the passenger door. Consider the following statements:

    I. The books were pushed against the door by a centrifugal force.
    II. The books were pushed against the door by a centripetal force.
    III. The forces acting on the books while they are crashing against the door are gravity, the normal force, and a force toward the right.

    Which is (are) true?

    A. I and III are true.
    B. II and III are true.
    C. III only is true.
    D. None is true.

*Use the following information for questions 19–23:*

A stopper is swung on a string, one end of which is fixed at a point P. The diagram shows the stopper and string from the top, and the stopper is swinging counterclockwise at constant speed.

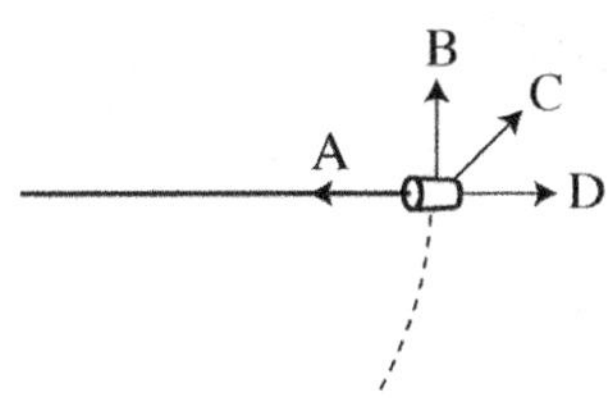


19. Which arrow best shows the direction of the velocity vector?
    A. A
    B. B
    C. C
    D. D

20. Which arrow best shows the direction of the acceleration vector?
    A. A
    B. B
    C. C
    D. D

21. Which arrow best shows the direction of the net force?
    A. A
    B. B
    C. C
    D. D

22. What provides the centripetal force?
    A. Gravity.
    B. Tension.
    C. Friction.
    D. Normal.

23. Which arrow best shows the direction the stopper would go if the string were to break at the moment shown in the diagram?
A. A
B. B
C. C
D. D

*Use the following information for questions 24–27:*

A '79 Buick Regal (1200 kg) is being driven at a constant speed 3 m/s and turning to the right on a curve of road which has effective radius 4 m.

24. Consider the car as viewed from the top. Ignore the gravitational and normal forces (which are vertical and add to zero anyway). Which is the best force diagram?

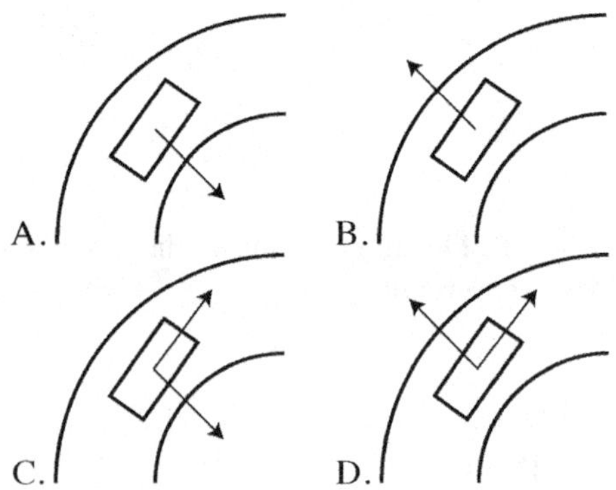


25. What is the acceleration of the Buick?
A. 0 $m/s^2$
B. 0.75 $m/s^2$
C. 2.25 $m/s^2$
D. 9.8 $m/s^2$

26. What is the net force on the Buick?
A. 0 N
B. 900 N
C. 2700 N
D. 11760 N

27. What force provides the centripetal force on the Buick?
A. Gravity.
B. Tension.
C. Friction.
D. Normal.

*Use the following information for questions 28–29:*

A bicycle wheel (mass 3 kg, radius 0.5 m) is spinning at a constant angular speed. It is situated horizontally, and a beetle (5 g) is sitting on the rim. The beetle is traveling at a speed 2 m/s. Use $g$ = 10 m/s$^2$.

28. In the following diagrams, the wheel is viewed almost from the side, so that "down" is toward the bottom of the page. Which is the best force diagram?

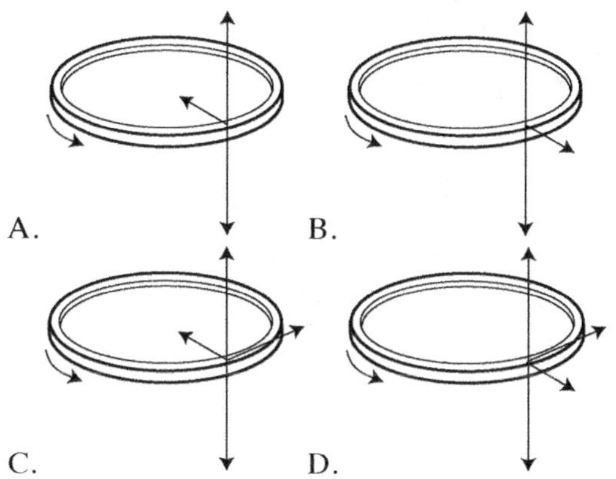


29. What is the acceleration of the beetle?
    A. 0 m/s$^2$
    B. 2 m/s$^2$
    C. 8 m/s$^2$
    D. 10 m/s$^2$

30. The figure shows a ball (from the top view) rolling on a table with a partial hoop. Which arrow best describes the path of the ball after it leaves the hoop?

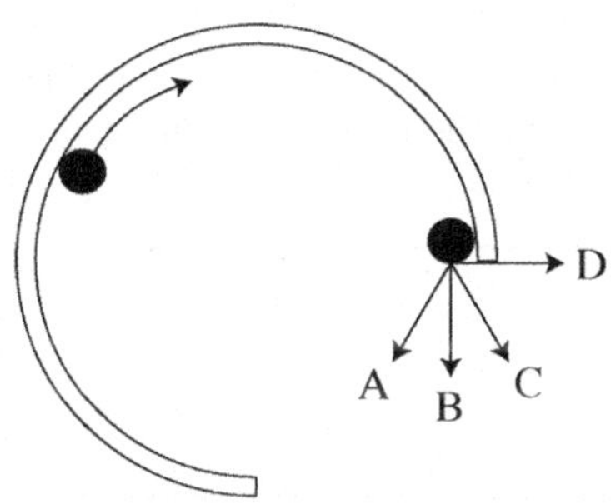


    A. A
    B. B
    C. C
    D. D

## All Sections

*Use the following information for questions 31 and 32:*

A winch pulls a crate of apples (mass $M$) up an incline (making an angle $\alpha$ with the horizontal). (See figure.) The tension exerted by the winch is $T$. At the bottom of the incline the crate begins at rest at $t = 0$. Assume there is no friction.

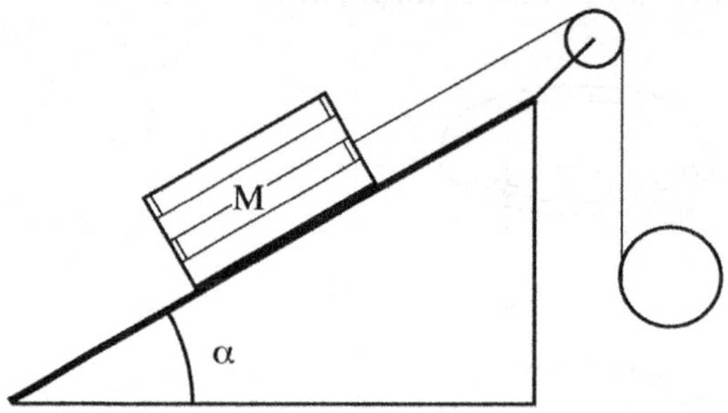


31. What is the normal force of the incline on the crate?
   A. $Mg$
   B. $Mg\cos\alpha$
   C. $Mg\sin\alpha$
   D. $Mg\sin\alpha - T$

32. What is the acceleration of the crate?
   A. 0
   B. $g$
   C. $N/M - g\cos\alpha$
   D. $T/M - g\sin\alpha$

*Use the following information for questions 33 and 34:*

A runner (50 kg) is running around a track (see figure). The curved portions of the track are arcs of a circle, and the dimensions of the track are shown. The runner is running a constant speed 8 m/s. Use 10 m/s$^2$ for the acceleration due to gravity.

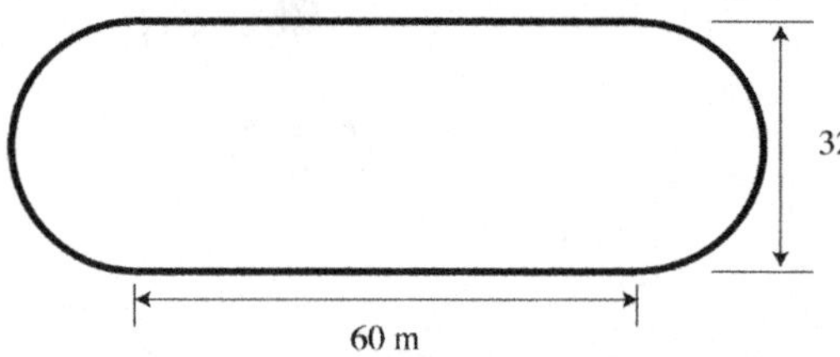


33. When the runner is on the curved portions of the track, what are the forces acting on her?
   A. Gravity, up; the normal force, down; and a force, inward.
   B. Gravity, up; the normal force, down; a force, inward; and a force, forward.
   C. Gravity, up; the normal force, down; and a force, outward.
   D. Gravity, up; the normal force, down; a force, outward; and a force, forward.

34. What is the net force on the runner on the curved portion of the track?
   A. 0 N
   B. 100 N
   C. 200 N
   D. 5000 N

*Use the following information for questions 35–37:*

A Ferris wheel (radius $R$) is turning in a counterclockwise direction at a given frequency ($f$).

35. How would the velocity of a chair on the Ferris wheel change if the frequency were doubled?
    A. It would stay the same.
    B. It would increase by a factor of 2.
    C. It would increase by a factor of 4.
    D. It would increase by a factor of 8.

36. How would the centripetal acceleration of a chair on the Ferris wheel change if the frequency were doubled?
    A. It would stay the same.
    B. It would increase by a factor of 2.
    C. It would increase by a factor of 4.
    D. It would increase by a factor of 8.

37. A person is sitting on a seat at the bottom of a Ferris wheel which is going counterclockwise and speeding up. Which arrow best shows the acceleration vector?

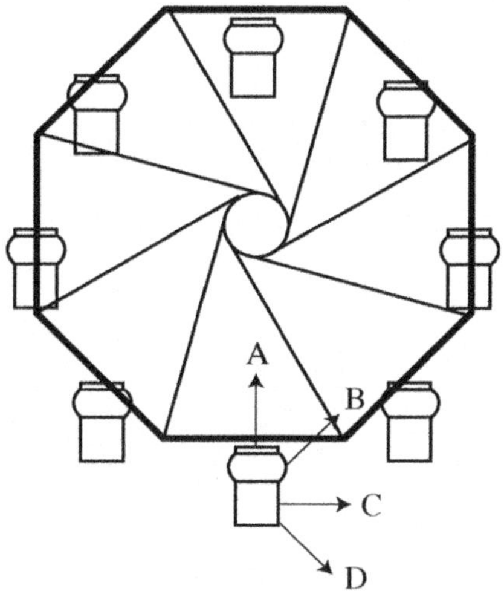


    A. A
    B. B
    C. C
    D. D

38. In a centrifuge, the bottoms of the centrifuge tubes are 10 cm away from the axis of rotation. The centrifuge spins with a frequency of 50 rps (revolutions per second). Which expression gives the velocity of a sample at the bottom of a spinning tube?
    A. $10\pi$ m/s
    B. $20\pi$ m/s
    C. $100\pi$ m/s
    D. $200\pi$ m/s

39. In the following diagram Jupiter is revolving about the Sun. If the gravity of the Sun were somehow cut off when Jupiter was at point P, what path would Jupiter take?

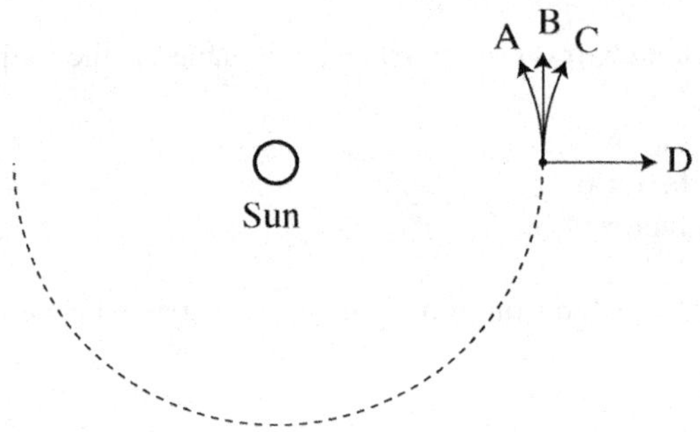


A. A
B. B
C. C
D. D

# Solutions

1. C.
The force diagram should include gravity, pointing down. Nothing else is touching the orange, except the problem mentions that wind exerts a horizontal force, so we add that. The force diagram is shown.

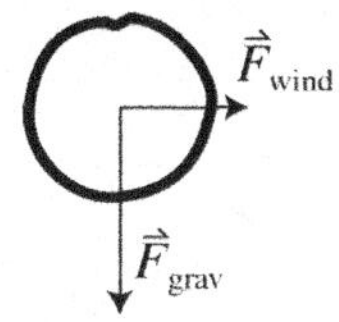


2. C.
The only vertical force is due to gravity, so $F_y = F_{grav} = mg = 30$ N.

3. C.
At the top of the orange's path we have $v_{2y} = 0$ m/s. Since this is vertical information, let's see what other vertical information we have. We have $a_y = -10$ m/s$^2$ and $v_{1y} = 5$ m/s. We want $\Delta t$, so we write $v_{2y} = v_{1y} + a_y\Delta t$ and obtain $\Delta t = 0.5$ s.

4. B.
We want $a_x$. The horizontal information we have is $v_{1x} = 0$ and $F_x = 6$ N. We don't have enough information for the equations of Chapter 2, but we can use $a_x = F_x/m = (6\text{ N})/(3\text{ kg}) = 2\text{ m/s}^2$.

5. B.
Since we know $a_x$, $v_{1x}$, and $\Delta t$, we can find $v_{2x} = v_{1x} + a_x\Delta t = 0 + (2\text{ m/s}^2)(0.5\text{ s}) = 1$ m/s.

6. B.
First we draw a force diagram. This problem has those key words "at constant velocity", which means there is a force balance on the shoe. The horizontal forces are equal in magnitude so that their vector sum is zero, so B is correct. If you chose C, then go back and read the section on the first law of motion.

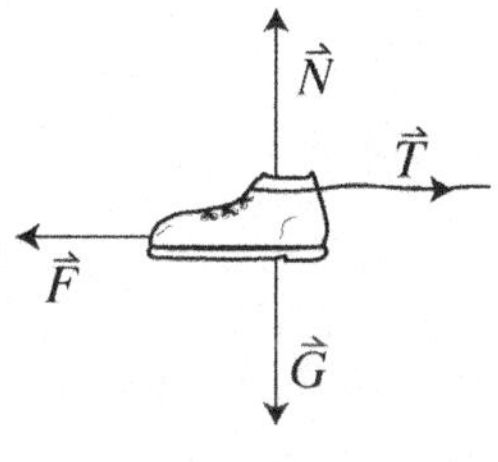

7. C.
First we draw a force diagram (see figure). In addition to gravity, we have two forces, normal and tension, due to two things touching the crate. There is no friction. The gravitational force vector can be separated into two components (see figure).

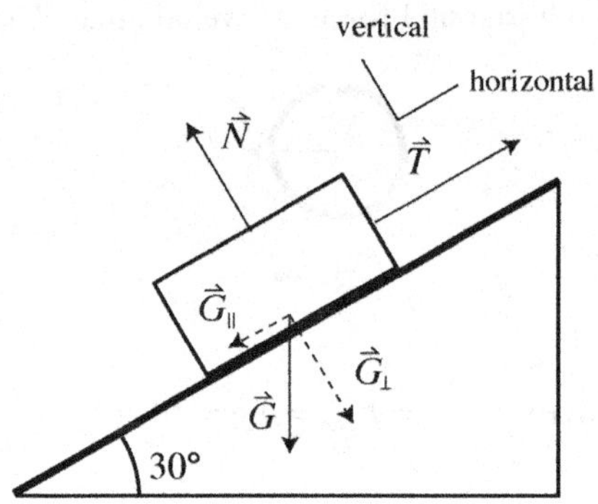


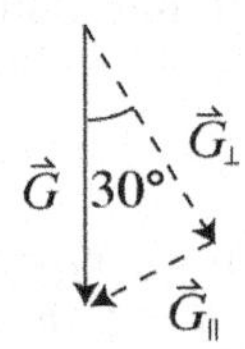


From trigonometry we know

$$\frac{G_\perp}{G} = \cos 30^\circ$$
$$G_\perp = G\cos 30^\circ$$

This gives us choice C.

8. B.
We get this from the same diagrams shown in the solution for 7.

9. C.
If we take the sum of all the perpendicular forces (in the text we called them "vertical"), we obtain $N - G_\perp = (F_{net})_y$. The negative sign denotes the "downward" direction of $G_\perp$. But the crate is moving at a constant velocity, which tells us that the acceleration is zero, and the net force is zero. Thus $N - G_\perp = 0$, giving choice C.

10. A.
As we noted, the net force is zero because of the information we have on the acceleration.

11. B.
This time we take the sum of parallel forces ("horizontal"), and we obtain $T - G_\parallel = (F_{net})_x = 0$. Again, we know the horizontal component of the net force is zero because the crate is moving at constant velocity. Thus the answer is B.

12. B.
There are two things touching the sled: the ground and the stick. So, in addition to gravity pointing down, we draw the normal force pointing up and the force due to the stick pointing down/right. There is no friction (which would act to the left). So B is the correct answer (see figure).

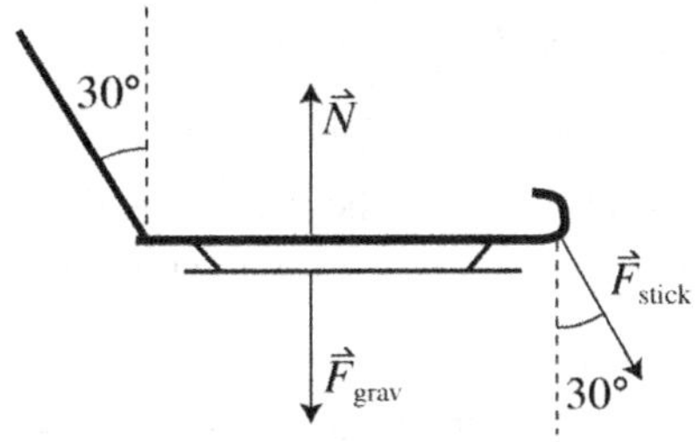


13. A.
We use the following diagram to obtain the vertical component of the stick's force.

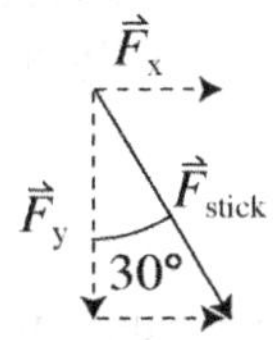


Using trigonometry, we obtain

$$\frac{F_x}{F_{stick}} = \sin 30°$$
$$F_x = F_{stick} \sin 30°$$
$$= (20\text{N})\left(\frac{1}{2}\right)$$
$$= 10\text{N}$$

14. B.
Using the same diagram as in solution 13, we obtain $F_y/F_{stick} = \cos 30°$, and $F_y = 17$ N.

15. D.
In order to obtain the normal force, we need to consider all the vertical forces. We can obtain the vertical component of the net force by looking at the force diagram, so we write

$$(F_{net})_y = N - mg - F_y$$

Here we have used $F_{grav} = mg$ and have chosen "up" to be positive. But we know that the sled is not moving up or down, so the vertical acceleration $a_y$ is zero. And from second law of motion, we know that $(F_{net})_y = ma_y = 0$. The above equation becomes

$$0 = N - mg - F_y$$

$$N = mg + F_y$$

$$= (5\text{ kg})(10\text{ m/s}^2) + 17\text{ N}$$

$$= 67\text{ N}$$

16. B.
In the last problem, we talked about the vertical component of the net force being zero. We can tell from the force diagram that there is a net force, however, and this net force is the horizontal component $F_x$.

17. C.
The acceleration of the sled is given by $a_x = (F_{net})_x/m = 2.0\ m/s^2$

18. D.
The books are following a straight path. By turning the wheel, the driver pulls her car door into the path of the books, giving the impression that the books have a force on them.

19. B.
The direction of the velocity vector is always changing but not its magnitude. At the moment shown in the diagram, the velocity vector is pointing in the direction of the stopper's motion, that is, B.

20. A.
The tangential acceleration is zero, since the speed is constant. However, because the stopper is moving in a circle, the velocity vector is changing direction, and the acceleration vector points toward the center of the circle.

21. A.
Since the acceleration vector $\vec{a}$ points toward the center, we can conclude that the net force $\vec{F}_{net} = m\vec{a}$ points toward the center as well.

22. B.
The centripetal force is provided by the string, which is a force of tension.

23. B.
If the string were to break, then there would no longer be a force to affect the velocity vector. The velocity vector would be constant (first law of motion), so B is correct.

24. A.
Because the car moves in a circle, we know there is a centripetal acceleration and a centripetal force. This narrows the choices to A and C. There is no reason to assume there is a force acting forward on the car, especially since the tangential acceleration is zero (because of the car's constant speed). Hence A is correct. If you chose C, perhaps you were thinking that motion in the forward direction implies there must be a force in the forward direction. Not so.

25. C.
The acceleration is given by $a_{cent} = v^2/r = (3\ m/s)^2/(4\ m) = 2.25\ m/s^2$.

26. C.
Once we know the acceleration of the car, we necessarily know the net force on the car: $F_{net} = ma = (1200\ kg)(2.25\ m/s^2) = 2700\ N$.

27. C.
To see that the centripetal force is due to friction, consider what would happen if there were no friction between the tires and the road. The car would simply slide straight forward into the other lane.

28. A.
Gravity and the normal force add to zero. We know there is a net force toward the center of the wheel, because the beetle is moving in a circle, so this narrows our choices to A and C. Choice C includes a force in the forward direction, but since the wheel is rotating at constant speed, there is no tangential acceleration and no reason to assume there is a tangential force.

29. C.
The acceleration can be calculated $a = v^2/r = (2\ \text{m/s})^2/(0.5\ \text{m}) = 8\ \text{m/s}^2$.

30. B.
Once there is no longer a centripetal force, there is no longer a centripetal acceleration. According to the first law of motion, the velocity vector would be constant.

31. B.
We draw a diagram, including the three forces on the crate. We choose "horizontal" and "vertical" to be parallel and perpendicular to the surface, respectively, and we divide gravity into components (see figure).

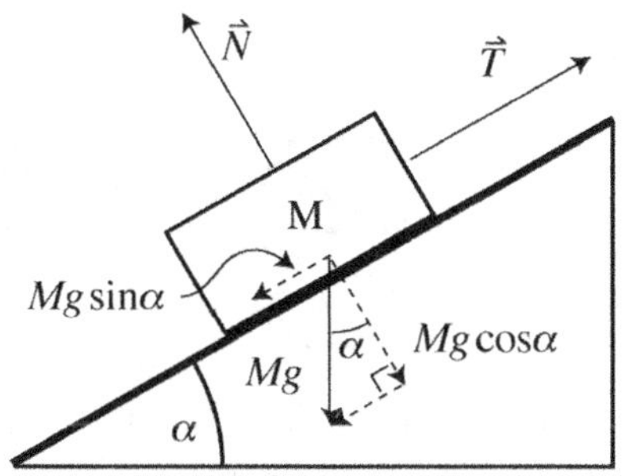


Since the normal force is "vertical", let us consider the vertical component of $F_{net} = a$. From the diagram we obtain

$$(F_{net})_y = N - Mg\cos\alpha$$

But we know that $a_y = 0$, because the crate is not moving up or down. This implies $(F_{net})_y = 0$, and we have

$$0 = N - Mg\cos\alpha$$

$$N = Mg\cos\alpha$$

32. D.
If we consider the "horizontal" component of $F_{net} = Ma$, then we can obtain from the diagram

$$(F_{net})_x = T - Mg\sin\alpha$$

where we have taken the positive direction to be up the incline. The acceleration is

$$a_x = \frac{(F_{net})_x}{M}$$

$$= \frac{T}{M} - g\sin\alpha$$

33. A.
The force inward is due to friction between the runner's shoes and the track. It may seem surprising that there is not a force forward, since the runner is actively running. There is a horizontal acceleration, and thus net force, only at the beginning of the race when the runner is accelerating. When she is going at a constant speed, she expends effort pushing down with the foot in contact with the track and moving it back fast enough.

34. C.
The net force is given by $F_{net} = ma$, and the only acceleration is centripetal, so we have $a = a_{cent} = v^2/r$.

35. B.
The frequency is the number of revolutions per unit of time. Each revolution represents a trip of length $2\pi R$. So the velocity is the total distance per time, that is, $v = 2\pi Rf$. If $f$ is doubled, then $v$ is doubled.

36. C.
On the other hand the centripetal acceleration is $a_{cent} = v^2/r$, so $a_{cent}$ is increased by a factor of 4.

37. B.
The person experiences a centripetal acceleration because he is moving in a circle, and he experiences a tangential acceleration because he is speeding up.

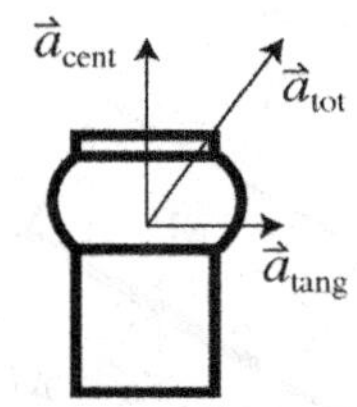


The tangential acceleration is in the same direction he is going. It would point the opposite way if he were slowing down. Adding the two vectors together gives the result B (see figure).

38. A.
In one second the centrifuge sample undergoes 50 revolutions, and each revolution is the circumference $2\pi r = 2\pi(0.1\text{ m})$. Thus in one second the sample travels $50\,(0.2\pi\text{ m})$, giving a velocity $10\pi$ m/s.

39. B.
If there is no gravity, then there is no net force, and Jupiter follows a straight line at constant speed, maintaining the same velocity vector as it had when the gravity was cut off.

# Biology Review 1
# Introduction

## A. Philosophy of the Biology Review Section

This review section is is designed to help you understand major concepts in biology and to use this knowledge to solve problems. Each chapter contains ACT-style questions with answers and full explanations.

The ACT questions are written to test your knowledge of basic concepts in biology and problem solving abilities; they do not stress rote memorization of facts. The questions are based on short passages. Four to eight questions will be asked for each passage. These questions may require you to interpret data from graphs, tables or figures.

This introductory chapter will familiarize you with the concepts that form the foundations of biology. These key ideas are implicit in every topic of biology, and we recommend that you keep them in the back of your mind at all times.

## B. Biology

We'll start with a simple definition of biology: *biology is the study of life*. This includes examination of the processes that govern how life is maintained and reproduced, and observation of how living things interact with each other and with the environment. Although this sounds rather simple, it involves studying many different and diverse areas, from the molecular level to the global scale.

There are two distinct perspectives from which one can study biological processes: *in vivo* or *in vitro*. *In vivo* means research is done in the body itself. *In vitro* means literally "in glass," as an experiment that is done in a test tube, or outside of the body. *In vivo* experiments, naturally, include all the factors that can influence a process; therefore, the reactions being studied represent a "real life" situation. However, all these factors and reactions can make for a very complex system that is not easily understood. *In vitro* studies, on the other hand, allow researchers to manipulate the system to study only one or two isolated factors. This helps to determine exactly what these factors do.

Although each approach has its merits and drawbacks, both perspectives are needed and work synergistically. For example, the study of AIDS has benefited from both *in vivo* and *in vitro* studies. The course of the illness, from infection with HIV to full blown AIDS, has been well characterized in humans through the use of *in vivo* studies. But the exact mechanism of how the virus invades a cell is known through *in vitro* studies. Drugs designed to fight the virus are first tested in vitro using HIV infected cells grown in the laboratory. Drugs that prove effective *in vitro* are then tested *in vivo*.

## C. The Scientific Method

As with all sciences, biological research involves methodically searching for information. The procedure associated with this search is called the *scientific method*. It involves

1. asking questions, which are then followed by one or more *hypotheses* (educated guesses or hunches that answer or explain the question).
2. making predictions from the hypothesis, usually in the form of "if....then" statements (*if* the influenza virus causes the flu, *then* those exposed to it will become ill).
3. testing the predictions through experimentation, observation, model building, etc., including appropriate controls with which to compare the results.
4. repeating the investigations and devising new ways to further test the hypothesis (this may include modification of the hypothesis based on the results of the tests).
5. reporting the results and drawing conclusions from them.

A *theory* is similar to a hypothesis in that it is subjected to the scientific method, but a theory usually explains a broad range of related phenomena, not a single one. Theories are well supported hypotheses, shown to be valid under many different circumstances.

In science, there is no real beginning or end. All hypotheses are based on previous work, and all results and conclusions can be expanded in the future. Often experiments raise more questions than they answer.

## D. Characteristics of Life

If biology is the study of life, then it is natural to ask: What is life? There is no one definition of life that can encompass all living things. However, there are certain characteristics that all living things share:

- *order and organization*: The basic unit of life is the cell. It is capable of performing all activities of life. For single-celled organisms, this is the limit of their organization. For multicellular organisms, cells may be arranged in tissues, tissues arranged in organs, and organs arranged in systems. Living things are further organized into populations, communities, ecosystems and, ultimately, the biosphere (the combined regions of the earth in which organisms can live). A corollary to this topic is the organization and structure of the genetic material: all living things have nucleic acids as the storage mechanism for genetic information.
- *growth and development*: All living things grow and develop during their life. This may be as simple as a bacterium that increases in size or as complex as a fertilized egg that develops into an elaborate, multicellular organism.
- *reproduction*: Organisms must reproduce in order for the species to survive. The exact mode of reproduction may be different (for example, *asexual* verses *sexual* reproduction), but the outcome is the same: an increase in the number of organisms.
- *energy metabolism*: All living things require energy to survive. Various processes are necessary in order to supply this energy. First of all, organisms must gain nutrition: those that can make their own food are called *autotrophs*, all others are called *heterotrophs*. The nutrients must be converted into energy, which includes the process of *respiration*. Finally, waste products must be eliminated.
- *stimuli response and homeostasis*: The ability to respond to the environment, whether it be the external or the internal environment, is a vital function in all living things. Organisms must regulate their life processes based on what is happening in the environment.
- *evolution*: The ability to change, to mutate, is an important characteristic of life. Were it not for this ability, the vast diversity of life would not exist on this planet. For that matter, no life would exist at all.

## E. Structure and function

Another topic which underlies all aspects of biology is the correlation between structure and function. If the structure is known, often the function can be determined, and vice versa. For example, knowing the amino acid sequence of a protein (the structure) can help predict how that protein functions in a cell. Or, as another example, if an animal has the ability to see (the function), then it must have the structures necessary to support that function (eyes, optic nerves, a region of the brain to interpret nerve impulses, etc.).

## F. Using this review section

Remember: the ACT stresses your problem solving skills and your knowledge of basic biological concepts. Therefore, you will be better prepared for the test by understanding these concepts rather than by just memorizing various facts and terms. These review sections will help you in this endeavor. Study to understand the topics, not just to memorize them.

# Chapter 1 Problems

**Passage**

Karposi's sarcoma (KS) is a cancer marked by purple tumors on the skin. Although extremely rare in the population, it is found in approximately 25% of gay men infected with HIV, the virus that causes AIDS. It has been suggested that KS itself is caused by a virus, the human herpesvirus 8 (HHV8). To help confirm this hypothesis, the following observations were made:

*Observation 1*:

At regular intervals over a two-year period, samples of blood were taken from gay men who tested positive for HIV at the beginning of the study. In 38 of the subjects, no KS developed, although 18% had antibodies against HHV8 in their blood samples. However, in 40 subjects who developed KS, 80% had HHV8 antibodies. In 11 of these men, HHV8 antibodies were detected from the beginning of the study, but 21 of the men developed HHV8 antibodies during the study. In these subjects, the antibodies appeared several months before the onset of KS.

*Observation 2*:

Individuals not known to have been infected with HIV were tested for antibodies against HHV8. In 141 blood donors, only 1% were found to test positive for the antibodies. In addition, 300 hemophiliacs who had regular blood transfusions were also examined. Three percent had HHV8 antibodies.

*Observation 3*:

In 176 patients who had syphilis, a sexually transmitted disease, 36 (or 20%) had antibodies to HHV8.

1. What conclusion can be drawn from these observations?
   A. Although HHV8 is associated with KS, it has yet to be proven that it causes KS.
   B. In observation 1, the fact that some individuals tested positive for HHV8 but did not develop KS proves this virus does not cause the disease.
   C. Individuals with syphilis will also develop KS.
   D. Hemophiliacs are at no risk of developing KS.

2. Which is the best hypothesis for the mode of transmission of HHV8?
   A. blood transfusions
   B. casual contact
   C. sexual transmission
   D. airborne particles

3. In another study, it was found that KS is prevalent in transplant patients taking drugs which suppress their immune system to prevent rejection of the new organ. This, along with the above evidence, indicates:
   A. a virus other than HHV8 causes KS.
   B. a weakened immune system is probably necessary for KS to develop.
   C. both A and B
   D. neither A nor B

4. Many researchers argue that viruses are not alive. Which of the following supports this hypothesis?
   I. Some viruses carry their genetic information in the form of RNA, not DNA.
   II. Viruses cannot reproduce by themselves.
   III. Viruses do not evolve; that is, they do not mutate or change.
   A. I only
   B. II only
   C. I and II only
   D. II and III only

# Solutions

1. A.

No direct evidence is presented to confirm that HHV8 causes KS, although it appears to be associated with the disease. Answer B is not the best choice as observation 1 was based on a 2-year study: those who tested positive for HHV8 may have developed KS at a later date. A statement about all syphilis patients developing KS is not valid, even if HHV8 is truly the cause of the disease: only a certain percentage would be predicted to develop the disease. And, conversely, it cannot be stated that all hemophiliacs are safe from the disease either.

2. C.

KS is prevalent in individuals with HIV and syphilis, two sexually transmitted diseases. Although HIV can also be contracted through blood transfusions, it appears HHV8 cannot due the hemophiliac data in observation 2. Casual contact and airborne transmission can also be ruled out as HHV8 antibodies are rarely found in HIV negative blood donors.

3. B.

No evidence is given that transplant patients had or didn't have HHV8, so the virus as a cause of KS cannot be ruled out based on this statement. However, since these transplant patients have a weakened immune system, as is found in AIDS, it appears KS thrives under this condition.

4. B.

Although some may argue that one characteristic of life is the storage of genetic information in the form of DNA, numerous viruses do contain DNA. Also, one outstanding feature of many viruses is their rapid mutation rate. However, all viruses need the cellular machinery to replicate: they cannot reproduce on their own.

# Biology Review 2
# Biological Molecules

## A. Introduction

The study of *biochemistry* specifically explores the structure and function of important molecules found in living organisms. All important biomolecules are *organic*, that is, based on the carbon atom. Such molecules are often called *macromolecules*, because they are relatively large in comparison with the molecules studied by traditional, *inorganic* chemistry. Biological macromolecules are often *polymers* -- molecules formed by the stepwise addition of smaller subunits (*monomers*). Four major classes of biological molecules have been identified, each with unique structural properties and different roles.

## B. Carbohydrates

*Carbohydrates* get their name from the fact that they are composed of only three types of atoms in particular combinations. The term literally means "carbon and water", belying their atomic composition: [C(H2O)]n. This means that only carbon, hydrogen, and oxygen are present, and there is usually twice as much hydrogen as oxygen or carbon. In animals, the major function of carbohydrates is to provide energy for the organism. The fundamental carbohydrate subunit is the *monosaccharide* (see Figure 2.1). Monosaccharides can exist alone, or can be polymerized into larger *disaccharides* (see Figure 2.2) and *polysaccharides*.

- *Monosaccharides:* These are the simplest carbohydrate subunits found in nature. Along with disaccharides, they have a sweet taste and have thus been referred to historically and nutritionally as *sugars* or *simple carbohydrates*. Many nutritional monosaccharides are six-carbon compounds, such as *glucose* (the body's favorite fuel molecule), *fructose* ("fruit sugar"), and *galactose*. Other important monosaccharides are the five-carbon sugars *ribose* and *deoxyribose*, part of the nucleotides that compose DNA and RNA.
- *Disaccharides*: Composed of two monosaccharide units joined by a *glycosidic* bond, these are also recognized as sugars or simple carbohydrates nutritionally. *Sucrose*, or common table sugar, consists of one glucose and one fructose subunit. *Lactose* ("milk sugar") is made up of one glucose and one galactose subunit. *Maltose* ("malt sugar") is composed of two glucose subunits. All of these molecules must be broken down into their constituent monosaccharides before they can be utilized by the body.
- *Polysaccharides*: Polysaccharides are made up of many, often hundreds, of monosaccharide subunits. In nature, all of the important polysaccharides are glucose polymers, differing only in their physical arrangement and the type of bonds that join the subunits. Because they do not taste sweet, they are referred to nutritionally as complex carbohydrates, and include such compounds as starch and fiber. *Starch* is an energy storage molecule found in plants, and often makes up a large part of the human diet in the form of grains and vegetables. Because it is a polymer of glucose, starch is broken down into glucose subunits to be used as fuel in our bodies. Fiber, or *cellulose*, is a structural polysaccharide, composing the cell walls of plants. Due to the nature of the glycosidic bonds joining the glucose subunits, however, most animals (including humans) are unable to digest it. *Glycogen* is very similar to starch, and is sometimes referred to as "animal starch". Animals often store excess glucose in this form in their livers and muscles as an energy reserve.

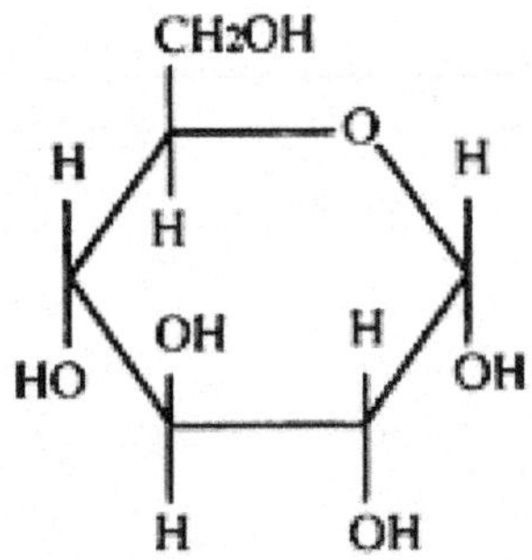


**Figure 2.1: Glucose, a monosaccharide**

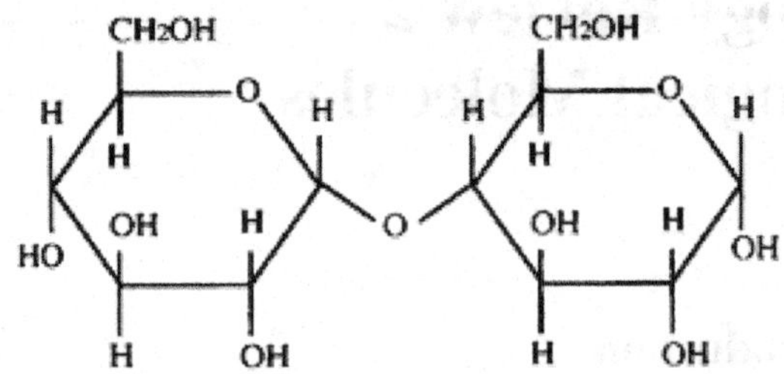


**Figure 2.2: Maltose, a disaccharide**

## C. Lipids

*Lipids* are macromolecules grouped together for different reasons than are carbohydrates. They do not share any particularly constant chemical structure, but they do share an important physical property brought about by their basic chemical composition: lipids are biological molecules that do not dissolve appreciably in water. This is because they contain nonpolar covalent bonds, and are largely composed of hydrocarbon chains or rings. Since the body is a very watery place, lipids face a challenge, as they are not able to dissolve, and must be handled similarly with regard to their transportation and usage. Several types of lipids exist.

- *Triglycerides:* Triglycerides are composed of one molecule of the trialcohol *glycerol* covalently attached to three *fatty acid* molecules, hydrocarbon chains of varying lengths bonded through a terminal carboxylic acid group (see Figure 2.3). Traditionally called *fats* and *oils*, the major role of triglycerides in the body is long-term energy storage. Fats tend to be solid at room temperature because the fatty acid chains are *saturated*, which means they do not contain carbon-carbon double bonds. Chains containing double bonds are called *unsaturated*. The more unsaturated a fatty acid chain, the more liquid the triglyceride. Thus oils are often *polyunsaturated* triglycerides.

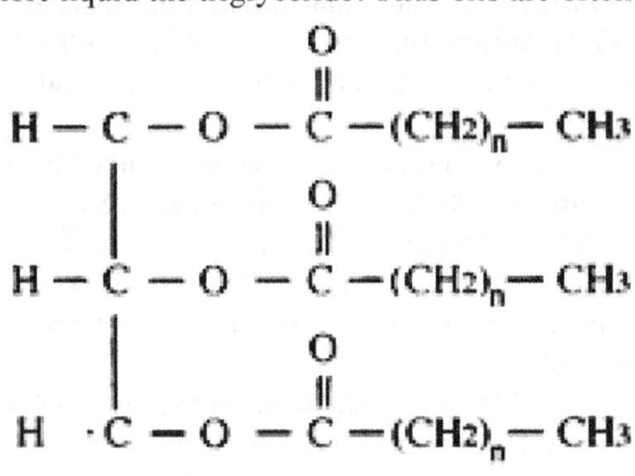


**Figure 2.3: A saturated triglyceride**

- *Phospholipids:* Phospholipids are a class of related compounds that structurally resemble triglycerides. In place of one of the fatty acids bonded to glycerol, however, is a hydrophilic molecule containing a phosphate group. This gives phospholipids a chemical "split personality". A portion of the molecule is *hydrophobic*, and unable to dissolve in water, while another portion is strongly charged and *hydrophilic*. This interesting combination of properties allows phospholipids to form the structures of plasma membranes and lipoproteins. They also act as emulsifying agents, allowing other lipids to dissolve more easily in the body.
- *Steroids:* Steroids are lipids that do not structurally resemble triglycerides, but are composed of a series of nonpolar rings (see Figure 2.4). *Cholesterol* is the most well known and prevalent steroid compound in the body. Cholesterol plays a role in the structure of cell membranes, as well as serving as the starting compound from which many others are synthesized. Other important steroids include the sex hormones and vitamin D.

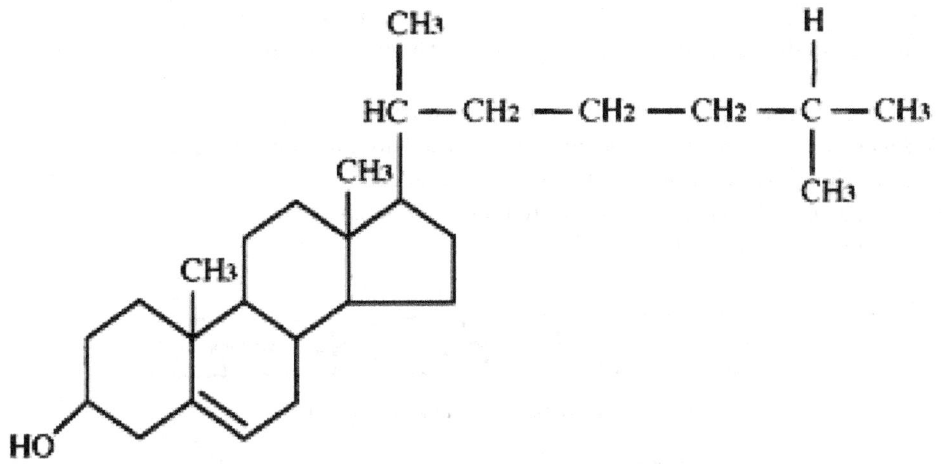

**Figure 2.4: Cholesterol, a steroid lipid**

## D. Proteins

*Proteins* are polymeric macromolecules made up of subunits of *amino acids* (see Figure 2.5). As you might expect, amino acids all contain an amino group and a carboxylic acid group; what differentiates them is a variable portion referred to as the *R group*. In a sense, proteins are structurally simple, since every one consists of a number of amino acids linked by *peptide bonds*. There are twenty different amino acids, however, which can be linked together in any order, and a typical protein contains anywhere from 30 to 1,000 amino acids. Thus, a remarkably vast diversity of different proteins is possible, and this is exactly what we find. There are probably close to 100,000 different proteins in the human body, each with a different function. The remarkable diversity of protein function is made possible by the fact that once a chain of amino acids is linked together (also called a *polypeptide*), it undergoes additional folding so that the final protein molecule exists in a particular three-dimensional conformation. It is this shape that allows it to function in a unique way. We can identify four levels of protein structure (see Figure 2.6).

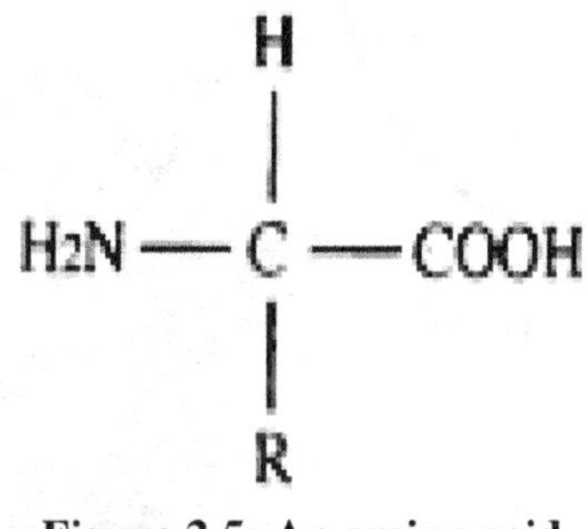

**Figure 2.5: An amino acid**

- *Primary structure:* A protein's primary structure simply refers to the linear order of amino acids it contains.
- *Secondary structure:* The secondary structure of a protein comes about due to local interactions, usually hydrogen bonds between atoms of adjacent amino and acid groups. Common secondary structures include the *alpha-helix* and the *beta-pleated sheet.*
- *Tertiary structure:* A protein's tertiary structure refers to its ultimate three dimensional shape. It folds uniquely due to long range interactions between the R groups of the amino acids. Such interactions include hydrogen bonding, electrostatic interactions, and hydrophobic interactions. It is the tertiary structure that is responsible for the protein's function.
- *Quaternary structure:* Not all proteins have a quaternary structure; only those that consist of multiple polypeptide chains. Quaternary folding refers to the interactions between multiple chains of amino acids to achieve a protein that can only function in this complex state.

Proteins perform a vast array of functions, acting as enzymes, antibodies, structural components, hormones, and a wide variety of other functional entities. Well-known proteins include:

- *Hemoglobin*, which helps carry oxygen in the blood;
- *Collagen and Keratin*, major components of skin, hair, and connective tissues;
- *Insulin*, a hormone that regulates blood glucose levels;
- *Pepsin*, an enzyme that digests other proteins in the stomach;
- and others too numerous to list!

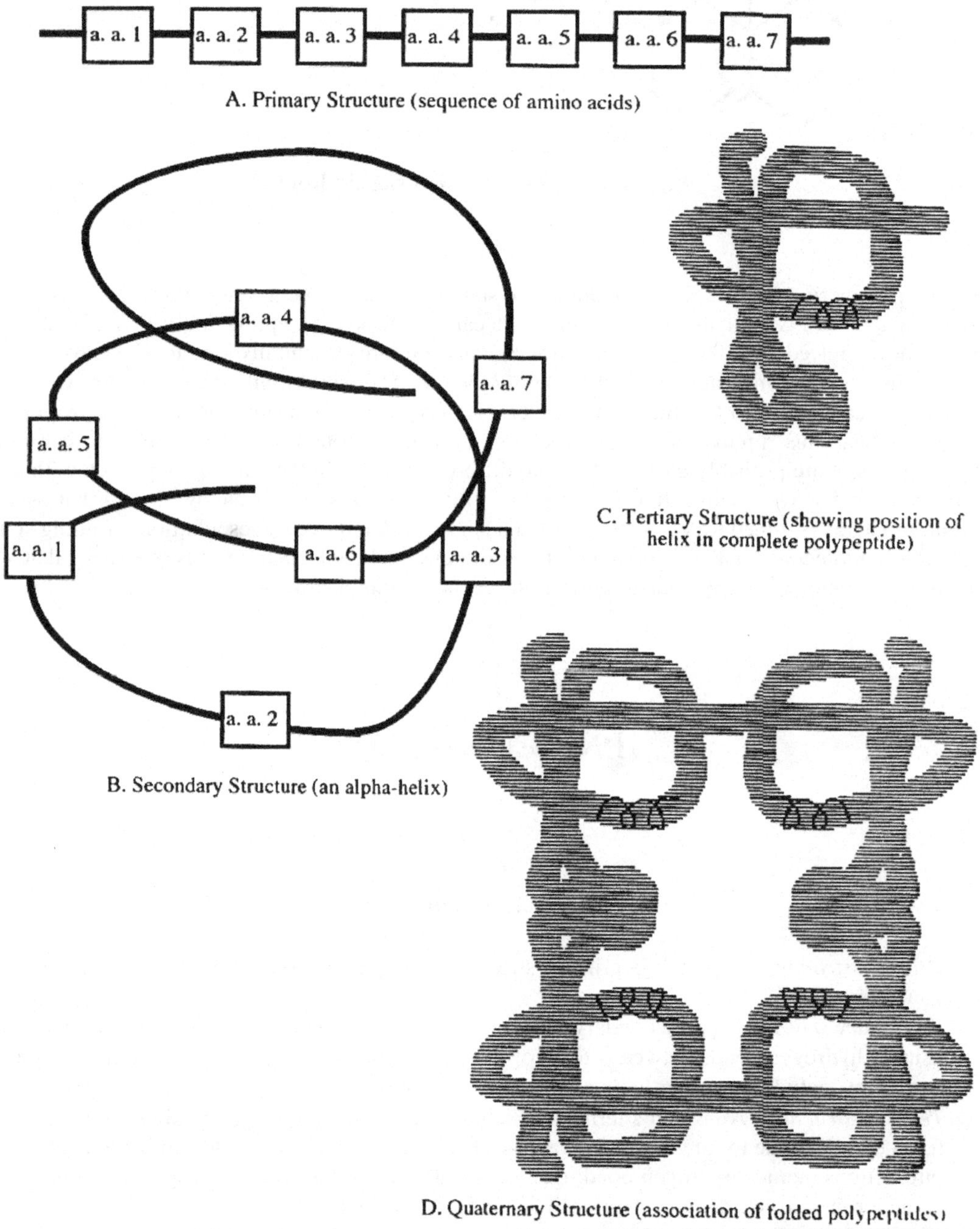


**Figure 2.6: Hierarchical folding of a hypothetical protein. aa: amino acid.**

## E. Nucleic Acids

Nucleic acids are also macromolecular polymers made up of a particular type of subunit, in this case, the *nucleotide* (see Figure 2.7). Nucleotides are more complex than the other subunits we have considered. Each one is made up of:

- a five-carbon sugar
- a nitrogenous base
- a phosphate group

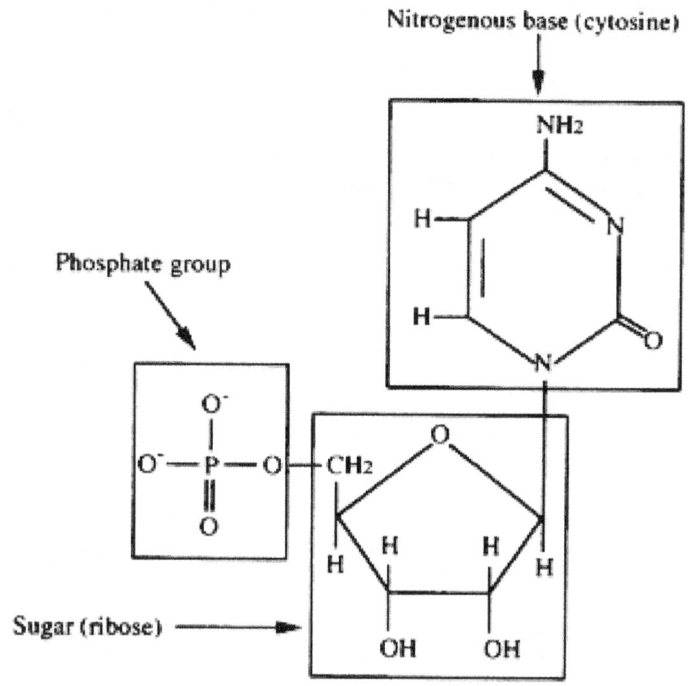


**Figure 2.7: Cytidine monophosphate (CMP), a ribonucleotide**

Two general types of nucleotides are recognized, depending upon which sugar they contain. Therefore, two major types of nucleic acids can be constructed, depending upon which kind of nucleotide is used.

- *DNA (deoxyribonucleic acid)* is made of nucleotides that contain the sugar *deoxyribose*.
- *RNA (ribonucleic acid)* is made of nucleotides that contain the sugar *ribose*.

Deoxyribonucleotides are of four types, depending upon which of four possible nitrogenous bases they contain. The four bases are:

- *Adenine*
- *Guanine*
- *Cytosine*
- *Thymine*

Ribonucleotides are of four types, depending upon which of four possible nitrogenous bases they contain. The four bases are

- *Adenine*
- *Guanine*
- *Cytosine*
- *Uracil*

As you can see, the bases are similar in DNA and RNA, with only one exception: the thymine of DNA is replaced by uracil in RNA. Due to their chemical structures, adenine and guanine are referred to as *purines*, while cytosine, thymine, and uracil are called *pyrimidines*.

Nucleotides are joined to one another by *phosphodiester bonds*, creating long chains. As with proteins, what makes one DNA molecule different from another is the sequence of bases that makes up the primary structure. This sequence carries encoded information, and is the basis for the "genetic code". In fact, the major functions of DNA and RNA all deal with the storage, transmission, and usage of genetic information.

It turns out that DNA normally exists in the form of a double helix in nature. Two antiparallel strands of nucleotides are held together by interactions between the bases to form a structure that resembles a twisted ladder. On the other hand, RNA is usually single-stranded.

# Biology Review 2 Problems

1. Which of the following sugars does not need to undergo any enzymatic processing before it can be absorbed by the human digestive system?
   A. Lactose
   B. Maltose
   C. Galactose
   D. Sucrose

2. Which of the following is true of the differences between DNA and RNA?
   A. DNA contains a five-carbon sugar, while RNA contains a six-carbon sugar.
   B. DNA contains phosphate groups, while RNA does not.
   C. RNA contains the purine base adenine in place of the purine base guanine contained in DNA.
   D. RNA contains a different five-carbon sugar than DNA.

3. Which of the following statements is true regarding proteins?
   A. All proteins exhibit quaternary folding.
   B. The primary structure of a protein refers simply to the linear order of amino acids it contains.
   C. Not all proteins contain amino acids, only the ones that have extensive three-dimensional folding.
   D. All proteins are composed of one polypeptide chain.

4. Which of the following compounds would one expect to be liquid (with the lowest density) at room temperature?
   A. A saturated triglyceride
   B. A monounsaturated triglyceride
   C. A polyunsaturated triglyceride
   D. None of the above, as all would be solids at room temperature.

5. Which of the following carbohydrates is a polymer of the monosaccharide glucose?
   A. Cellulose
   B. Lactose
   C. Ribose
   D. None of the above

# Solutions

1. C.
The human digestive system can only absorb sugars in the form of monosaccharides. Galactose is the only monosaccharide listed (glucose and fructose are the others). Lactose, maltose, and sucrose are all disaccharides, and must be enzymatically split before absorption.

2. D.
One of the major differences between DNA and RNA is that RNA contains the five-carbon sugar ribose, while DNA contains the five-carbon sugar deoxyribose. Both DNA and RNA contain phosphate groups, and both contain adenine and guanine. (RNA contains the pyrimidine base uracil instead of the pyrimidine base thymine.)

3. B.
Protein primary structure refers only to the linear order of amino acids. Secondary and tertiary folding are characteristics of all proteins, and involve interactions between various amino acids and ultimately cause specific three-dimensional folding. Both A and D are incorrect for related reasons; some proteins consist of more than one polypeptide chain, and it is only these proteins that exhibit quaternary folding (the association of multiple polypeptides). All proteins contain amino acids, by definition, so C cannot be correct.

4. C.
The major factor that determines whether or not a triglyceride will be solid or liquid at room temperature is the level of saturation of the fatty acids it contains. A completely saturated triglyceride is completely "filled" with hydrogen atoms, and is therefore linear in shape. This allows the molecules to pack closely and easily together, which results in a solid phase physical structure. We would call this type of molecule a "fat" or "saturated fat". A point of unsaturation occurs where a carbon-carbon double bond exists, and hydrogen atoms could potentially be added by hydrogenation. The more points of unsaturation, the more "kinks", or bends, the fatty acid will have. This makes packing together difficult, and we obtain a liquid, usually called an "oil".

5. A.
Cellulose is the only polysaccharide, or complex carbohydrate, listed. Lactose is a disaccharide, and ribose is a monosaccharide.

# Biology Review 3
# Enzymes and Energy

## A. Introduction

The study of energy transformations, or *thermodynamics*, is covered in the chemistry. However, it is useful to remind ourselves that the laws of thermodynamics apply as much to living organisms and cells as they do to inanimate objects. Cells therefore had to evolve methods of obtaining and processing energy that are in accordance with the general principles of energy transformations. It is a general chemical principle that in order for a chemical reaction to proceed, a certain amount of energy, the *activation energy*, must be absorbed by the reactants to break the bonds already in place. One way of providing activation energy is simply to add heat to the reactants. This is what we do when we light a match to start a fire, which allows a combustion reaction to proceed. Living systems, however, cannot withstand the high temperatures necessary to overcome the activation energy barriers for biochemical reactions. Another method must exist to allow cells to facilitate and control chemical reactions. *Enzymes* are biological catalysts that facilitate reactions by lowering the necessary activation energy (see Figure 3.1).

## B. Structure and Function of Enzymes

What types of molecules are enzymes and how do they work? Enzymes are almost always large protein molecules, folded into a particular three-dimensional configuration (see Chapter 2). Recently, some RNA molecules have been found to have enzymatic functions, but the vast majority of enzymes are proteins. The protein is folded so that a particular portion of the molecule, the *active site*, is accessible and forms a surface that attracts and aligns the reactant(s) in a favorable configuration. The reactant(s) are referred to as the enzyme's *substrate*, and when they associate with the enzyme, the intended reaction is able to proceed efficiently at the relatively low temperature of the cell. The amino acids that comprise the active site are close together in space, but may be far apart in the primary structure of the protein. Thus factors that disturb the overall folding of the protein may decrease or totally destroy the enzyme's ability to function.

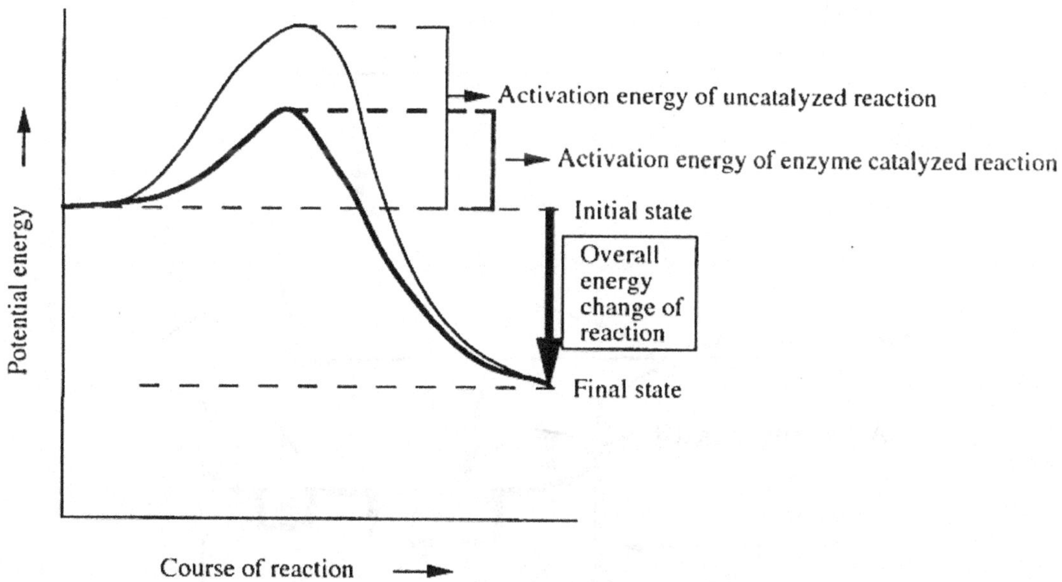


**Figure 3.1: Activation energies of catalyzed and uncatalyzed reactions**

The way the enzyme interacts with the substrate is only now becoming completely clear. Initially, the association was envisioned as a "lock and key" model, in which the shape of the active site matched the shape of the substrate exactly. This would account for their ability to come together easily. A more modern idea is the "induced fit" model, in which the active site and substrate have an affinity for each other, but the binding of the substrate may change the conformation of the active site, inducing a better fit and perhaps straining the bonds that will be broken in the substrate (see Figure 3.2).

Enzymes have great *specificity*, which means that one enzyme can catalyze only one reaction or a set of related reactions. This is of great benefit because it allows the cell to control different reactions independently, by regulating the activity or quantity of the enzyme involved. Furthermore, the enzyme is not permanently altered in any way by participating in the reaction, and so is "recyclable", being used over and over again. Therefore, enzymes typically do not need to be manufactured in large quantities.

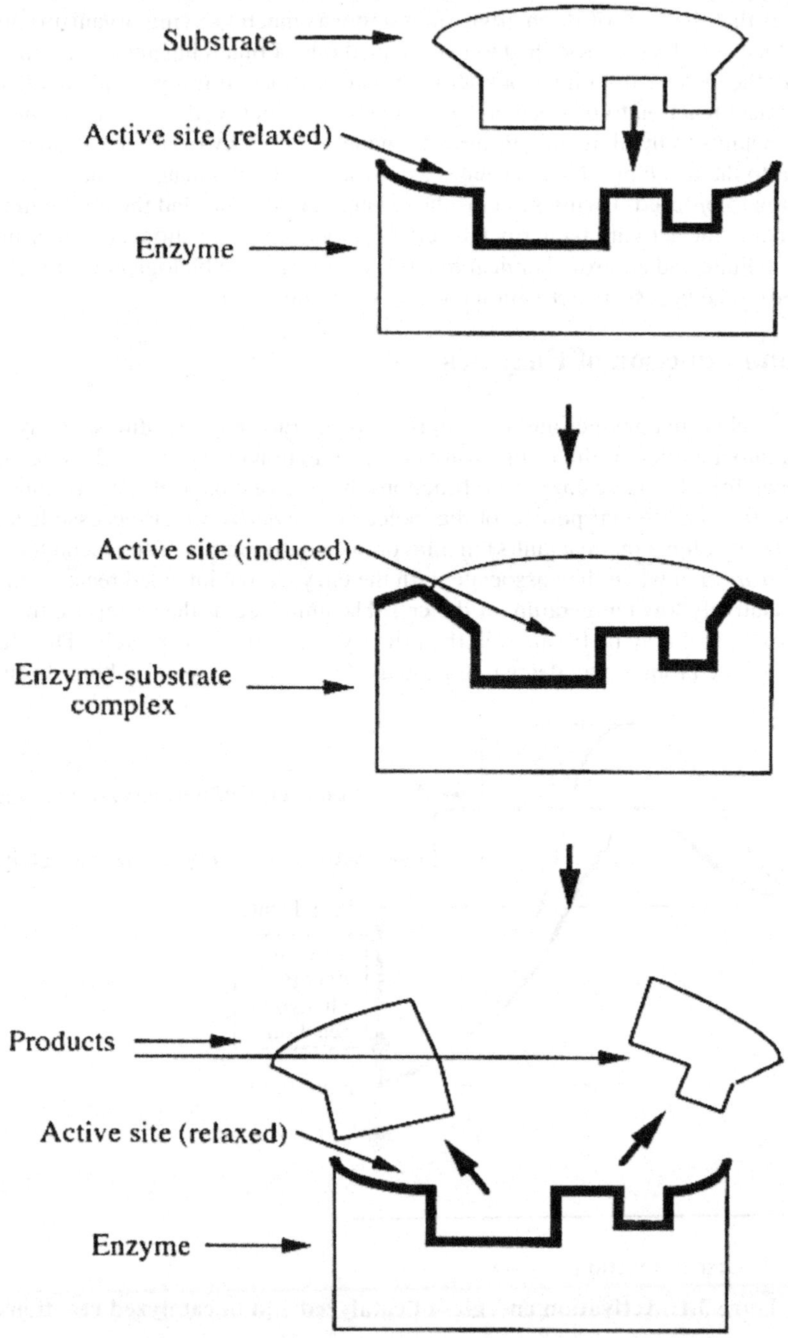


**Figure 3.2: A model of enzyme action (induced fit)**

## C. Factors Affecting Enzyme Activity

Sometimes it takes more than just the presence of the substrate and the enzyme to allow a reaction to take place. Many enzymes require the presence of a *cofactor* to function. Simple cofactors are usually ionic minerals. For example, many enzymes require magnesium or zinc ions, for which they have binding sites, in order to act as catalysts. If the cofactor is not present, the enzyme will not work or will have reduced activity.

Sometimes the cofactor is a nonprotein organic molecule, called a *coenzyme*. The coenzyme plays a central role in the catalyzed reaction, often by accepting or providing electrons, and becoming *reduced* or *oxidized*, respectively, in the process. The vitamins of the B-complex group or their derivatives serve as coenzymes in the catabolic reactions that oxidize food molecules to release energy.

Environmental conditions also play a major role in determining an enzyme's effectiveness. In particular, temperature and pH influence enzyme activity enormously, and this is the major reason these parameters are so tightly controlled in humans. The temperature and pH of our bodies cannot vary considerably without dire consequences. Why are enzymes affected so profoundly by these environmental conditions?

Enzyme activity, in general, is reduced as the temperature drops and is increased as the temperature rises. If our body temperature becomes colder than normal, enzymes will function, but at slower and slower rates as the temperature falls. Ultimately, these rates will be too slow to sustain life. Raising the temperature will increase enzyme activity, to a certain point. If the temperature is too high, the interactions that maintain the shape of the enzyme will be disrupted, changing the overall shape of the protein and destroying, often irreversibly, the functionality of the active site. When an enzyme loses its activity due to disruption of its three-dimensional configuration, that enzyme has become *denatured*. Both low and high pH values can also denature enzymes by disrupting the interactions that hold them in their proper orientation, in this case by affecting the charges of various R groups and subsequently changing their affinities for each other. Every enzyme has an optimum pH and temperature at which it functions best; any deviation from these values will reduce its ability to catalyze reactions.

## D. Control of Enzyme Activity

While ultimately the control of enzymes is accomplished genetically, several other processes affect when and how efficiently an enzyme works, so the cell carries out the desired reactions at the right times and rates. A classic example of an enzyme control mechanism is called *feedback inhibition*. Many enzymes, in addition to their active sites, contain binding sites for other molecules. These are referred to as *allosteric sites*, and enzymes that contain them are *allosteric enzymes*. If the allosteric site has an affinity for the product of the enzyme-catalyzed reaction, a feedback loop will automatically control how much of the product is produced. If product concentrations are low, the cell requires the reaction to proceed; little of the product exists to bind to the allosteric site, and the enzyme functions normally. As the reaction progresses, and enough of the product is made, it binds to the allosteric site. When this happens, the overall shape of the enzyme changes so that the active site is either hidden or disrupted, and the reaction will take place at a reduced rate or not at all. When product levels drop again, the product dissociates from the allosteric site, exposing the active site and allowing the reaction to resume. This process acts like a "thermostat", and is a common cellular strategy for regulating enzyme activity.

Sometimes molecules are present that act to inhibit an enzyme's function, and these are appropriately called *enzyme inhibitors*. *Competitive* inhibitors resemble the substrate, and thus compete with it for binding to the active site. *Non-competitive* inhibitors bind to an allosteric site, causing the enzyme's shape to change and affecting its function. Therefore, competitive inhibitors are less effective as the substrate concentration rises, while non-competitive inhibitors work regardless of the substrate concentration.

## E. ATP as the Energy Currency of the Cell

Reactions that are *endergonic* require the input of energy. Cells must perform many endergonic reactions in order to remain alive; protein synthesis and DNA replication are just two examples. While enzymes are required to catalyze these reactions, they cannot provide the energy to drive them. Therefore, all endergonic reactions must be coupled to energy-releasing, or *exergonic*, reactions in order to proceed. While cells can ingest energy containing molecules in many forms (monosaccharides, fatty acids, amino acids, etc.), the energy contained in these molecules must be harvested and stored in a usable form by the cell. In almost all cases, this usable form of energy is the molecule *ATP (adenosine triphosphate)*, the triphosphate form of the DNA nucleotide adenosine (see Figure 3.3). When the bond linking the terminal phosphate to the rest of the molecule is *hydrolyzed* (broken), *ADP (adenosine diphosphate)* and free phosphate are the products, and energy is released. Thus, this is an exergonic reaction. Many enzymes that catalyze endergonic reactions associate with ATP, and often contain *ATPase* (ATP hydrolyzing) activity in addition to their other catalytic capabilities. Thus, it is ATP that directly provides the energy required by enzymes to catalyze endergonic reactions. This intimate association of exergonic ATP hydrolysis with endergonic reactions is referred to as *coupling* of reactions. The next chapter addresses the process by which cells transform the chemical energy of food molecules into energy stored as ATP.

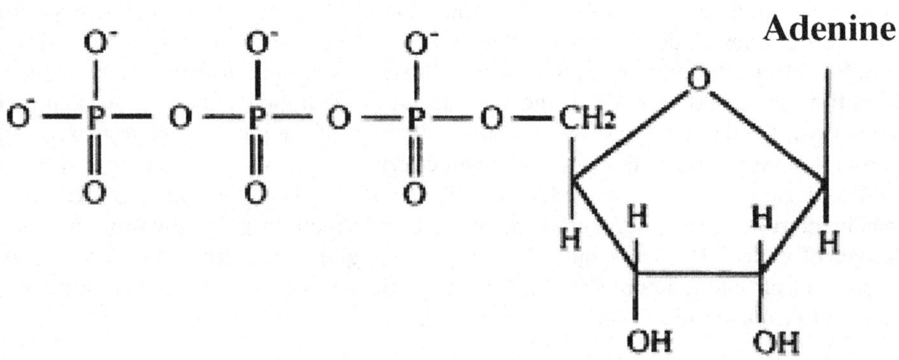


**Figure 3.3: Adenosine triphosphate (ATP)**

# Biology Review 3 Problems

**Passage**

The reagent iodine potassium iodide (IKI) can be used to detect the presence of starch in a solution. IKI is normally light yellow in color; in the presence of starch it turns a deep blue. IKI can therefore be used to test for the presence and activity of the enzyme amylase, which breaks starch into maltose disaccharide units. (Maltose does not affect the color of IKI). Thus, if starch is initially present and mixed with IKI, the deep blue color created will begin to lighten and disappear if amylase is present as it begins to break down the starch to maltose. Using the same concentration of enzyme, the longer it takes for the blue color to disappear, the lower the amylase activity.

Amylase is usually present in vertebrates in two forms that work in different parts of the digestive tract. Salivary amylase, secreted in the saliva by the salivary glands, begins to break down starch in the mouth, which has a slightly acidic pH. Pancreatic amylase is manufactured by the pancreas and released into the small intestine, where it similarly breaks down remaining starch molecules to maltose.

The following tables show the results of an experiment designed to test the activities of one form of amylase at various pH's and temperatures.

| pH | Time for blue color to disappear (in minutes) |
|---|---|
| 3 | 10 |
| 6 | 1 |
| 9 | 5 |
| 12 | 30 |

| Temperature (°C) | Time for blue color to disappear (minutes) |
|---|---|
| 15 | 10 |
| 30 | 5 |
| 37 | 1 |
| 60 | blue color never disappears |

1. Which of the following is the enzyme being tested?
   A. Human salivary amylase
   B. Shark salivary amylase
   C. Human pancreatic amylase
   D. Shark pancreatic amylase

2. What is the most likely explanation for the observation that the blue color never disappears at 60° C?
   A. The chemical bonds in starch are stabilized by the heat so that it cannot break down even though the enzyme is highly active.
   B. Heat causes the IKI to become unable to stain the starch.
   C. The amylase has become denatured at this temperature.
   D. All of the above are reasonable explanations.

3. What is the optimal temperature and pH for the enzyme being tested?
   A. 15° C, pH 12
   B. 37° C, pH 6
   C. 37° C, pH 12
   D. 15° C, pH 6

4. The breakdown of starch is an exergonic reaction, which would occur spontaneously at temperatures of about 200° C. Amylase allows the reaction to proceed at physiological temperatures by:
   A. Increasing the activation energy for the reaction.
   B. Increasing the potential energy of the reactants.
   C. Changing the amount of energy released by the reaction.
   D. Lowering the activation energy of the reaction.

5. Which of the following statements is true of the action of amylase at its optimum pH and temperature?
   A. The enzyme is irreversibly changed, so that one enzyme molecule can only catalyze the reaction once.
   B. It is denatured, and the enzyme's activity is the highest possible.
   C. The active site of amylase consists of every amino acid in the protein.
   D. This pH and temperature represents the physiological conditions under which it functions in nature.

# Solutions

1. A.
From the information given, we can tell that the optimal temperature and pH for the enzyme in question is 37 degrees C, pH 6 (the values at which the reaction proceeds to completion most quickly). As with the previous passage, the high temperature rules out any shark enzymes (choices B and D), and the mildly acidic pH could only exist in the mouth, not the small intestine (which is mildly alkaline).

2. C.
A human enzyme would likely become denatured at 60° C, as we saw in the last passage, and this would account for the lack of activity (the failure of the blue color to disappear implies that the starch remains present forever, since the enzyme cannot break it down). Choice A does not make sense, since heat will always tend to destabilize chemical bonds. Choice B could not be correct, because the IKI is still active; it is responsible for the blue color. Logically, this leaves out choice D, too.

3. B.
The explanation is the same as in question 1.

4. D.
This question is simply testing your general knowledge of how an enzyme functions in the context of this passage. Namely, an enzyme always functions by lowering the activation energy of a reaction (usually by providing a surface at the active site that puts the reactant(s) in a proper orientation and proximity). A is incorrect since it states the opposite of this idea. Choice B reflects what would happen if we were to facilitate a reaction by adding heat; the whole point of enzymes is to facilitate the reaction at a temperature that would not be harmful to life. With regard to choice C, an enzyme or any catalyst can never alter the change in free energy during a reaction, it can only affect the rate of that reaction! (This is a basic principle of thermodynamics and enzyme function).

5. D.
Again, this question is simply testing your general knowledge of how an enzyme functions in the context of this passage. The optimal pH and temperature for any enzyme will reflect the environment in which it normally functions; that is, in fact, why the optimal values are optimal! Choice B must be incorrect, because when an enzyme is denatured, it has no activity, the exact opposite of optimal functioning. Two other ideas you are expected to know regarding the general mechanism of enzyme action are the following: the active site of an enzyme is typically composed of only a few amino acids, and the enzyme is "recyclable", or not irreversibly changed by participating in the reaction. These ideas invalidate choices A and C.

# Biology Review 4
# Cellular Metabolism

## A. Introduction

All living organisms, as we have already discussed, require energy in order to survive. Animals obtain this energy by ingesting compounds that contain potential chemical energy (carbohydrates, fats, and proteins), and metabolizing them so that energy is released and stored in the form of ATP (see Chapter 3). Energy is almost always used directly in the form of ATP. What processes occur at the cellular level to accomplish these energy transformations and create ATP? The net sum of all reactions that take place in a cell or organism is called *metabolism*, but often the term metabolism is used to refer to the *catabolic* (breaking down) reactions that make energy available to the cell. These are the reactions on which we will focus in this chapter.

## B. Types of Metabolism

While in theory there are many pathways cells could utilize to metabolize their food, in nature, especially in animals, only a few pathways are used. While rather complex, these pathways vary little from organism to organism. Two major pathways are generally available to heterotrophic animal cells during metabolism: *aerobic respiration* and *fermentation*. Aerobic respiration is an oxygen requiring series of reactions, and is necessary in all vertebrates. Fermentation does not require oxygen, and while most vertebrates can perform fermentation reactions, it serves them mainly in emergencies when extra energy is needed. Only organisms such as yeast and some bacteria can live entirely by engaging in fermentation.

While many molecules may be metabolized (monosaccharides, amino acids, fatty acids, etc.), most cells prefer glucose as their source of fuel. It is therefore convenient to examine glucose metabolism as a model for the overall process, while keeping in mind that other molecules may also be used. Whichever pathways are ultimately utilized to harvest energy from glucose, they always involve the oxidation of glucose and always begin with a series of reactions called *glycolysis*. If the aerobic pathway is used, glycolysis is followed by two processes: the *Krebs cycle*, a cyclic series of reactions, and the *electron transport chain (ETC)*, where much ATP is synthesized. Let's look at each of these processes in more detail.

## C. Understanding Glycolysis

Glycolysis, which literally means "splitting sugar", is a series of nine reactions that partially oxidize glucose and harvest two molecules of ATP. Figure 4.1 shows all nine of the reactions, but it is not necessary to attempt to memorize them all in this context. We are most interested in understanding the major events and the ultimate products of glycolysis.

Glycolysis occurs in the cytoplasm of animal cells, and it is important to remember three major aspects of the process:

- Glucose, a six-carbon compound, is ultimately broken down into 2 molecules of pyruvic acid, a three-carbon compound.
- Two molecules of ATP are produced when glycolysis is complete.
- Two molecules of the coenzyme *$NAD^+$ (nicotinamide adenine dinucleotide)* are reduced to *NADH* when glycolysis is complete.

A few comments may be helpful at this point.

- Glycolysis is the first step in glucose metabolism in all vertebrates and almost all living cells.
- The ATP made during the process is generated by *substrate level phosphorylation*, which simply means a phosphate group attaches to ADP directly from one of the reactants in the pathway to make ATP.

- $NAD^+$ is a common coenzyme (see Chapter 3) that acts as an electron shuttle. It contains more potential energy when it is reduced to NADH than in its oxidized form, so the generation of NADH represents the temporary storage of energy.

## D. The Anaerobic Option: Fermentation

Glycolysis occurs as the first step in glucose metabolism regardless of whether the cell is performing fermentation or aerobic respiration. It is important to note that glycolysis can never

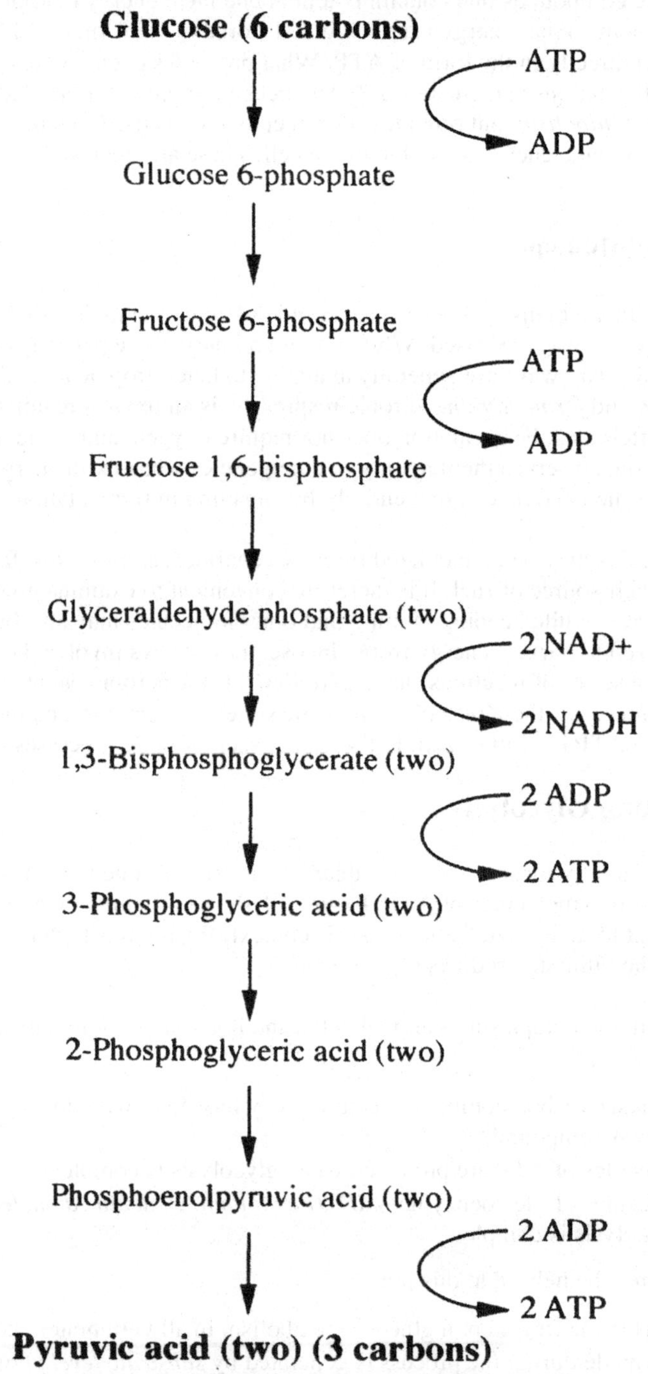


**Figure 4.1: A summary of glycolysis**

occur alone. It must be coupled to other reactions to be useful. Since fermentation is a comparatively simple process, we will examine it first. If a cell performs fermentation, no oxygen is required, but the pyruvic acid generated in glycolysis must be further processed. While countless variations exist, two major types of fermentation are common (see Figure 4.2):

- In *ethanol fermentation*, pyruvic acid is broken down into ethanol (a two-carbon compound) and carbon dioxide ($CO_2$). This type of fermentation is especially prevalent in yeast, and is utilized in many commercial processes, including the baking of bread and the making of alcoholic beverages.
- In *lactic acid* fermentation, the atoms of pyruvic acid are rearranged to form lactic acid (another three-carbon compound). This type of fermentation is carried out by vertebrates, usually in their muscle tissues. During heavy exertion, not enough oxygen may be available to supply ATP needs via the aerobic pathway.

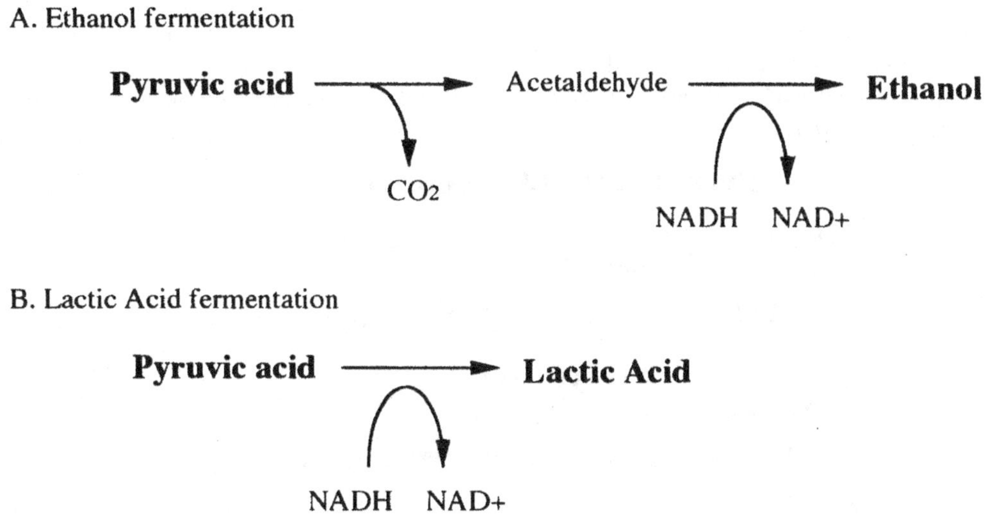


**Figure 4.2: a summary of fermentation reactions**

It is important to note in examining Figure 4.2 that no additional ATP is produced by either of the fermentation reactions. Why, therefore, must they occur? The answer is simple: the two molecules of NADH formed during glycolysis are oxidized back to $NAD^+$ during the fermentation reactions. If this did not occur, all of the $NAD^+$ in a cell would quickly be used up, and glycolysis could no longer continue. This explains why glycolysis can never "stand alone".

## E. The Aerobic Option: Cellular Respiration

Vertebrates must perform *aerobic cellular respiration* in order to obtain enough ATP to live. This means that they need a constant supply of oxygen (a major participant in the reactions), and produce carbon dioxide as a waste. This explains the need for breathing, or *physiological* respiration. The reactions of aerobic respiration take place in the *mitochondria*, so, after glycolysis, the pyruvic acid to be processed must be transported to this organelle. We should note here that the mitochondrion (singular) is surrounded by two membranes; this creates an *intermembrane space*. In addition, the inner membrane is folded to increase its surface area, and the folds are referred to as *cristae*. The inner, liquid portion of the mitochondria is called the *matrix*.

## F. The Krebs Cycle

Pyruvic acid must now enter a cyclic series of reaction, the *Krebs cycle*, also known as the *citric acid cycle* or the *TCA (tricarboxylic acid) cycle*. The reactions of the Krebs cycle take place in the mitochondrial matrix, and are summarized in Figure 4.3. Again, it is inadvisable to attempt to memorize all of these reactions. Before pyruvic acid can enter the cycle, it must undergo some initial preparatory steps:

- Pyruvic acid is oxidized, releasing one molecule of $CO_2$ and a two-carbon acetyl group.
- The acetyl group is attached to a molecule called *Coenzyme A*, to form *acetyl CoA*.
- In the process, a molecule of $NAD^+$ is reduced to NADH.

It is important to note that carbon atoms from pyruvic acid can enter the Krebs cycle only in the form of acetyl CoA.

Next, the acetyl group is donated to the four-carbon molecule *oxaloacetic acid* to form *citric acid*, a six-carbon compound. The citric acid is subsequently broken down, in a series of steps, back to oxaloacetic acid. As with glycolysis, there are three main points to keep in mind regarding the Krebs cycle. Specifically, for each turn of the cycle:

- Two carbons enter the cycle as an acetyl group, and two carbons are released as carbon dioxide.
- One molecule of ATP is harvested.
- Three molecules of $NAD^+$ are reduced to NADH, and one molecule of another coenzyme, *FAD*, is reduced to *$FADH_2$*.

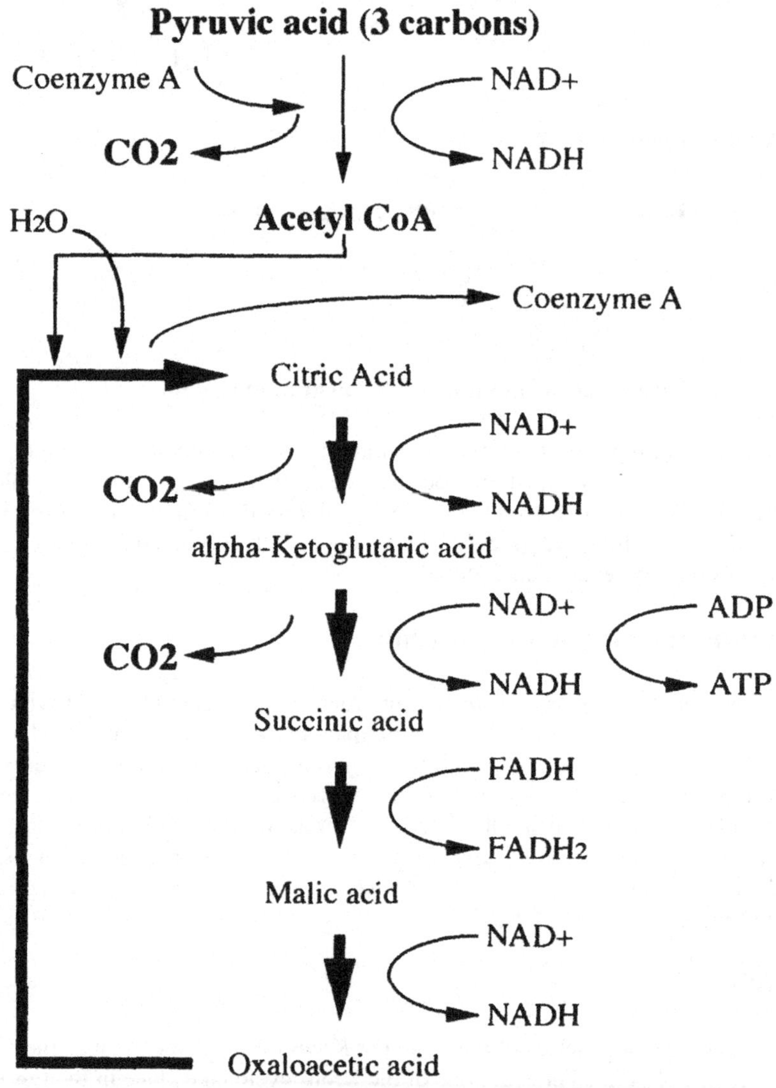


**Figure 4.3: Highlights of the Krebs cycle**

It is important to remember that for every glucose molecule we started with, two molecules of pyruvic acid were created. Since we have only been keeping track of one pyruvic acid molecule as it travels through the Krebs cycle, we must multiply our totals by two if wish to know the yield per molecule of glucose. If we consider the preparatory events with those of the Krebs cycle itself, we can summarize as follows:

- For each three-carbon molecule of pyruvic acid that enter the mitochondria, three molecules of carbon dioxide are released, for a total of six carbon dioxides. This is the source of the $CO_2$ we exhale.
- A total of two molecules of ATP are harvested.
- Overall, eight molecules of $NAD^+$ are reduced to NADH, and two FAD molecules are reduced to $FADH_2$.

## G. The Electron Transport Chain: ATP Harvesting

After the Krebs cycle, the original glucose molecule has been completely oxidized, which means that its potential energy has been released. One glucose molecule theoretically contains enough potential energy to manufacture close to one hundred molecules of ATP. It is clear that we have not stored a significant amount of the released energy in this form yet. Where is the energy?

The energy is temporarily residing in the reduced coenzyme molecules NADH and $FADH_2$. During the final stages of respiration, *the electron transport chain*, these coenzymes will donate electrons to a series, or chain, of carriers, and will become oxidized in the process, back to $NAD^+$ and FAD (the coenzymes can then go back and participate in glycolysis and the Krebs cycle). The carriers are physically located in the inner mitochondrial membrane, and include *cytochrome* proteins, among other molecules. As electrons are passed down the chain, energy is released; the final electron acceptor is molecular oxygen (O2), which is reduced and converted to water as it accepts electrons (this is where the oxygen we inhale is actually used). The energy released is temporarily stored and ultimately used to make ATP by *oxidative phosphorylation*. The *chemiosmotic model* explains how ATP is produced this way (see Figure 4.4). The model highlights three major points:

- The reduced coenzymes NADH and $FADH_2$ ultimately react with oxygen by donating electrons through a series of intermediaries. This reaction causes oxidation of the coenzymes to $NAD^+$ and FAD, and reduction of oxygen to water. A large amount of energy is released in the process. If the intermediate carriers did not exist, all of the energy would be released at once, which would be difficult for the cell to control and manage.
- As energy is slowly released, it is used to pump *protons* ($H^+$ ions) from the matrix into the intermembrane space by active transport. This establishes an *electrochemical proton gradient* which stores the energy that has been released. This gradient can be used to do work.
- The enzyme *ATP synthase* is embedded in the inner membrane, and protons diffuse through a channel in this protein back into the matrix. As the protons "fall" through the enzyme, the energy they release is used to do work: the *phosphorylation* (addition of a phosphate group) of ADP to make ATP.

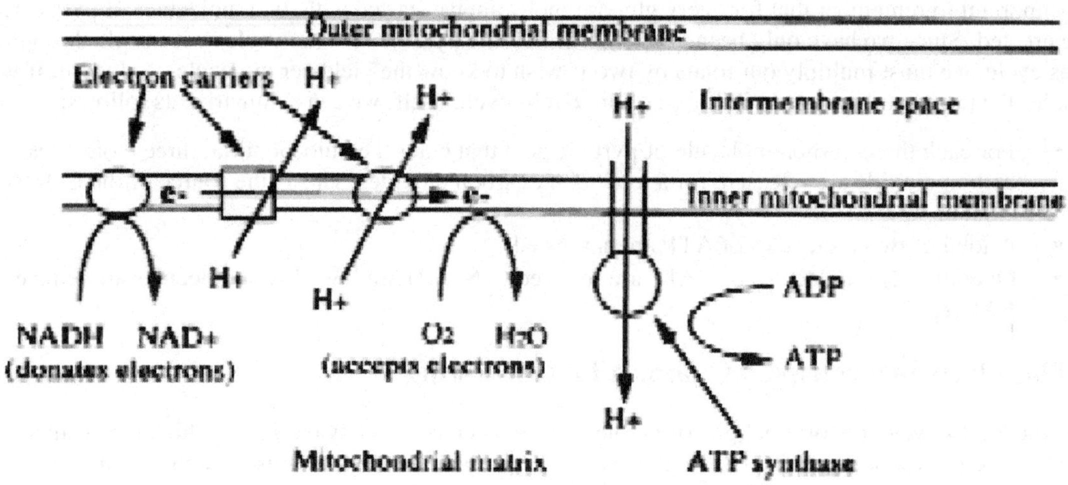


**Figure 4.4: The electron transport chain and oxidative phosphorylation**

When all is said and done, 32 ATP molecules are harvested by the electron transport chain from all of the reduced cofactors generated through glycolysis and the Krebs cycle. If we add this to the 2 ATP molecules obtained from glycolysis and the 2 produced in the Krebs cycle, we come up with a grand total of 38 molecules of ATP harvested. However, the net amount is 36 ATP molecules because some energy is used to transport pyruvic acid into the mitochondria.

## H. Overall Energy Harvest

Given that glucose, if burned in a calorimeter, releases 686 kilocalories/mole, and ATP hydrolysis usually yields approximately 7 kilocalories/mole, we can calculate the efficiency of aerobic respiration. If we harvest 36 ATP molecules and multiply by 7 kilocalories/mole, we have obtained 252 kilocalories/mole of glucose burned. This represents an efficiency of 252/686, or approximately 37%. While far from perfect, it is certainly preferable to the efficiency of fermentation, which by the same logic is approximately 2%. Where does the rest of the energy go? The first law of thermodynamics tells us that energy can never be created or destroyed, so it must have been transformed into another type of energy. In fact, it was converted to heat, and is in effect "wasted" energy. Birds and mammals, however, have figured out another use for this energy so that it is not completely wasted: it is the major source of internal heat used to maintain a relatively high body temperatures.

## I. Summary of Aerobic Respiration

After looking at all of the details of respiration, we can formulate a net equation that takes them all into account. That equation is:

glucose + oxygen + ADP + P —> carbon dioxide + water + ATP

We can make this a proper, balanced chemical equation:

$$C_6H_{12}O_6 + 6O_2 + 36ADP + 36P \longrightarrow 6CO_2 + 6H_2O + 36\ ATP$$

# Biology Review 4 Problems

**Passage**

Many chemicals that are poisons exert their toxic effects by interfering with some aspect of aerobic respiration, usually involving the electron transport chain of the mitochondria. Three such poisons are cyanide, 2,4-dinitrophenol, and the antibiotic oligomycin.

Cyanide is a potent and deadly human poison. It causes its effects by binding to one of the electron carriers and inhibiting the passage of electrons to oxygen, so that electron transport, proton-pumping, and ATP synthesis stop virtually instantaneously.

2,4-dinitrophenol is also a deadly poison to humans. It is an example of the general class of poisons known as "uncouplers", which allow protons to pass back from the intermembrane space to the matrix without passing through the ATP synthase enzyme. Electron transport and proton pumping continue, but ATP is not made. Such uncouplers are also called ionophores.

Oligomycin, an antibiotic, is not deadly to humans, but does interfere with respiration, which is how it kills bacteria, and accounts for its side effects in humans. It is representative of a group of poisons that directly inhibit ATP synthase by blocking the passageway for protons. As with the uncouplers, electron transport and proton pumping continue, but ATP is not made (although for a different reason).

Use the information above, the following observations, and your knowledge of respiration when answering the questions.

*Observation 1*

When a person breathes a particular deadly poison (compound A), the following effects are observed. Clinically, body temperature quickly increases, causing profuse sweating, and ultimate death. At the biochemical level it is noted that normal to greater than normal amounts of oxygen are used, and normal to greater amounts of carbon dioxide are produced. The pH of the mitochondrial intermembrane space does not change appreciably; if it does, it may increase slightly.

*Observation 2*

When a person ingests a particular toxin (compound B), the following effects are observed. At the biochemical level, it is noted that the Krebs cycle continues to function, producing NADH, and that the NADH is oxidized back to NAD+ as it donates electrons to the electron transport chain. Strikingly, the pH of the mitochondrial intermembrane space is noted to drop significantly.

1. The toxin ingested that is referred to as compound A in observation 1 is likely to be:
   A. Cyanide.
   B. 2,4-dinitrophenol.
   C. Oligomycin.
   D. The information presented does not allow differentiation of the three.

2. The toxin ingested that is referred to as compound B in observation 2 is likely to be:
   A. Cyanide.
   B. 2,4-dinitrophenol.
   C. Oligomycin.
   D. The information presented does not allow differentiation of the three.

3. During cyanide poisoning, which of the following molecules would increase its concentration dramatically?
   A. NAD+
   B. NADH
   C. Carbon dioxide
   D. FAD

4. In the 1950s, certain weak uncoupling agents were used to promote weight loss. They worked very well; in fact, they worked so well at "burning calories" that many people died from using them and they were pulled from the market. Which of the following statements explains how uncouplers could cause weight loss?
   A. Uncouplers allow ATP to be made but prevent its transport out of the mitochondria, thus uncoupling its manufacture from its use.
   B. Uncouplers increase the metabolic rate, and allow caloric energy temporarily stored in reduced coenzymes to remain "unharvested" as ATP; it is simply released as heat.
   C. Uncouplers prevent oxygen from accepting electrons, so that ATP is not made, and energy is not available to digest and absorb food.
   D. Uncouplers, due to their toxic effects, cause appetite suppression, and lower the overall metabolic rate.

5. Glycolysis could continue to operate in all of the following poisoning situations except:
   A. Cyanide poisoning.
   B. Poisoning from a fatal uncoupler.
   C. Oligomycin intoxication.
   D. None of the above; glycolysis is not affected by the events of the electron transport chain.

# Solutions

1. B.
The important information to remember when answering both questions 1 and 2 is the following: First, with cyanide poisoning, electron transport stops completely; this causes a buildup of NADH, and the ultimate shutdown of all previous processes, including the Krebs cycle and glycolysis, due to the unavailability of NAD+. Cyanide poisoning would therefore not allow any oxygen to be used up or carbon dioxide to be manufactured. In addition, no energy is released, as all processes are stopped. Since ATP synthase is still functioning, the proton gradient will eventually disappear, causing the pH in the intermembrane space to rise dramatically, as equilibrium with the matrix is reached. With 2,4-dinitrophenol, or any uncoupler, electron transport continues, and so do oxygen consumption and carbon dioxide evolution. Since the membrane is leaky to protons, they will flow back to the matrix without making ATP; the energy released will be dissipated as heat, which accounts for the increased body temperature. To the extent the proton gradient is relieved, the pH will rise in the intermembrane space. With an ATP synthase inhibitor like oligomycin, again, oxygen will continue to be used and carbon dioxide produced, as electron transport will continue. Since protons cannot re-enter the matrix, however, the gradient will be come steeper and steeper, and the pH in the intermembrane space will drop dramatically as it becomes very acidic.

So, the high body temperature of the person from observation 1, coupled with continued oxygen consumption and the relatively unchanged or alkaline pH of the intermembrane space, identifies the poisoning agent as an uncoupler.

2. C.
Refer to the information from question 1. The dropping pH in the intermembrane space alone allows identification of this toxin as an ATP synthase inhibitor.

3. B.
Again, refer to the information from question 1. NADH will build up as electron transport is discontinued, as it has no place to give up electrons. NAD+ and FAD, the oxidized forms of coenzymes, will eventually decrease in amount until there are none left. Carbon dioxide will ultimately stop being produced, so its concentration will not change very much.

4. B.
As we have seen from the solution to question 1, uncouplers function by allowing the inner mitochondrial membrane to become permeable to protons, and we know from question six that energy is released but not used (harvested as ATP); it is simply dissipated as heat. Choice A is incorrect because uncouplers decrease the formation of ATP; choice C is incorrect because it describes a cyanide-like poison that shuts down electron transport. Choice D makes little sense; if anything, an uncoupler would cause the metabolic rate to rise.

5. A.
Again referring to the information from question 1, cyanide poisoning is the only situation we mentioned where glycolysis (and the Krebs cycle) will ultimately cease to function, due to the unavailability of oxidized coenzymes.

# Biology Review 5
# DNA Structure and Function

## A. Introduction

Perhaps the most important molecule in all of biology is *DNA, deoxyribonucleic acid.*

All living things contain DNA as the storage unit of genetic information. The DNA molecule in every organism has the exact same structure and function. How it is copied by the cell (*replication*) and how it is interpreted (*transcription* and *translation*) may differ in some details; however, the basics are the same in every living entity. We will concentrate on eukaryotic cells in this chapter.

## B. The Function of DNA

Earlier, in Chapter 1, we considered the relationship between structure and function. We will now see how the structure of DNA directly relates to its function. In brief, the functions of DNA are

- to carry the genetic information of the organism;
- to control the development of the cell and the organism;
- to direct the function of the cell, including its reproduction and metabolism.

Since DNA has the same structure in all organisms, and it dictates the function of the cell, it is natural to wonder what makes species different. It is not the basic structure of the DNA, but rather the exact arrangement of the components of DNA that determines this difference. In fact, this arrangement not only accounts for the difference among species, but also for the uniqueness of individuals in the same species.

## C. The Chemical Structure of DNA

By the early 1950s, it was well known that DNA was a polymer made up of monomers called *nucleotides*. Each nucleotide consists of three chemical groups (see Figure 5.1, see also Chapter 2):

- a 5 carbon sugar, *deoxyribose*;
- a nitrogen rich *base* attached to the first carbon of the sugar;
- a *phosphate group* attached to the fifth carbon of the sugar.

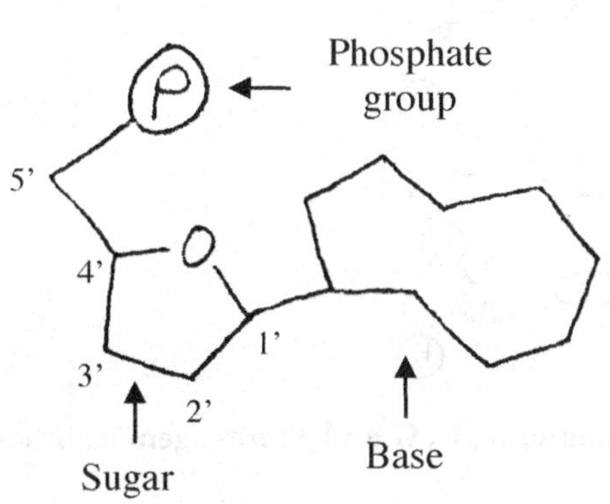


**Figure 5.1: General structure of a nucleotide**

Four types of nucleotides exist, and each differ only at the nitrogenous base:

- *adenine* (*A*),
- *cytosine* (*C*),
- *guanine* (*G*) and
- *thymine* (*T*).

In 1953, James Watson and Francis Crick proposed a model for the structure of DNA by considering a wide variety of data from other researchers. In particular, two observations became crucial to their model:

- Erwin Chargaff had discovered that, in every molecule of DNA, the amount of *A* was always equal to the amount of *T*, and the amount of *C* was always equal to the amount of *G*.
- Rosiland Franklin had obtained X ray diffraction data that showed DNA exists in a *double helix*, similar in structure to a winding staircase.

From this information, Watson and Crick were able to correctly determine the structure of the polymer (see Figure 5.2).

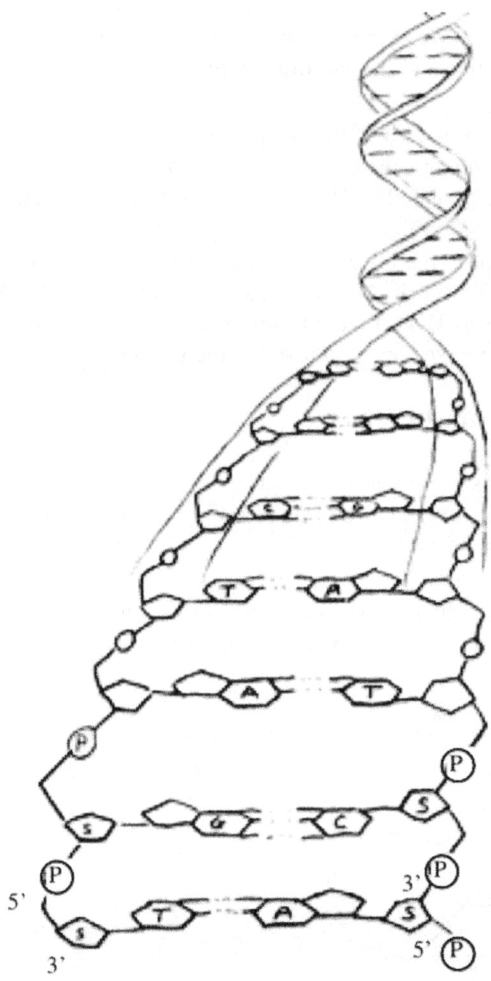


**Figure 5.2: The structure of DNA. S: sugar; P: phosphate; A, C, G and T: nitrogenous bases**

The monomers of DNA are linked together via the sugar and the phosphate groups in the nucleotides. This is often called the *sugar-phosphate backbone*, and the bond that forms between the monomeric nucleotides is called *a phosphodiester bond*. Remember how the phosphate group is attached to the fifth carbon on the deoxyribose? The phosphate from one nucleotide attaches to the sugar of another nucleotide at the third carbon. This actually gives DNA a direction, like north and south, except we call it 5' and 3' (pronounced "5 prime" and "3 prime," see Figure 5.2). We'll talk more about this attachment a little later.

One strand (polymer) of DNA is usually found attached to another strand of DNA, forming a double stranded molecule. The bases on each strand bond to form a pair. This pairing follows a strict rule: *A* always pairs with *T*, and *C* always pairs with *G* (this is what Chargaff saw). We call this *complementary base pairing*. When the strands of DNA come together to form a double strand, the structure twists around itself, creating a double helix (see Figure 5.2). By knowing the *sequence* (arrangements of nucleotides) of one strand, the sequence of the second strand can be determined. It is this sequence that determines the structure and function of the cell and the organism.

The base pairs are held together by *hydrogen bonds*, weak bonds that form between hydrogens and oxygens or hydrogens and nitrogens. The G-C base pairs form 3 hydrogen bonds while the A-T base pairs form only 2. This makes the G-C pairs inherently stronger and more stable.

Another interesting feature of the double helix is that, in order for the bases to pair correctly, the two strands of DNA must run in opposite directions, or in an antiparallel fashion. This means that the 5' end of one strand pairs with the 3' end of the other strand (see Figure 5.2).

## D. Chromosome Structure

Before going any further, let's look at the structure of the DNA as it exists in a cell. If you could remove the DNA from one of your cells and stretch it out, it would be approximately one meter long. How can so much DNA fit into a microscopic cell? The answer is in the packaging. DNA is highly organized into a structure called a *chromosome*. The chromosome is made up of DNA and proteins. The DNA double helix is wrapped around proteins known as *histones*, and the histones form complexes called *nucleosomes* (see Figure 5.3). The nucleosomes are further packaged into *supercoiled loops* sometimes called *solinoids*. These are packaged into chromosomes.

Human cells contain 23 pairs of chromosomes (46 total). Certain genetic diseases can be diagnosed by examining the chromosome. This is what is done when a pregnant woman has an amniocentesis. The field of biology that studies chromosomes is called *cytogenetics*.

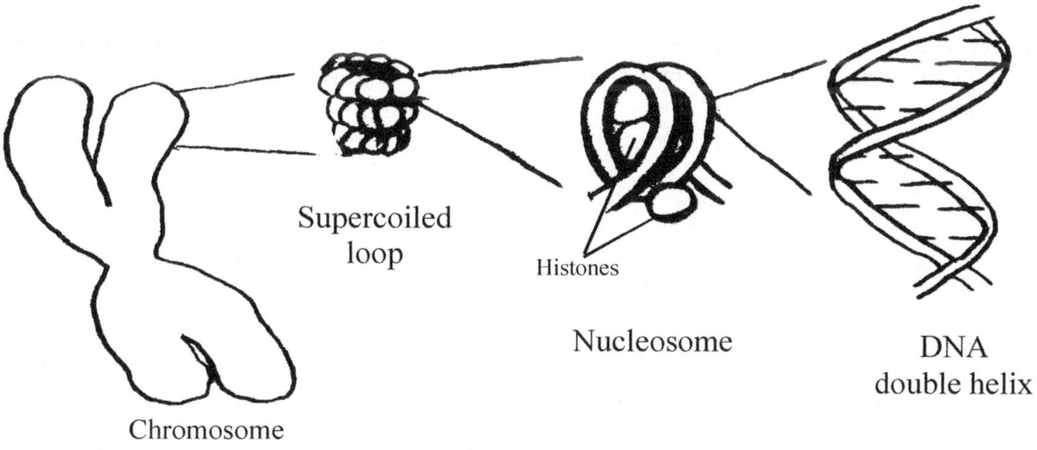


**Figure 5.3: Levels of DNA packaging**

## E. RNA Structure

This is an opportune time to examine the structure of *RNA (ribonucleic acid)*, a molecule closely related to DNA. RNA is also a polymer of nucleotides, but differs from DNA in three major respects:

- RNA is usually single stranded, meaning it is not usually found base paired with another RNA molecule;
- the sugar in an RNA nucleotide is *ribose*, not deoxyribose. The difference is found on the 2' carbon of the sugar: ribose has an OH group attached while deoxyribose has only an H;
- a nitrogenous base called *uracil* (*U*) substitutes for thymine (T) in RNA. Uracil is similar in structure to thymine, and will base pair with adenine (this will become important when we discuss transcription and translation).

## F. Replication

When a cell divides, the new cell (or *daughter* cell) must receive the same genetic information as the original cell, or the daughter cell will not function correctly. Therefore, before a cell divides, the DNA must be copied, or *replicated*, faithfully. It must also be transferred to the new cell. This entire process is known as *mitosis* and will be addressed in Chapter 12. The details of how DNA is replicated will be discussed here.

The beauty of the structure of DNA is that it simplifies its own replication. Due to the base pairing rules, each strand of the helix serves as a template to make a new strand. The copy is the complement of the template, and is identical to the strand originally bound to the template. This type of replication is referred to as *semiconservative*. Let's consider the details.

## G. The Replication Machinery

One of the most important concepts to remember is that polymers of nucleotides can be built only in one direction, in the 5' to 3' direction. The phosphate from one nucleotide binds to the 3' carbon of the sugar in the preceding nucleotide (refer to Figure 5.2). The process can never occur in reverse.

Before the DNA can be replicated, the two strands must dissociate (this is sometimes referred to as unzipping, as the DNA double helix resembles a zipper). The physical point at which the DNA is unzipped is referred to as the *replication fork* (see Figure 5.4).

The main enzyme involved in replication is called *DNA polymerase*. It binds to each single stranded DNA chain and builds a complementary strand by reading the template strand. We will now examine some of the details of this process.

## H. Leading and Lagging Strand Synthesis

Since DNA is synthesized from 5' to 3', the template DNA must be read from 3' to 5' (remember the antiparallel structure of the double helix). This causes a problem at the replication fork: one strand can be copied *continuously* as the fork extends, but the other strand must be copied *discontinuously*. This discontinuous replication occurs in the following manner (see Figure 5.4):

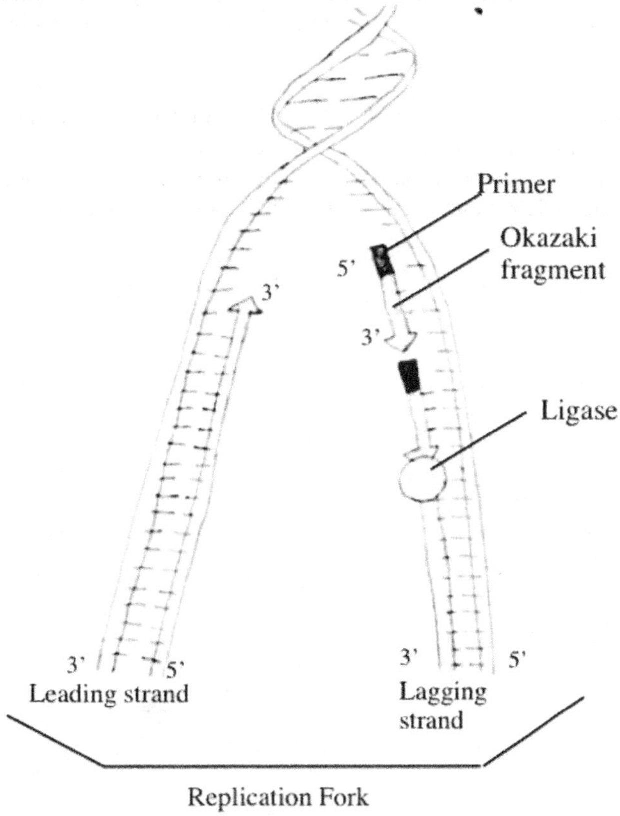


**Figure 5.4: DNA replication**

1. An RNA *primer* (a short stretch of nucleotides) attaches to one strand via complementary base pairing. The primer is synthesized by a *primase* enzyme.

2. The DNA polymerase begins synthesizing a new, complementary strand in the 5' to 3' direction. The strands made between RNA primers are known as *Okazaki fragments*.

3. The primers are degraded and the Okazaki fragments are linked together by the enzyme *DNA ligase*.

The strand that is synthesized continuously is called the *leading strand*. The other strand, synthesized discontinuously, is referred to as the *lagging strand*.

## I. Proofreading and Repair

Everyone makes mistakes, and so does DNA polymerase. When this enzyme puts the wrong nucleotide into the growing chain, the result is a *mutation* which changes the genetic make up of an organism. Therefore, mistakes cannot be tolerated.

Fortunately, DNA polymerase has another ability besides synthesizing DNA strands: it also proofreads its work. If an error is detected, the enzyme cuts out the incorrect nucleotide and replaces it with the correct one.

In addition to random errors during replication, DNA can be mutated by other factors, such as UV light and cancer causing agents (*carcinogens*). If these mutations are not fixed, the function of the cell will be altered. This may cause the cell to die, or to grow in an unregulated fashion (cancer). Many repair mechanisms exist in the cell to detect and correct mutations, and more often than not they work just fine.

# Chapter 5 Problems

**Passage** (Questions 1-4)

The Watson and Crick model of the structure of DNA eloquently suggested the method of DNA replication. The double helical nature, and the complementary base pairing, implied that one strand of DNA provided the information for making the other strand. Once the strands of DNA separated, each would be used as a template to make a new, complementary strand. Therefore, the two resulting strands of DNA would contain both an old strand and a new strand. This is known as the *semiconservative* theory of replication. The *conservative* theory stated that one double helix would contain only the newly synthesized DNA while the other would contain only the original strands.

In 1958, Matthew Meselson and Franklin Stahl confirmed the semiconcervative model. They grew the bacterium *E. coli* for many generations in a heavy isotope of nitrogen, $^{15}N$. The isotope was incorporated into the DNA.

The bacteria were then placed in medium that contained only the light isotope of nitrogen, $^{14}N$. The bacteria were sampled over a period of time. Their DNA was extracted and subjected to centrifugation techniques, which would separate the DNA based on density.

DNA containing solely $^{15}N$ was "heavy" whereas DNA containing solely $^{14}N$ was "light". After one round of division in the medium containing $^{14}N$, all the DNA was found to be intermediate in size, between the light and heavy types.

This "intermediate" DNA was further analyzed. The hydrogen bonds between the base pairs were broken so the DNA was single stranded. These strands were then centrifuged. Half of the single stranded DNA was in the heavy form, and half was in the light form.

1. If the replication theory suggested by Watson and Crick's model was correct, then the density of the double stranded DNA after 2 rounds of replication in $^{14}N$ medium would be:
   A. half heavy, half light
   B. half heavy, half intermediate
   C. half intermediate, half light
   D. all intermediate

2. $^{14}N$ and $^{15}N$ were good choices as isotopes because they would be incorporated into the DNA via:
   I. the deoxyribose sugar
   II. the phosphate group
   III. the bases
   A. I only
   B. II only
   C. III only
   D. I and III

3. If DNA replication was conservative, after 1 round of replication in $^{14}N$, the density of the double stranded DNA would be:
   A. all intermediate
   B. all heavy
   C. half intermediate, half heavy
   D. half heavy, half light

4. In this experiment, the Okazaki fragments would contain:
   A. only $^{14}N$
   B. only $^{15}N$
   C. neither $^{14}N$ nor $^{15}N$
   D. both $^{14}N$ and $^{15}N$

Questions 5 - 7 are independent of any passage and independent of each other.

5. Heat is often used to separate or denature double stranded DNA into single stranded DNA by breaking the hydrogen bonds between base pairs. Which of the following statements is true?
   A. G-C base pairs would require higher temperatures to break than A-T base pairs.
   B. A-T base pairs would require higher temperatures to break than G-C base pairs.
   C. All long DNA chains should denature at the same temperature.
   D. DNA-DNA double strands would require higher temperatures to denature that DNA-RNA double strands containing the same sequence.

6. Part of the process of purifying DNA from isolated chromosomes must involve:
   A. adding DNase, an enzyme that breaks apart DNA.
   B. adding proteases, enzymes that break apart proteins.
   C. adding RNase, an enzyme that breaks apart RNA.
   D. adding detergents, chemicals that break apart lipids.

7. An RNA molecule was synthesized to complementary base pair with a DNA molecule. The sequence of the DNA was: 5' ATCCGCTAAG 3'. The RNA sequence should be:
   A. 5' CUUAGCGGAU 3'
   B. 5' UAGGCGAUUC 3'
   C. 5' CTTAGCGGAT 3'
   D. 5' TAGGCGATTC 3'

# Solutions

1. C.
After one round of replication, all the DNA was found to be intermediate in density, with one strand containing the light isotope and one containing the heavy isotope. A second round of replication would mean that these intermediate DNA double strands would now serve as the template strands. The resulting copies would contain the light isotope. Therefore, from the template strand containing the heavy isotope, the resulting double strand would be intermediate, and from the template containing the light isotope, the double strand would contain only the light isotope. This would result in half the DNA being intermediate and half being light.

2. C.
Only the bases in a nucleotide contain nitrogen.

3. D.
Conservative replication implies that the template strands remain together and the copied strands form the new double stranded DNA molecule. If this were true, then the template strands would contain only $^{15}N$, the heavy isotope, and the copy strands would contain only $^{14}N$, the light isotope.

4. A.
Okazaki fragments are generated during replication of the lagging strand. Therefore, they would contain only $^{14}N$.

5. A.
Since G-C base pairs contain three hydrogen bonds, whereas A-T base pairs contain only two, the G-C pairing would require more energy, and hence a higher temperature, to disrupt. Extending this idea, long chains of DNA require different temperatures to denature depending on how many G-C pairs they have relative to A-T base pairs. Since there is no difference in the number of hydrogen bonds between a DNA-DNA double strand and a DNA-RNA double strand with the same sequence (A-U base pairs only contain two, just like A-T), there would be no difference in the temperature necessary to denature them.

6. B.
Chromosomes contain DNA and proteins. It is necessary to keep the DNA intact during isolation for further experiments, so adding DNase would not be wise. Since chromosomes do not contain RNA or lipids, reagents used to degrade these components are not necessary. However, to purify the DNA, proteins must be removed. Proteases are often used to accomplish this.

7. A.
Since RNA contains *U* instead of *T*, the answers using *T* can be eliminated automatically. This question stresses the property of base pairing that requires the strands to be in the opposite orientation (antiparallel). Therefore, although the sequence UAGGCGAUUC (response B) matches the DNA sequence as written, the 5' and 3' ends are not in the correct orientation. If the sequence was reversed, as in answer A, the two strands would complementary base pair.

# Chemistry Review 1
# Introduction

The science portion of the ACT consists of questions that test the basic concepts in chemistry. The concepts are tested with respect to application, problem solving, and analytical thinking. These skills, which are measured by the test, are considered essential in the study of college level material. The questions asked are based on passages.

The ACT tests your ability to understand new concepts based on high school science courses. Sometimes the passages contain topics and data that you have not seen before. Nevertheless, you do not need advanced courses in chemistry to be successful in the ACT. All you need is to understand the basic concepts in chemistry. You should be familiar with the common equations and constants that are used in the high school level of studies. For example, simple formulas such as the density-mass equation or the gas equations. More importantly, you should know how to use the equations in both numerical and conceptual situations. Advanced equations will be provided to you in the passages, if necessary.

When studying this section, you should not focus just on facts. You should grasp the reasoning behind the facts and think conceptually about the various ideas in chemistry.

# Matter

## A. INTRODUCTION

A firm grasp of the basic ideas of division of matter is important for the understanding of physical sciences. These basic ideas presented here are not only used in chemistry and physics, but in many diverse fields such as medicine, engineering, astronomy, geology, and so on. In this chapter, we will discuss ideas about atoms and molecules, and related aspects such as moles, Avogadro number, percentage composition, atomic mass, atomic weight, and subatomic particles.

## B. ATOMS

Atoms are the basic units of elements and compounds. In normal chemical reactions, atoms retain their identity. In this section, we will present a quick review of some of the basic terms and concepts such as elements, compounds, and mixtures.

### Elements

An element is defined as matter that is made of only one type of atom. Elements are the basic building blocks of more complex matter. Some examples of elements include hydrogen (H), helium (He), potassium (K), carbon (C), and mercury (Hg).

### Compounds

A compound is matter formed by the combination of two or more elements in fixed ratios. Let's consider an example. Hydrogen peroxide ($H_2O_2$) is a compound composed of two elements, hydrogen and oxygen, in a fixed ratio.

## Mixtures

A combination of different elements, or a combination of elements and compounds, or a combination of different compounds is called a mixture. For example, an aqueous solution of potassium hydroxide ($KOH + H_2O$). In this example, the two components are potassium hydroxide and water.

Though these definitions illustrate basic ideas, you need to understand them fully; otherwise it will be almost impossible to decipher higher concepts that are based on these simple ideas. The ACT tests your understanding of basic concepts by incorporating simple ideas into passages. So in order to succeed on the test, you need to thoroughly understand the basics.

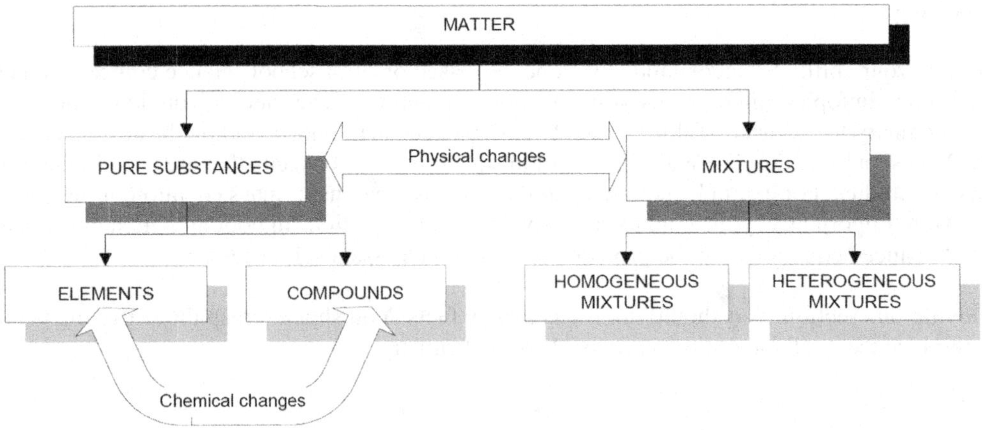


## C. DALTON'S ATOMIC THEORY

In 1803, John Dalton proposed the atomic theory of matter. The main postulates of his atomic theory can be summarized as follows:

1) Matter is composed of indivisible particles - atoms.

2) An element is composed of only one kind of atom. These atoms in a particular element have the same properties such as mass, size, or even shape.

3) A compound is composed of two or more elements combined in fixed ratios or proportions.

4) In a chemical reaction, the atoms in the reactants recombine, resulting in products which represent the combination of atoms present in the reactants. In the process, atoms are neither created, nor destroyed. So a chemical reaction is essentially a rearrangement of atoms.

### Ramifications of Dalton's Theory

The atomic theory put forward by Dalton is consistent with the law of conservation of mass. As the fourth postulate says, chemical reaction is just a rearrangement of atoms, and thus the total mass remains constant during a chemical reaction.

The postulates also account for the law of definite proportions. Compounds are made of elements in fixed or definite proportions. Since the atoms have fixed mass, compounds should have elements in a fixed ratio with respect to mass. Finally, these postulates predict what is known as the law of multiple proportions. According to this law, if two elements form two or more different compounds, the ratio of the mass of one element of these compounds to a fixed mass of the other element is a simple whole number.

## D. THE GENERAL STRUCTURE OF THE ATOM

During the early twentieth century, scientists discovered that atoms can be divided into more basic particles. Their findings made it clear that atoms contain a central portion called the nucleus. The nucleus contains protons and neutrons. Protons are positively charged, and neutrons are neutral. Whirling about the nucleus are particles called electrons which are negatively charged. The electrons are relatively small in mass. Take a look at Table 1-1 for a size comparison.

**Table 1-1**

| *PARTICLE* | *ABSOLUTE CHARGE (Coulombs)* | *RELATIVE CHARGE* | *MASS (kg)* |
|---|---|---|---|
| Neutron | 0 | 0 | $1.675 \times 10^{-27}$ |
| Proton | $+1.6 \times 10^{-19}$ | +1 | $1.673 \times 10^{-27}$ |
| Electron | $-1.6 \times 10^{-19}$ | −1 | $9.11 \times 10^{-31}$ |

## E. ELECTRONS

As mentioned above, the late nineteenth century scientists conducted several experiments, and found that atoms are divisible. They conducted experiments with gas discharge tubes.

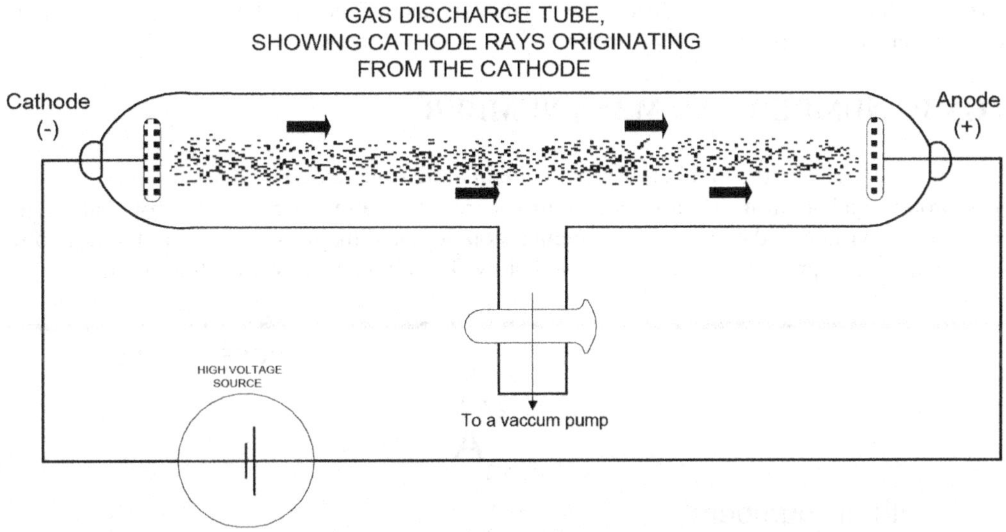


**Figure 1-1 Gas discharge tube**

A gas discharge tube is shown in Figure 1-1. The gas discharge tube is an evacuated glass tube and has two electrodes, a cathode (negative electrode) and an anode (positive electrode). The electrodes are connected to a high voltage source. Inside the tube, an electric discharge occurs between the electrodes. The discharge or 'rays' originate from the cathode and move toward the anode, and hence are called cathode rays. Using luminescent techniques, the cathode rays are made visible and it was found that these rays are deflected away from negatively charged plates. The scientist J. J. Thompson concluded that the cathode ray consists of negatively charged particles (electrons).

## Charge of Electrons

R. A. Millikan conducted the famous oil drop experiments and came to several conclusions: The charge of an electron is $-1.602 \times 10^{-19}$ C. From the charge-to-mass ratio, the mass of an electron was also calculated.

$$\frac{\text{Charge}}{\text{Mass}} = -1.76 \times 10^{8} \text{ Coulombs / gram}$$

$$\text{mass} = \frac{-1.6 \times 10^{-19}}{-1.76 \times 10^{8}} = 9.11 \times 10^{-23} \text{ g} = 9.11 \times 10^{-31} \text{ kg}$$

## F. PROTONS

Protons are positively charged nuclear particles. The charge of a proton is (positive electronic charge) $+1.6 \times 10^{-19}$ C. The net positive charge of the nucleus is due to the presence of the protons. A proton is about 1800 times more massive than an electron.

## G. NEUTRONS

Neutrons have mass comparable to that of protons, but neutrons are devoid of any electric charge. We will talk more about neutrons and their whereabouts when we study radioactivity.

Now a natural question is whether electrons, protons, and neutrons are the most fundamental particles. The answer is no. These fundamental particles are made of more fundamental particles called quarks. But, we don't have to go that far for the ACT. Just be aware that such sub-fundamental particles exist, fundamental particles being electrons, protons, and neutrons.

## H. ATOMIC NUMBER AND MASS NUMBER

The atomic number denotes the number of protons in an atom's nucleus. The mass number denotes the total number of protons and neutrons. Protons and neutrons are often called nucleons. By convention, the atomic number is usually written to the left of the elemental notation, and the mass number to the left above the elemental notation as represented by the example below. The element shown is aluminum.

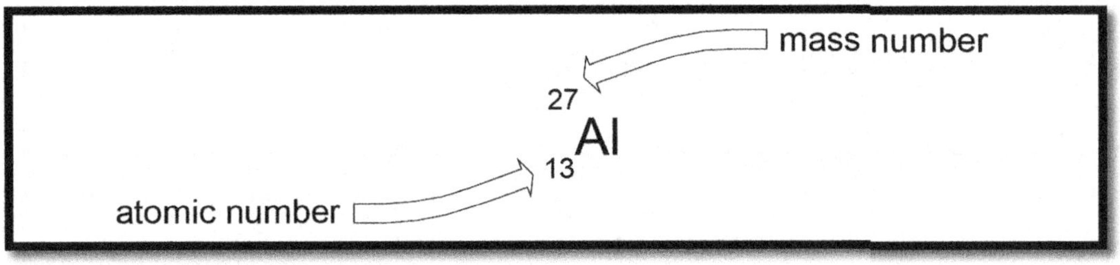


Some atoms have the same atomic number, but different mass numbers. This means different number of neutrons. Such atoms are called isotopes.

## I. ATOMIC WEIGHT

The atomic weight of an element is the average weight of all the isotopic masses of the element, calculated on the basis of their relative abundance in nature. The atomic weights are set on a "carbon-12" scale. This is the standard weight scale that is used worldwide to express atomic weights. Exploring this further, we can say that 12 atomic mass units (amu) make up the mass of one atom of ${}^{12}_{6}C$ isotope. In other words, one amu is equal to 1/12 the mass of one carbon-12 atom. We can also say that the atomic weight of carbon-12 is 12 amu. Even though it is popular to use the term atomic weight, atomic mass is a more appropriate term since we are really talking about the mass rather than the weight.

## J. MOLECULES

A molecule is a set or group of atoms which are chemically bonded. It can be represented by a molecular formula. A molecular formula represents the different kinds of atoms that are present in a molecule, along with the actual ratio of its atomic combination in forming that molecule.

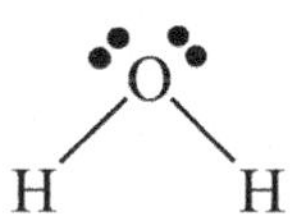

A molecule of $H_2O$ (water) contains two hydrogen atoms bonded to one atom of oxygen.

NaOH Sodium Hydroxide

A molecule of sodium hydroxide contains one sodium atom, one oxygen atom, and one hydrogen atom. The point is that molecular formula represents the molecules and the actual ratio of the atoms present in them.

## Molecular Weight

Molecular weight represents the sum of the atomic weights of all the atoms in that molecule. Molecular weight is also known as formula weight.

**Example 1-1**

Calculate the molecular weight of sulfuric acid ($H_2SO_4$).

2 hydrogens $2 \times 1 = 2$
1 sulfur $1 \times 32 = 32$
4 oxygens $4 \times 16 = \underline{64}$
98 g/mol

**Example 1- 2**

Calculate the molecular weight of carbon dioxide ($CO_2$).

1 carbon $1 \times 12 = 12$
2 oxygens $2 \times 16 = \underline{32}$
44 g/mol

## Empirical Formula

Empirical formula of a molecule represents the simplest ratio of the atoms present in the molecule. For example, acetylene has the molecular formula $C_2H_2$. The empirical formula of acetylene is CH. In essence, empirical formula gives the simplest ratio of atoms in a molecule.

**Problem 1-1**

Write the empirical formula of the following molecules.

1. $H_2O_2$
2) $C_2H_6$
3) $H_2O$

**Answers:**

1) HO
2) $CH_3$
3) $H_2O$

Notice that sometimes the empirical formula is the same as the molecular formula as in the case of water ($H_2O$).

## K. THE CONCEPT OF MOLE

The quantity of a given substance that contains as many units or molecules as the number of atoms in 12 grams of carbon-12 is called a **mole**. For example, one mole of glucose contains the same number of glucose molecules as in 12 grams of carbon-12. The mass of one mole of a substance is called its *molar mass*. The number of atoms in 12 grams of carbon-12 is represented by the *Avogadro number* ($6.023 \times 10^{23}$). So one mole of any substance contains Avogadro number of units in them.

To understand this concept thoroughly, try different possible scenarios where the Avogadro number can be used. Here are some. One mole of hydrogen atoms contains Avogadro number of hydrogen atoms. One mole of hydrogen molecules contains Avogadro number of hydrogen molecules. A mole of water contains $6.023 \times 10^{23}$ water molecules. These are all different ways of expressing the same concept.

The molar mass of a substance is equal to the molecular weight of that substance. The molecular weight (formula weight) of water is 18 amu. Since this is the molar mass, we can express it as 18 grams/mol.

**Example 1-3**

Calculate the mass of one molecule of sodium hydroxide (NaOH).

**Answer:** We know that the formula weight of sodium hydroxide is 40 g/mol.

| | |
|---|---|
| Sodium | 23 |
| Oxygen | 16 |
| Hydrogen | 1 |
| | 40 g/mol |

We also know that one mole of NaOH contains Avogadro number of molecules. So the mass of the NaOH molecule can be found by the following method:
Mass of one molecule of sodium hydroxide

$$= \frac{40\text{g}}{6.02 \times 10^{23}} = 6.64 \times 10^{-23}\text{g}$$

**Example 1-4**

Calculate the number of moles in 109.5 grams of hydrogen chloride.

**Answer:** Just like the last example, we have to find the molar mass of the molecule. The molar mass is 35.5 + 1 = 36.5 g/mol.

$$\text{Number of moles} = 109.5\ \text{g HCI} \times \frac{1\ \text{mol HCI}}{36.5\ \text{g HCI}} = 3\ \text{moles HCI}$$

So, 109.5 grams of HCl correspond to 3 moles of HCl.

You should be able to do these types of conversions back and forth, from grams to moles and moles to grams.

Try the next problem to see whether you have mastered the idea.

**Problem 1-2**

Calculate the number of grams in 8 moles of sulfur dioxide.

**Answer:** If your answer is close to 512.8 g, you solved the problem correctly.

## L. COMPOSITION BY PERCENTAGE

The ACT often contains percentage composition problems. Percentage composition is the percentage contribution (by weight) of each element to the total mass. Let's explore this idea by looking at some examples.

**Example 1-5**

Calcium carbonate ($CaCO_3$), commonly known as limestone is used in the preparation of a variety of compounds. Calculate the percentage composition of each element in calcium carbonate.

**Solution:**

| # of atoms per molecule | molar weight of the atoms | total mass of the element per mol |
|---|---|---|
| 1 Calcium | 1 x 40.1 g | 40.1 g |
| 1 Carbon | 1 x 12.0 g | 12.0 g |
| 3 Oxygen | 3 x 16.0 g | 48.0 g |
| | | 100.1 g |

The percentage composition of each element can be found as follows:

$$\%\ \text{of calcium} = \frac{40.1}{100.1} \times 100 = 40.1\%\ \text{calcium}$$

$$\%\ \text{of carbon} = \frac{12}{100.1} \times 100 = 12\%\ \text{carbon}$$

$$\%\ \text{of oxygen} = \frac{48}{100.1} \times 100 = 47.9\%\ \text{oxygen}$$

**Predicting Formulas from Percentage Compositions**

You should be able to predict the formula of a compound on the basis of a given data of percentage compositions. Study the next example to understand how it is done.

**Example 1-6**

A carbon compound contains 27.27% carbon and 72.73% oxygen by mass. Predict the simplest ratio or formula of the compound.

**Solution:**

The best way to approach this problem is to consider that we have 100 grams of this compound. Logically it should contain 27.27 grams of carbon and 72.73 grams of oxygen. With that in mind, we can calculate the number of moles of each element or atom. After that we can obtain the simple ratio.

**Step 1**

# of moles of carbon atoms equals 27.27/12 = 2.275 moles of carbon atoms
# of moles of oxygen atoms equals 72.73/16 = 4.546 moles of oxygen atoms

**Step 2**

Divide every number of moles with the smallest number of moles that you got in Step 1. Here the smaller one is 2.2725. So divide the number of moles of carbon atoms and the number of moles of oxygen atoms by 2.2725. That will give you the simplest ratio between them.

$$\text{Carbon: } 2.2725/2.2725 = 1$$

$$\text{Oxygen: } 4.546/2.2725 \approx 2$$

Since the ratio of carbon to oxygen is 1:2, the compound is $CO_2$.

**Example 1-7**

Calculate the mass of sulfur in 150 grams of $H_2SO_4$.

**Solution:**

The easiest way to calculate this is to find the percentage composition of sulfur. Then, use that percentage to find the mass of sulfur in the given amount of substance.

**Step 1**

$$\% \text{ of sulfur} = 32.1/98 \times 100, \text{ roughly } 33\%$$

**Step 2**

The mass of sulfur present in 150 grams of sulfuric acid is

$$150 \times 33\%/100 = 49.5 \text{ g}$$

## M. DENSITY

Density is defined as the mass per unit volume.

$$\text{Density} = \frac{\text{mass}}{\text{volume}}$$

This property can be used to identify a compound or an element, since the density of a pure substance is a constant. Since density relates mass and volume, it can be used to find the volume occupied by a given mass, or if the volume is known, we can find the mass.

Density of water is 1.0 g/ml.

Let's explore some calculations involving density. You'll see that these calculations have tremendous laboratory significance.

**Example 1-8**

The density of carbon tetrachloride is about 1.6 g/ml at 20°C. Calculate the volume occupied by 320 g of $CCl_4$.

**Solution:**

$$\text{Density} = \frac{\text{mass}}{\text{volume}}$$

$$\text{Volume} = \frac{\text{mass}}{\text{density}} = \frac{320\text{ g}}{1.6\text{g / ml}} = 200\text{ ml}$$

So, 320 grams of carbon tetrachloride will occupy a volume of 200 ml.

**Example 1-9**

A 20 ml sample of mercury has a mass of 271 g. Calculate the density of mercury.
Solution:

$$\text{Density} = \frac{\text{mass}}{\text{volume}} = \frac{271\text{ g}}{20\text{ ml}} = 13.55\text{ g / ml}$$

This question tests your knowledge of a basic equation. Though the equation is simple, you should be able to manipulate this equation so that you can connect this piece of information with other facts and formulas that are given in your test question or passage.

# Chemistry Review 1 Problems

1. A student preparing a solution for an experiment measured the weight of the sample solute to be used. If she is supposed to use 2 moles of calcium hydroxide, she must use:
A. 57.1 grams.
B. 74.1 grams.
C. 114.2 grams.
D. 148.2 grams.

2. Which of the following best represents the total number of ions present in a sample of NaCl weighing 102 g?
A. $6.0 \times 10^{23}$ ions
B. $10.5 \times 10^{23}$ ions
C. $12.0 \times 10^{23}$ ions
D. $21 \times 10^{23}$ ions

3. Experiments conducted with gas discharge tubes during the late $19^{th}$ century resulted in many important conclusions in atomic chemistry. Among those, one of the most important was the discovery of rays known as cathode rays. Cathode rays when passed near negative plates will most likely:
A. bend toward the negative plate.
B. bend away from the negative plate.
C. will not bend, because they are uncharged.
D. will not bend, because they are high energy radiations.

4. Which of the following is true regarding a typical atom?
A. Neutrons and electrons have the same mass.
B. The mass of neutrons is much less than that of electrons.
C. Neutrons and protons together make the nucleus electrically neutral.
D. Protons are more massive than electrons.

5. The empirical formula of butane is:
A. $CH_3$
B. $C_2H_5$
C. $C_4H_{10}$
D. $CH_2$

6. The mass of one mol of a substance is numerically equal to:
A. mass number.
B. Avogadro number.
C. molecular weight.
D. 22.4.

7. If m represents the number of moles of a substance, M represents the molar mass of the substance, and d represents the density of the substance, which of the following expressions equals to the volume of the sample substance?
A. mM/d
B. m/dM
C. d/mM
D. m/d

8. The mass of oxygen in 96 grams of sulfur dioxide is closest to:
A. 16 g.
B. 24 g.
C. 32 g.
D. 47 g.

9. Choose the best value that corresponds to the percentage composition of chlorine in carbon tetrachloride?
A. 92%
B. 90%
C. 86%
D. None of the above

10. A student researcher analyzing the identity of the by-product of a reaction found that the compound contained 63.6% nitrogen and 36.4% oxygen. What is the most likely formula of this compound?
A. NO
B. $NO_2$
C. $N_2O$
D. $N_2O_3$

Questions 11-14 are based on the following passage.

**Passage**

The following reactions were conducted in a lab for studies related to reaction kinetics. Consider the reactions shown:

Reaction 1

$$CH_4 + 2O_2 \rightarrow CO_2 + 2H_2O$$

$$2C_2H_6 + 7O_2 \rightarrow 4CO_2 + 6H_2O$$

Reaction 2

11. Reaction 2 is best described by which of the following?
A. A metathesis reaction
B. A combustion reaction
C. An endothermic reaction
D. A decomposition reaction

12. In Reaction 2, if 54 g of water was formed, how much ethane and oxygen must have reacted?
A. 30 g of ethane and 224 g of oxygen
B. 60 g of ethane and 224 g of oxygen
C. 30 g of ethane and 112 g of oxygen
D. 60 g of ethane and 112 g of oxygen

13. Which of the following is the actual formula of dextrose, if the empirical proportion is 1:2:1, where the proportion number 2 is that of hydrogen. Dextrose has a molecular weight of 180 g/mol? (Dextrose molecule is composed of carbon, hydrogen, and oxygen)
A. CHO
B. $C_6H_{12}O_6$
C. $C_{12}H_6O_3$
D. $CH_2O$

14. Which of the following equals the number of hydrogen atoms in 40 grams of methane?
A. $1.5 \times 10^{24}$
B. $2.4 \times 10^{24}$
C. $6.0 \times 10^{24}$
D. $6.023 \times 10^{23}$

# Solutions

1. The answer is D. This question tests your understanding of the concept of mole. The student is preparing a solution using calcium hydroxide. The formula of calcium hydroxide is $Ca(OH)_2$. The formula weight of calcium hydroxide is 74.1 grams/mol. Here, the student added two moles of it. So the answer is 74.1 x 2 = 148.2 g.

2. The answer is D. The sample weighs 56 g. The formula weight of $O_2$ is 32 g/mol. So the number of moles of oxygen present is roughly 1.75. Remember that 1 mol contains Avogadro number of particles. Since 1.75 moles of oxygen is present, there are $1.75 \times 6.0 \times 10^{23}$ molecules of oxygen. Are we done at this point? No! The question asks for the total number of ions present in the sample. Since each oxygen molecule is composed of oxygen atoms, the total number of atoms in this sample equals $2 \times 1.75 \times 6.0 \times 10^{23} = 21 \times 10^{23}$.

3. The answer is B. Gas discharge tubes are vacuum tubes with electrodes connected to very high voltages. Cathode rays are emitted from the cathode of the tube. A cathode ray consists of a beam of electrons. Since electrons are negatively charged, cathode rays will bend away from a negatively charged plate.

4. The answer is D. This question asks for a true statement from the choices. Let's look at each of those choices. Choice A is incorrect, because neutrons and electrons do not have the same mass. Choice B claims that the mass of a neutron is less than that of an electron. That is not true either. Protons are positive and neutrons are neutral. They cannot together make the nucleus neutral. In fact, the nucleus is positively charged, not electrically neutral. So Choice C is wrong. The last choice makes perfect sense. A proton is more massive than an electron.

5. The answer is B. The actual formula of butane is $C_4H_{10}$. But the question is not about butane's actual formula. The question asks for the empirical formula of butane. Empirical formula is the most simplified ratio of the actual formula. So the correct answer is $C_2H_5$.

6. The answer is C. Let's look at the question carefully. If you did not get the right answer, carefully think about it again. The question is simply testing your understanding of a definition. The mass of a substance in one mole of it is numerically equal to its molecular weight.

7. The answer is A. Density is equal to mass per volume. Further more, the mass is equal to the product of number moles and molar mass. Equating these two equations we can get the answer. Using the letter notations given in the question, the expression of volume is given by Choice A.

8. The answer is D. The formula of sulfur dioxide is $SO_2$. The formula weight of $SO_2$ is 64 g/mol. This sample of $SO_2$ contains 1.5 moles of it. If we calculate the percentage mass of the oxygen, we can see that the oxygen's mass contribution is 50% in $SO_2$. So regardless of the amount of $SO_2$ present, half of its mass is due to the presence of oxygen and the other half is due to the presence of sulfur. Here we have 96 g of $SO_2$. Now, one half of 96 g gives 48 g.

9. The answer is A. The formula of carbon tetrachloride is $CCl_4$. The formula weight of $CCl_4$ is 154 g/mol. The percentage composition of Cl can be calculated as follows:

$$\%\text{ of chlorine} = \frac{\text{mass of chlorine}}{\text{mass of } CCl_4} \times 100\% = \frac{142}{154} \times 100\% \approx 92\%$$

10. The answer is C. In the question, the percentage compositions of nitrogen and oxygen are given. With that information, we have to find the formula of that compound. Just as discussed in our review, assume that we have 100 grams of this compound. If there are 100 grams, logically it should contain 63.6 grams of nitrogen and 36.4 grams of oxygen.

Step 1

$$\text{\# of moles of nitrogen} = 63.6/14 \approx 4.5 \text{ moles of nitrogen}$$
$$\text{\# of moles of oxygen} = 36.4/16 \approx 2.275 \text{ moles of oxygen}$$

Step 2
Divide every number of moles with the smallest number of moles calculated in Step 1. Here the smaller one is 2.275. Divide both mole numbers by 2.275. This will give you the simplest ratio between them.

$$\text{Nitrogen } 4.5/2.275 \approx 2 \qquad \text{Oxygen } 2.275/2.275 \approx 1$$

Since the ratio of nitrogen to oxygen is 2:1, the compound is $N_2O$.

11. The answer is B. Reaction 2 is a typical example of a combustion reaction.

12. The answer is C.

$$2C_2H_6 + 7O_2 \longrightarrow 4CO_2 + 6H_2O$$

Reaction 2

According to Reaction 2, two moles of ethane will react with 7 moles of oxygen to form 4 moles of carbon dioxide and 6 moles of water. In the question, it is given that 54 g of water was formed. This corresponds to 3 moles of water. From the equation, we can say that 1 mol of ethane must have reacted to form 3 moles of water. We can also say that 3.5 moles of oxygen must have reacted in the formation of 3 moles of water. The last step is the conversion of moles to grams. The corresponding quantities in grams are 30 g of ethane and 112 g of oxygen.

13. The answer is B. From the question, we can say that the empirical formula of dextrose is $CH_2O$. The molecular weight of dextrose is also given. Let's first calculate the weight of the empirical formula. Maybe the empirical formula represents the actual formula. The weight of the empirical formula is 30 g/mol. But the question says it is 180 g/mol. In this case, the molecular formula is not the same as the empirical formula. By dividing the actual weight by the empirical weight, we will get the coefficient with which we have to multiply the empirical formula to get the actual formula. The coefficient is 180/30 = 6. So the molecular formula is $C_6H_{12}O_6$. Have you heard of glucose? Well, dextrose is the same thing. If you knew that, you could have picked the answer right away. But more importantly, it is the process that you have to learn from this question.

14. The answer is C. The formula of methane is $CH_4$. The molecular weight is 16 g/mol. We have 40 g of methane. That means we have 2.5 moles of methane. So the number of hydrogen atoms can be calculated.

$$6.0 \times 10^{23} \times 2.5 \text{ moles} \times 4 \text{ hydrogen atoms/mol} = 6.0 \times 10^{24} \text{ atoms}$$

# Chemistry Review 2
# Chemical Reactions

## A. INTRODUCTION

Chemical reaction is a process at the molecular or ionic level by which one or more types of substances are transformed into one or more new types of substances by different modes of combination. In this chapter, we will explore the different types of chemical reactions including oxidation-reduction reactions. We will also learn how to balance equations.

## B. CHEMICAL REACTIONS

A chemical reaction can be represented by a chemical equation. In a chemical equation representing an irreversible reaction, the substances that react (reactants) are written on the left side, while the resulting substances (products) are written on the right side of an arrow.

$$2\,H_2 \quad + \quad O_2 \longrightarrow 2\,H_2O$$

In the reaction shown above, two molecules of hydrogen react with one molecule of oxygen, forming two molecules of water. Balancing of equations will be covered a little later in this chapter.

## C. TYPES OF CHEMICAL REACTIONS

There are five types of chemical reactions.

1) Combination reaction
2) Combustion reaction
3) Decomposition reaction
4) Displacement reaction or single-replacement reaction
5) Metathesis reaction or double-replacement reaction

### Combination Reaction

A reaction involving the formation of a compound from two or more substances is called a combination reaction.

Some representative combination reactions

$$2Na\,(s) \quad + \quad Cl_2\,(g) \longrightarrow 2\,NaCl\,(s)$$

$$SO_2\,(g) \quad + \quad H_2O\,(l) \longrightarrow H_2SO_3\,(aq)$$

### Combustion Reaction

A combustion reaction involves the reaction of substances with oxygen, and it is usually accompanied by the release of large amounts of heat. Combustion reactions are thus highly exothermic.

Some representative combustion reactions

$$C\,(s) \quad + \quad O_2\,(g) \longrightarrow \quad O\ _2\,(g)$$

$$2\,H_2S\,(g) \; + \; 3\,O_2\,(g) \longrightarrow 2\,SO_2\,(g) \; + \; 2\,H_2O\,(g)$$

## Decomposition Reaction

A decomposition reaction is a process in which one compound decomposes or splits to form two or more simpler compounds and/or elements.

A representative decomposition reaction

$$CaCO_3\,(s) \longrightarrow CaO\,(s) \; + \; CO_2\,(g)$$

## Displacement Reaction (single-replacement reaction)

In a *single-replacement reaction*, an element reacts with a compound, and results in the displacement of an element or group from the compound. An example of a single-replacement reaction is shown.

A single-replacement reaction

$$Zn\,(s) + CuCl_2\,(aq) \rightarrow ZnCl_2\,(aq) + Cu\,(s)$$

In this reaction, Zn substitutes for Cu.

## Metathesis Reaction (double-replacement reaction)

A *metathesis (double-replacement) reaction* involves the exchange of two groups or two ions among the reactants. Remember that in a single-replacement reaction, there is only one group or ion being switched. A metathesis reaction can often result in an insoluble product from soluble reactants, and the insoluble compound formed is called a precipitate.

A metathesis reaction

$$AgNO_3\,(aq) \; + \; NaCl\,(aq) \longrightarrow AgCl\,(s) \; + \; NaNO_3\,(aq)$$

Note that this reaction involves the formation of a precipitate of AgCl.

With respect to the types of reactions, your objective should be to understand the basis behind the categorization.

## D. BALANCING SIMPLE EQUATIONS

A chemical equation is said to be balanced if all the atoms present in the reactants appear in the same numbers among the products. Here is an example.

**Example 2-1**

Balance the following equation.

$$Fe + O_2 \longrightarrow Fe_2O_3$$

**Solution:**

Start by balancing the oxygen atoms. There are two oxygen atoms on the reactant side and three oxygen atoms on the product side. To balance this, put 3 as the coefficient of oxygen on the reactant side. When we write '3 $O_2$,' that means we have 6 oxygen atoms on the reactant side. To make the same number of oxygen atoms on the product side, let's put 2 as the coefficient of $Fe_2O_3$. Now the oxygen atoms seem to be balanced.

Let's take a look at Fe. Since the coefficient of $Fe_2O_3$ is 2, we have 4 atoms of Fe on the product side. We can balance this by writing 4 as the coefficient of Fe on the reactant side. So the balanced equation is as follows:

$$4\,Fe + 3\,O_2 \longrightarrow 2\,Fe_2O_3$$

**Problem 2-1**

Balance the following equations:

**(a)** $Cu + AgNO_3 \longrightarrow Ag + Cu(NO_3)_2$

**(b)** $Fe_2O_3 + CO \longrightarrow Fe + CO_2$

**(c)** $H_2SO_4 + NaOH \longrightarrow Na_2SO_4 + H_2O$

**(d)** $Ba(OH)_2 + HCl \longrightarrow BaCl_2 + H_2O$

Answers:

**(a)** $Cu + 2\,AgNO_3 \longrightarrow 2\,Ag + Cu(NO_3)_2$

**(b)** $Fe_2O_3 + 3\,CO \longrightarrow 2\,Fe + 3\,CO_2$

**(c)** $H_2SO_4 + 2\,NaOH \longrightarrow Na_2SO_4 + 2\,H_2O$

**(d)** $Ba(OH)_2 + 2\,HCl \longrightarrow BaCl_2 + 2\,H_2O$

## E. OXIDATION NUMBER

Electrons are exchanged during oxidation-reduction reactions. The behavior of atoms or ions in terms of the number of electrons transferred is expressed as the oxidation state (oxidation number). We can define oxidation number as the charge of an atom or ion, based on a set of standard rules. If the given species is an ion containing a single atom, then its oxidation state is its charge itself. Let's analyze this by looking at a few examples.

In NaCl, the oxidation state of sodium is +1 and the oxidation state of chlorine is –1. Generally, the elements at the top right corner of the periodic table are assigned negative oxidation numbers. Some of the elements on the right side of the periodic table can have positive or negative oxidation numbers depending upon the atom to which the given element is bonded to. The elements in the middle and the left portions of the periodic table have almost exclusively positive oxidation numbers.

**Table 1-1**

**General guidelines for assigning oxidation numbers**

1. The elemental natural state oxidation number of any atom is zero. For example, the oxidation number of oxygen atom in $O_2$ is zero.
2. The sum of the oxidation numbers of the atoms in a compound should be zero.
3. The sum of the oxidation numbers of the atoms in an ionic species (a species with a net charge) should equal the net charge of the ionic species.
4. The oxidation number of a given ion containing a single atom is its charge itself.

**Oxidation numbers of some common elements**

1. The common oxidation number of Group IA metals is +1. E.g., lithium, sodium, potassium.
2. The common oxidation number of Group IIA metals is +2. E.g., beryllium, magnesium, calcium.
3. The common oxidation number of Group IIIA is +3. E.g., aluminum, boron.
4. The common oxidation number of Group IVA is +4. +2 is also seen in some compounds such as CO.
5. The common oxidation numbers of Group V A are +5 and –3.
6. The common oxidation number of Group VIA is –2.
7. The common oxidation number of Group VIIA is –1.
8. The common oxidation number of H is +1. In some metal hydrides, hydrogen shows an oxidation number of –1.

The above list of common oxidation numbers is not comprehensive. Nevertheless, it gives you a basic and essential picture about assigning oxidation numbers in common compounds and ionic species. Most elements can have multiple oxidation states, depending on the element or ionic species they are bonded to. You have to always follow the general guidelines in Table 1-1, and check whether the items listed are satisfied.

Now that we have learned the theory of assigning oxidation numbers, let's do an example to see how it works.

**Example 2-2**

What is the oxidation number of sulfur in sulfuric acid?

**Solution:**

Sulfuric acid is $H_2SO_4$. The oxidation number of hydrogen is +1. But we have two hydrogens which add up to a charge of +2. Since the total charge of this molecule should be zero, we can say that the charge of sulfate ion is –2. We also know that the oxidation number of oxygen is –2. But there are four oxygens in a sulfate ion. So the charge adds to –8. Now let's solve this algebraically.

$ON^{Sulfur}$ - Oxidation number of sulfur
$ON^{Oxygen}$ - Oxidation number of oxygen
$ON^{Sulfur} + 4\,(ON^{Oxygen}) = -2$
$ON^{Sulfur} + 4\,(-2) = -2$
$ON^{Sulfur} = +6$

So the oxidation number of sulfur in sulfuric acid is +6.

**Problem 2-2**

Calculate the oxidation state of the element indicated in each of the following problems.

(a) What is the oxidation state of hydrogen in $MgH_2$?
(b) What is the oxidation state of chlorine in $ClO_3^-$?
(c) What is the oxidation state of oxygen in $Na_2O_2$?
(d) What is the oxidation state of nitrogen in $NH_3$?
(e) What is the oxidation state of oxygen in $O_2$?
(f) What is the oxidation state of bromine in $HBrO_2$?
(g) What is the oxidation state of manganese in $KMnO_4$?

**Answers:**

(a) –1
(b) +5
(c) –1
(d) –3
(e) 0
(f) +3
(g) +7

## F. OXIDATION-REDUCTION REACTIONS

Oxidation-reduction (redox) reactions involve the transfer of electrons from one compound or species to another. In this section, we will discuss oxidation-reduction reactions and learn how to balance them.

*Oxidation* is the process by which an atom or species loses its electrons. In *reduction*, an atom or species gains electrons. Let's first consider oxidation and reduction separately.

Consider the conversion of iron from its neutral elemental state to its ionic form.

$$Fe \longrightarrow Fe^{2+} + 2\,e^-$$

Notice that in the process iron lost electrons. The process is oxidation.

Consider another example. An example involving the conversion of bromine to its ionic form.

$$Br_2 + 2\,e^- \longrightarrow 2\,Br^-$$

Notice that in the process bromine gained electrons. The process is reduction.

Now let's go one step forward. Consider the next reaction.

$$2\ Fe + 3\ Br_2 \longrightarrow 2\ FeBr_3$$

This reaction is a typical example of an oxidation-reduction reaction. The oxidation number of iron on the reactant side is 0. The oxidation number of bromine on the reactant side is also 0. What are the oxidation numbers of iron and bromine in $FeBr_3$? Well, we know that bromine has an oxidation number of −1. So the oxidation number of iron in $FeBr_3$ is +3. Thus iron is oxidized and bromine is reduced. The species that gets oxidized is called the reducing agent. The species that gets reduced is called the *oxidizing agent*. In this reaction, evidently iron acts as the *reducing agent*, and bromine acts as the oxidizing agent.

*Oxidation results in an increase in the oxidation number. In the process of oxidation, electrons are lost.*

*Reduction results in a decrease in the oxidation number. In the process, electrons are gained.*

## Balancing Oxidation-Reduction Reactions

Balancing of an oxidation-reduction reaction is a little more complex than balancing a simple reaction. The main rule that you have to follow when balancing oxidation-reduction reactions is that the absolute value of the increase in oxidation number of all the atoms that are oxidized should equal the absolute value of the decrease in oxidation number of all the atoms that are reduced. Balancing oxidation-reduction reaction is sometimes time-consuming and quite often frustrating. We will look at two methods of balancing oxidation-reduction reactions.

## Method A

1. Write the unbalanced equation.
2. Find the oxidation numbers of the atoms that undergo change in oxidation states and write on top of each atom the corresponding oxidation number.
3. By this process, we will be able to see which atoms are getting oxidized and reduced.
4. Compare and indicate the change in oxidation numbers from the reactant side and the product side, and write down the change in the oxidation numbers.
5. Make the necessary changes by writing coefficients that will equalize the changes in the oxidation numbers. In other words, the net decrease in the oxidation numbers should equal the net increase in the oxidation numbers. Add water if necessary.
6. Do a final check on whether all the atoms and charges balance out.

## Method B

1. Write the unbalanced equation.
2. Separate the two half-reactions, write them out, and balance any of the atoms. From this point onward, we balance the reactions separately. Do the obvious or the simple balancing by inspection if possible.
3. Balance the oxygen atoms by adding water on the appropriate side of the half-reaction.
4. Balance the hydrogen atoms by adding $H^+$ on the appropriate side of the half-reaction.
5. Add sufficient number of electrons so that the charges are balanced.
6. Once the half-reactions are balanced, combine the half-reactions and cancel out any common terms that appear on both sides of the equation to get the refined and balanced oxidation-reduction equation.

**Example 2-3**

Balance the following oxidation-reduction reaction.

$$Zn + NO_3^- \longrightarrow Zn^{2+} + N_2O$$

**Method A**

**1.** $Zn + NO_3^- \longrightarrow Zn^{2+} + N_2O$

**2.** $\overset{0}{Zn} + \overset{+5}{NO_3^-} \longrightarrow \overset{+2}{Zn^{2+}} + \overset{+1}{N_2O}$

**3.** $\overset{0}{Zn} + 2\ \overset{+5}{NO_3^-} \longrightarrow \overset{+2}{Zn^{2+}} + \overset{+1}{N_2O}$

**4.** 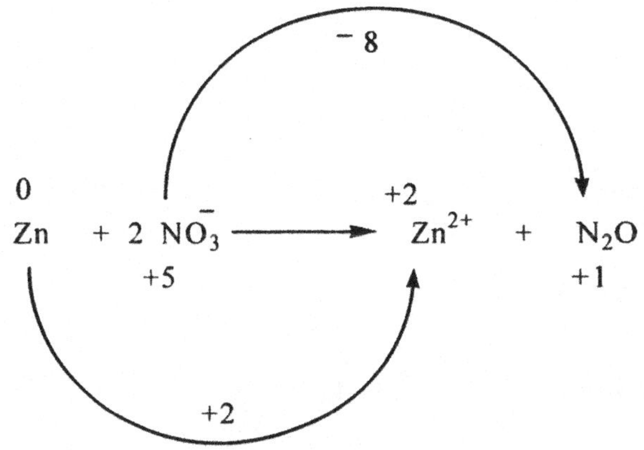


**5.** 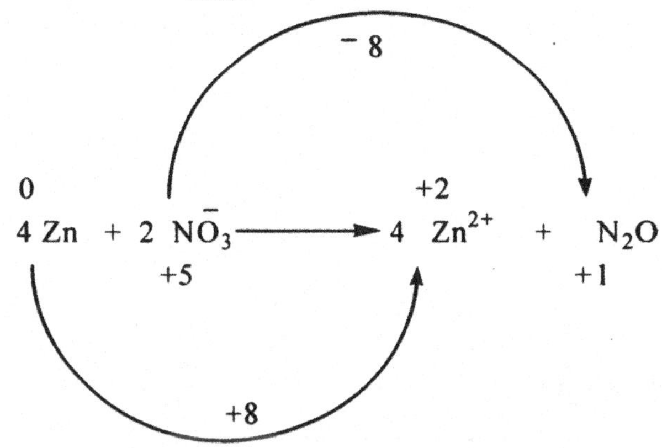


**6.** $10\ H^+ + 4\ Zn + 2\ NO_3^- \longrightarrow 4\ Zn^{2+} + N_2O + 5\ H_2O$

**Method B**

$$Zn + NO_3^- \longrightarrow Zn^{2+} + N_2O$$

Half-reaction I

**1.** $Zn \longrightarrow Zn^{2+}$

**2.** $Zn \longrightarrow Zn^{2+} + 2e^-$

Half-reaction II

**1.** $NO_3^- \longrightarrow N_2O$

**2.** $2\,NO_3^- \longrightarrow N_2O$

**3.** $2\,NO_3^- \longrightarrow N_2O + 5\,H_2O$

**4.** $10\,H^+ + 2\,NO_3^- \longrightarrow N_2O + 5\,H_2O$

**5.** $8e + 10\,H^+ + 2\,NO_3^- \longrightarrow N_2O + 5\,H_2O$

Before we combine the equations, the electrons need to be balanced out. So multiplying the balanced half-reaction # (I) by 4, we get

$$4\,Zn \longrightarrow 4\,Zn^{2+} + 8e^-$$

Combining the two half reactions

$$4\,Zn \longrightarrow 4\,Zn^{2+} + \cancel{8e^-}$$

$$\cancel{8e^-} + 10\,H^+ + 2\,NO_3^- \longrightarrow N_2O + 5\,H_2O$$

$$10\,H^+ + 4\,Zn + 2\,NO_3^- \longrightarrow 4\,Zn^{2+} + N_2O + 5\,H_2O$$

## G. CALCULATIONS INVOLVING CHEMICAL REACTIONS

In this section, we will look at some chemical equations and find what we can do with the information represented in a chemical equation. The equation shown below represents the reaction of methane and oxygen to form carbon dioxide and water.

$$CH_4 \quad + \quad 2\,O_2 \longrightarrow CO_2 \quad + \quad 2\,H_2O$$

From this balanced equation we can infer many things. Let's consider a few.

1) One molecule of methane reacts with 2 molecules of oxygen to form 1 molecule of carbon dioxide and 2 molecules of water.

2) We can also say that 1 mole of methane reacts with 2 moles of oxygen to form 1 mole of carbon dioxide and 2 moles of water.

3) Since one mole contains Avogadro number of molecules, we can say that $6.023 \times 10^{23}$ molecules of methane reacts with $1.2046 \times 10^{24}$ (= $2 \times 6.023 \times 10^{23}$) molecules of oxygen to form $6.023 \times 10^{23}$ molecules of carbon dioxide and $1.2046 \times 10^{24}$ molecules of water.

4) We can confidently say that 16 g of methane reacts with 64 g of oxygen to form 44 g of carbon dioxide and 36 g of water.

**Example 2-4**

Calculate the number of moles of water produced when 5.25 moles of methane undergo the reaction depicted below. (Assume there is plenty of oxygen for the reaction)

$$CH_4 \quad + \quad 2\,O_2 \longrightarrow CO_2 \quad + \quad 2\,H_2O$$

**Solution:**

From the equation, it is clear that for every mole of methane, 2 moles of water are formed. So without any elaborate calculations, you should be able to come up with the correct answer. It is very much like a ratio problem. Since there are 5.25 moles of methane, 10.5 moles of water will be formed.

Number of moles of water formed = 5.25 x 2 = 10.5 moles.

**Example 2-5**

110 g of $CO_2$ were formed as a result of the reaction shown below. How many grams of oxygen must have reacted to form that much carbon dioxide?

$$CH_4 \quad + \quad 2\,O_2 \longrightarrow CO_2 \quad + \quad 2\,H_2O$$

**Solution:**

According to the equation, 2 moles of oxygen result in 1 mole of carbon dioxide. For example, if 2 moles of carbon dioxide were formed, 4 moles of oxygen must have reacted. Here, the amount of carbon dioxide formed is given in terms of grams. So the first step is to convert the grams to moles.

$$\text{Moles of carbon dioxide} = \frac{110\text{ g}}{44\text{ g / mol}} = 2.5\text{ moles}$$

Hence, 5 moles of oxygen must have reacted to form 2.5 moles of carbon dioxide. But the question asks for this quantity in grams. So the final step is to convert moles to grams.

$$\text{The amount of required oxygen} = 5 \text{ moles} \times \frac{32 \text{ g}}{1 \text{ mol}} \approx 160 \text{ g}$$

## H. THE CONCEPT OF LIMITING REAGENT

So far we have been considering reactions in which all the reactants exist in adequate quantities. In this section, we will consider what happens when the amount of one of the reactants available is less than the amount required to complete the reaction. When such a condition exists, we call that reactant or reagent the limiting reagent.

We will further explore this scenario through the following examples.

**Example 2-6**

A reaction mixture contains 60.75 g magnesium and 146 g hydrogen chloride. Predict the limiting reagent if the reaction occurs as shown.

$$Mg + 2\,HCl \longrightarrow H_2 + MgCl_2$$

**Solution:**

First, we have to convert the grams of the substances to moles. Then make the comparison to see which one is the limiting reagent. By now, you should be comfortable with the conversion of moles to grams and vice versa. The number of moles of magnesium present is 2.5 moles. According to the equation, 1 mole of magnesium reacts with 2 moles of hydrogen chloride. For magnesium to completely react, there should be at least 5 moles of hydrogen chloride present. If you calculate the number of moles of hydrogen chloride present, you will get 4 moles. This amount of hydrogen chloride is not enough to completely react with the amount of magnesium present. So the limiting reagent is hydrogen chloride.

## I. PERCENT YIELD

If we know the chemical equation and the amounts of reactants, we can calculate the theoretical yield of that reaction. But in reality, the yield depends on many other factors also. Most of the time in synthesis reactions, even in your own lab experiments, you probably noticed that the actual yield is lower than the theoretical yield. The percent yield denotes the amount of actual yield in terms of the theoretical yield. The formula to find the percent yield is given below:

$$\text{Percent yield} = \frac{\text{actual yield of the product}}{\text{theoretical yield of the product}} \times 100\%$$

**Example 2-7**

$$NaNO_2 + HCl \longrightarrow HNO_2 + NaCl$$

A student conducted the above reaction in a lab as a part of her research assignment. She used 138 g of sodium nitrite, with excess of hydrogen chloride. What is the percent yield of $HNO_2$, if the actual yield of $HNO_2$ was 61.1 g?

**Solution:**

First, you should find the number of moles of $NaNO_2$. Since she used 138 g, the number of moles of $NaNO_2$ is 2 (Mol.wt of $NaNO_2$ is 69 g/mol). Since the ratio of formation of $HNO_2$ is 1:1 with respect to $NaNO_2$, theoretically 2 moles of $HNO_2$ should be formed. Two moles of $HNO_2$ correspond to 94 g. But actually, only 61.1 g of $HNO_2$ was formed. Now it is just a matter of plug and chug in the percent yield formula.

$$\text{The percentage yield of HNO}_2 = \frac{61.1 \text{ grams}}{94 \text{ grams}} \times 100\% = 65\%$$

In this experiment, the actual yield was not high (i.e., only 65% of the theoretically predicted yield) as expected.

**Example 2-8**

Match the following reactions with the appropriate type of reaction

| | |
|---|---|
| **1.** $2KClO_3 \longrightarrow 2KCl + 3O_2$ | **A.** Combination reaction |
| **2.** $2KCl + Cl_2 \longrightarrow 2KCl$ | **B.** Double-replacement reaction |
| **3.** $HNO_3 + NaOH \longrightarrow NaNO_3 + H_2O$ | **C.** Decomposition reaction |

**Solution:**

| | |
|---|---|
| $2KClO_3 \longrightarrow 2KCl + 3O_2$ | **C.** Decomposition reaction |
| $2KCl + Cl_2 \longrightarrow 2KCl$ | **A.** Combination reaction |
| $HNO_3 + NaOH \longrightarrow NaNO_3 + H_2O$ | **B.** Double-replacement reaction |

# Chemistry Review 2 Problems

1. The reaction shown here can be best classified as a:

$$2HI + PbCl_2 \rightarrow PbI_2 + 2HCl$$

A. combustion reaction.
B. combination reaction.
C. single-replacement reaction.
D. double-replacement reaction.

2. All the following reactant-combinations are neutralization reactions, except:

A. $HNO_2$ + NaOH
B. KOH + HCl
C. $Al^+$ + 3OH
D. $H_3PO_4$ + NaOH

3. An unbalanced equation is shown. What is the coefficient of aluminum hydroxide in the final balanced equation?

$$H_2SO_4 + Al(OH)_3 \rightarrow H_2O + Al_2(SO_4)_3$$

A. 1
B. 2
C. 3
D. 7

4. How many grams of sodium chloride are required to synthesize 73 grams of hydrogen chloride, if the reaction involves sodium chloride and sulfuric acid?

$$NaCl + H_2SO_4 \rightarrow Na_2SO_4 + HCl$$

A. 58.5 grams
B. 117 grams
C. 175.5 grams
D. 234 grams

5. In some reactions, you will often encounter ions in aqueous solutions which are not actually involved in the reaction. Such ions are best termed:

A. cations.
B. anions.
C. salt-bridge ions.
D. spectator ions.

6. Some substances can act as a base or an acid. Such substances are called:

A. aliphatic substances.
B. amphibasic substances.
C. lyophilic substances.
D. amphoteric substances.

7. Predict the coefficient of iron in the balanced equation for a reaction in which iron reacts with oxygen to form $Fe_2O_3$.

A. 1
B. 2
C. 3
D. 4

8. The number of molecules are the same in which of the following pairs?

A. 32 grams of $O_2$ and 32 grams of $SO_2$
B. 49 grams of $NO_2$ and 40 grams of NaOH
C. 20 grams of HF and 36 grams of $H_2O$
D. 60 grams of $C_2H_6$ and 156 grams of $C_6H_6$

Questions 9-15 are based on the following passage.

## Passage 1

Information regarding the amounts of substances that actually react to form the products are of extreme help. Analyzing reactions and having the balanced equations help chemists to determine the correct and optimum proportions of reactants to be used. There are various other factors besides the amount of reactants, which determine the efficiency of reactions.

**Experiment 1**

$$2\ Al + Fe_2O_3 \rightarrow 2\ Fe + Al_2O_3$$

Student 1 used 40.5 g of aluminum and 80 g of $Fe_2O_3$ for Reaction I. A second student conducted the same reaction with a different amount for one of the reactants. The second student used 40.5 g of aluminum, and 90 grams of $Fe_2O_3$. A third student also conducted the same reaction with 54 g of aluminum and the same amount of $Fe_2O_3$ used by the first student.

**Experiment 2**

Reaction II involved the production of $Al_2O_3$ from aluminum hydroxide This was done by heating aluminum hydroxide.

$$2\ Al(OH)_3(s) \rightarrow Al_2O_3(s) + \text{product x}$$

**Experiment 3**

$$Fe_2O_3(s) + 3\ CO(g) \xrightarrow[\text{heat}]{} 2\ Fe(s) + 3\ CO_2(g)$$

9. What is the most likely identity of product x?

A. aluminum
B. water
C. hydrogen
D. oxygen

10. In Experiment 3, which of the following is the limiting reagent?

A. $Fe_2O_3$ because there is only 1 mole of it
B. CO because of its gas phase
C. Fe because of its insufficiency
D. Cannot be determined without more data

11. Based on the given information, which of the following are true regarding Experiment 1?

I. The second student had the highest yield for aluminum oxide.
II. The third student had a higher yield for aluminum oxide than the first student.
III. The first and the third students had the same yield for aluminum oxide.

A. I only
B. I and II only
C. I and III only
D. II and III only

12. Which of the following changes will further increase the overall yield for the reactions conducted in the Experiment 1?

A. Increasing the amount of aluminum used by Student 1
B. Increasing the amount of aluminum used by Student 2
C. Increasing the amount of $Fe_2O_3$ used by Student 2
D. None of the above

13. Roughly 80 g of $Fe_2O_3$ was present in Experiment 3, and upon completion of the reaction, it was measured that 22 g of $CO_2$ was formed. If this is true, how much CO must have reacted before reaching completion of the reaction?

A. 14 grams
B. 28 grams
C. 56 grams
D. 84 grams

14. In the previous question, which of the reactants acts as the limiting reagent?

A. $Fe_2O_3$
B. CO
C. $CO_2$
D. Cannot be determined

15. If 0.1 mol of CO was reacted with excess of $Fe_2O_3$, how many molecules of carbon dioxide will be produced?

A. $6.0 \times 10^{23}$ molecules
B. $6.0 \times 10^{22}$ molecules
C. $3.0 \times 10^{23}$ molecules
D. $18.0 \times 10^{23}$ molecules

# Solutions

1. The answer is D. This question tests your understanding of the general classification of reactions. The reaction is clearly not a combustion reaction. Combustion reactions occur in the presence of oxygen and involve release of heat. There is no indication of that in the given reaction. You might be tempted to select Choice C since the given reaction is a replacement reaction, but double-replacement reaction best describes the given reaction. In the reaction, the iodine is replaced with the chlorine, and the chlorine is replaced with the iodine.

2. The answer is C. Choices A, B, and D are all acid-base reactions, which are neutralization reactions. Choice C is actually the formation of a base ($AlOH_3$), and it is not a neutralization reaction.

3. The answer is B. Let's start balancing the equation from the sulfate point of view. You have 3 sulfate groups in aluminum sulfate. It has to come from the sulfuric acid. So the coefficient is 3 for the sulfuric acid. Since there are 2 aluminums in aluminum sulfate, we can assign a coefficient of 2 to the aluminum hydroxide. The rest of the hydrogens and oxygens can be balanced with water.

The balanced equation should look like this:

$$3\ H_2SO_4 + 2\ Al(OH)_3 \longrightarrow 6\ H_2O + Al_2(SO_4)_3$$

4. The answer is B. First, we have to check whether the given reaction is balanced. Without the balanced equation, the calculations will be incorrect. As a matter of fact, the given reaction is not balanced. The balanced equation is this:

$$2\ NaCl + H_2SO_4 \longrightarrow Na_2SO_4 + 2\ HCl$$

Now we can work out our magic calculations. The question specifies the grams of HCl synthesized. The 73 grams converted to moles should give 2 moles. According to the equation, 2 moles of NaCl are required to synthesize two moles of HCl. All you have to do now is to convert moles to grams. The correct value is 117 grams of NaCl.

5. The answer is D. This is a straightforward definition-type question. The ions that are present in the solution sometimes do not actually participate in the reaction. By writing the complete ionic equation of such a reaction, we can see that the spectator ions on both sides of the equation are in the same state or form. We can cancel them out and get the net equation of the reaction. Spectator ions are not usually depicted in the net ionic equation.

6. The answer is D. This is also a definition-type question. Amphoteric substances can exhibit both acidic and basic properties depending on the environment that they are subjected to.

7. The answer is D. The reaction mentioned in this question is the rusting of iron. The balanced reaction should look like this:

$$4\ Fe + 3\ O_2 \longrightarrow 2\ Fe_2O_3$$

8. The answer is D. The question tests your understanding of the concept of mole. In Choice A, we have 32 g of $O_2$ which corresponds to 1 mol of oxygen, and 32 g of $SO_2$ which corresponds to half a mol. Choice A is not correct. In Choice B, we have 49 g of $NO_2$ which corresponds to a little more than 1 mol, and 40 g of NaOH which corresponds to 1 mol. So this may be our answer. Choice C is not correct, because we have 20 g of HF which is 1 mol, and 36 g of water which is 2 moles. So Choice C is incorrect. Choice D has 60 g of $C_2H_6$ which is 2 moles, and 156 g of $C_6H_6$ is also 2 moles. So D is a better choice than B.

9. The answer is B. To answer this question, you have to use your intuition. The reaction occurs by means of the heat supplied to the aluminum hydroxide, and it is a dehydration reaction. So the most plausible answer is water.

10. The answer is D. The reactant coefficients in a balanced equation cannot be used to determine the identity of the limiting reagent. Without comparing the amounts of reactants that are actually present, we cannot determine the identity of the limiting reagent. CO being a gas has nothing to do with being the limiting reagent for the given reaction. Choice C is not logical, because it is a product. The best answer is that we cannot determine the identity of the limiting reagent, if any, without any data regarding the amount of reactants used.

11. The answer is C. Look at the items one by one. Item I says that the second student had the highest yield of products. In Experiment 1, the first student used 40.5 g of Al and 80 g of $Fe_2O_3$. This translates to 1.5 moles of Al and roughly 0.5 moles of $Fe_2O_3$. Here $Fe_2O_3$ is the limiting reagent. What was the second student doing? Well, the second student increased the amount of $Fe_2O_3$ used, keeping the same amount of Al. Since $Fe_2O_3$ is the limiting reagent, an increase of $Fe_2O_3$ will increase the product yield. So the second student must have had more yield than the first student. The first and the third students have the same yield for aluminum oxide, because the amount of limiting reagent present is the same in both cases. So items I and II are true.

12. The answer is C. By increasing the amount of aluminum used by the first student, there will not be an increase in the yield because $Fe_2O_3$ is the limiting reagent. Hence, Choice A is not correct. The same reasoning rules out Choice B. Increasing the amount of $Fe_2O_3$ by the second student will increase the yield further, because we actually have unreacted Al left according to the amounts used in the experiment. Hence, if we further increase the amount of $Fe_2O_3$ used, it will increase the yield until the Al runs out. Choice D is wrong.

13. The answer is A. In the question, we are given that 80 g of $Fe_2O_3$ is used in the reaction mixture. How much of this actually reacted is another story. That is to be found out from the yield data. The amount of $CO_2$ produced is 22 g. Based on the balanced equation, we can infer that for every 3 moles of CO reacted, 3 moles of $CO_2$ are produced. So CO and $CO_2$ have a 1:1 ratio. If 22 g or 0.5 moles of $CO_2$ is produced, 0.5 moles of CO must have reacted.

14. The answer is B. Based on the question, we can say that about 0.5 mol of $Fe_2O_3$ and 0.5 mol CO was present in the reacting mixture. Since the $Fe_2O_3$ to CO reacting ratio is 1:3, CO will completely react, and there will be some leftover unreacted $Fe_2O_3$ after the reaction is complete. So the limiting reagent is CO.

15. The answer is B. This question tests your understanding of the mole concept. Since we have excess amounts of $Fe_2O_3$, we do not have to worry about it. But, what you should consider is that the amount of CO present dictates the reaction. One mol of $CO_2$ contains Avogadro number ($6.02 \times 10^{23}$) of molecules. The reaction results in 0.1 mol of $CO_2$. So 0.1 mol of $CO_2$ contains $6.02 \times 10^{22}$ molecules.

# Chemistry Review 3
# Electronic Structure

## A. INTRODUCTION

In this chapter, we will discuss the electronic arrangement of atoms. We will also talk about quantum numbers, orbitals, various rules pertaining to electron-filling, and electronic configuration.

## B. ATOMIC STRUCTURE

The first ideas about electronic arrangement in atoms were primarily figured out from atomic emission spectra. In various experiments, atoms were made to be thermally or electrically excited, and this resulted in different kinds of bands or lines on photographic plates. Our understanding of atomic structure is based on these types of experiments. All elements have their characteristic line spectra with which they can be analyzed and identified.

### Electromagnetic Waves

Before we discuss the atomic structure, we will touch on the topic of electromagnetic radiation to have a better analytical understanding of the key ideas. All electromagnetic radiations travel with a constant speed of $3 \times 10^8$ m/s. The electromagnetic spectrum ranges from radio waves to gamma rays.

### The Wave Nature

Light has wave nature. It has electric and magnetic fields which are perpendicular to each other, and can travel through space. No medium is required. Because of its wave character, we can define light in terms of frequency and wave length. The distance between two adjacent crests or troughs, or any two adjacent identical points on a wave is called *wave length* ($\lambda$). *Frequency* (f) is the number of wavelengths passing through a point in unit time. Wavelength and frequency are related by the relation given below. Frequency is usually expressed in 1/second ($s^{-1}$), which is otherwise known as *hertz* (Hz).

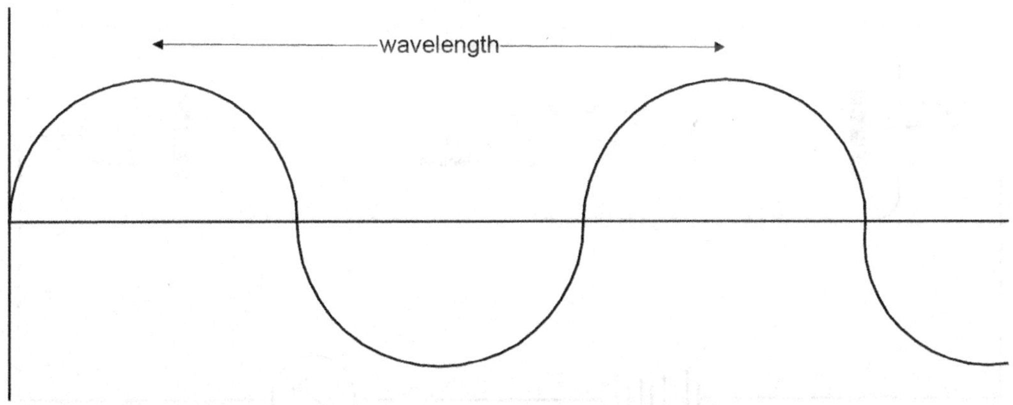


| Velocity = frequency x wavelength |
|---|
| $c = f\lambda$ |

## The Particle Nature and Quantum Theory

Light has particle nature. These particles or forms (packages) of energy are called quanta. A more modern term for such as particle of light is photon. The energy of a photon can be expressed in terms of the following formula.

$$\text{Energy} = h\,\text{f},$$

where $h$ is the Planck's constant, and f is the frequency.

Planck's constant = $6.63 \times 10^{-34}$ J.s

According to *Heisenberg's Uncertainty Principle*, we cannot determine both the momentum and the position of subatomic particles simultaneously. This is because we are using other particles (electromagnetic particles-like photons) of comparable energy to detect these subatomic particles, and by the time these other particles find the subatomic particles (say electrons), they are also disturbing the pathway of these electrons. In essence, the study of something extremely small and fast (about the magnitude of electrons) cannot be done without interference of its natural course or position.

## Photoelectric Effect

The Photoelectric Effect can be defined as the ejection of electrons from a metal surface when light rays strike on it. The ejected electrons are often called "photoelectrons." The ejection of electrons occurs only if the incident light has a certain minimum or *threshold frequency*. The required threshold frequency is a characteristic specific to each metal. Experimentally, it has been found that the photoelectrons emitted with maximum energy do not have the full energy equivalent supplied by the incident photon. This is because energy is required to break loose the electrons from the surface of the metal. The energy required for this is called "work function," which is characteristic of each metal. The photoelectrons can be accelerated to a positively charged plate, creating a flow of charges along a wire-photocurrent. The current can be measured by an ammeter connected to the wire.

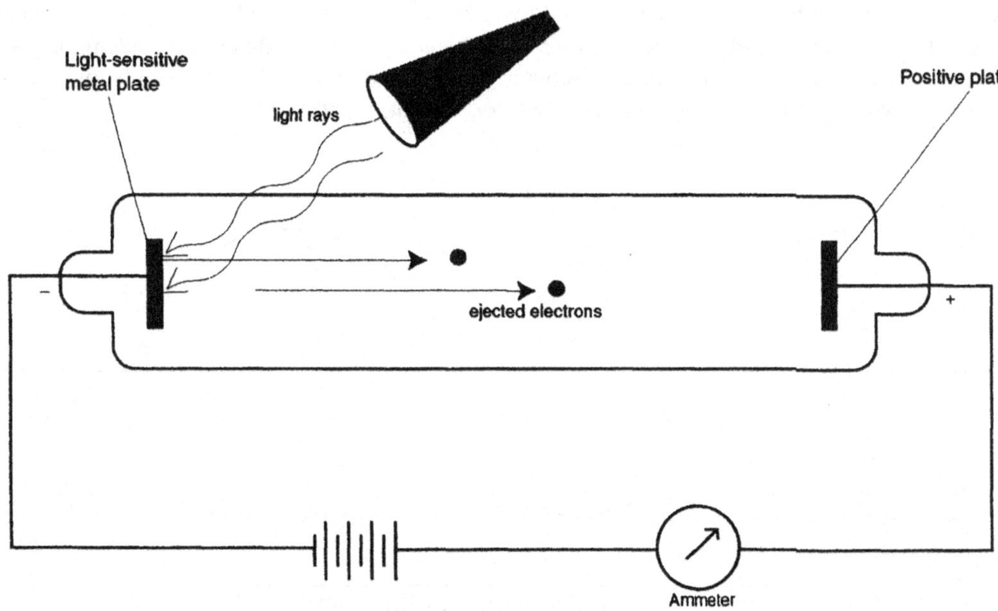

The maximum kinetic energy ($K_{max}$) of a photoelectron is given by the following equation:

$$K_{max} = 1/2\ mv_{max}^2 = hf - \varphi$$

In this equation, *m* is the mass of an electron, $v_{max}$ is the maximum velocity of the electrons, *h* is the Planck's constant, *f* is the frequency of the incident light, $\varphi$ (pronounced *phi*) is the "work function" of the metal. The entity *hf* represents the energy of the incident photon.

## Key Observations on Photoelectric Effect

1) The photoelectric effect exemplifies the particle nature of light.

2) Based on conservation of energy, no photoelectron can have energy more than that of an incident photon.

3) The energy of the photoelectrons is always less than that of the incident photons, because some energy (work function) is required to break the electrons loose.

4) The maximum energy of the photoelectrons is independent of the intensity of the incident light.

5) Electrons are not ejected no matter how high the intensity of the incident light is, unless the incident light has the energy corresponding to the threshold frequency characteristic of a particular metal.

## Atomic Emission Spectra

When we pass white light through a prism, dispersion of the light occurs resulting in *continuous spectrum* of wavelengths. Another type of spectrum results when heated gas emits light. This results in a *line spectrum*. Line spectrum contains only certain specific wavelengths of light. The wavelengths in the visible spectrum of hydrogen is given by the following formula:

$$\frac{1}{\lambda} = \frac{R}{hc}\left(\frac{1}{4} - \frac{1}{n^2}\right)$$

where λ is the wavelength of the light, *R* (Rydberg constant) = $2.18 \times 10^{-18}$ J, *h* (Planck's constant) = $6.63 \times 10^{-34}$ J.s, *c* (speed of light) = $3.0 \times 10^8$ m/s, and *n* is some whole number that is greater than 2 which corresponds to the orbit-number from which the electron is making the transition. For example, if the transition of an electron is from orbit number 4 to 2, the *n* value is 4.

## Bohr's Model of Hydrogen Atom

Niel Bohr's explanation of the hydrogen spectrum was a major breakthrough toward the understanding of atomic structure. The following are the postulates:

1) In each hydrogen atom, the electron revolves around the nucleus in one of the several stable orbits.

2) Each orbit has a definite radius and thus has a definite energy associated with it.

3) An electron in an orbit closest to the nucleus has the lowest energy, and if the electron is in the lowest orbit the atom is said to be in its *ground state*.

4) The electron in an atom may absorb discrete amounts of energy and move to another orbit with higher energy, and this state is called the *excited state*.

5) An electron in an excited atom can go back to a lower energy level and this process will result in the release of excess energy as light.

6) The amount of energy released or absorbed is equal to the difference between the energies of the initial and final orbits.

Based on Bohr's theory, light energy is emitted when an electron in a higher energy level ($E_{initial}$) jumps to a lower energy level ($E_{final}$). Based on the law of conservation of energy, the sum of energies of the emitted photon (*hf*) and the electron's final energy ($E_{final}$) should be equal to the electron's initial energy ($E_{initial}$). This can be represented mathematically as follows:

$$hf + E_{final} = E_{initial}$$

Transitions of the electron in the hydrogen atom result in different spectral lines.

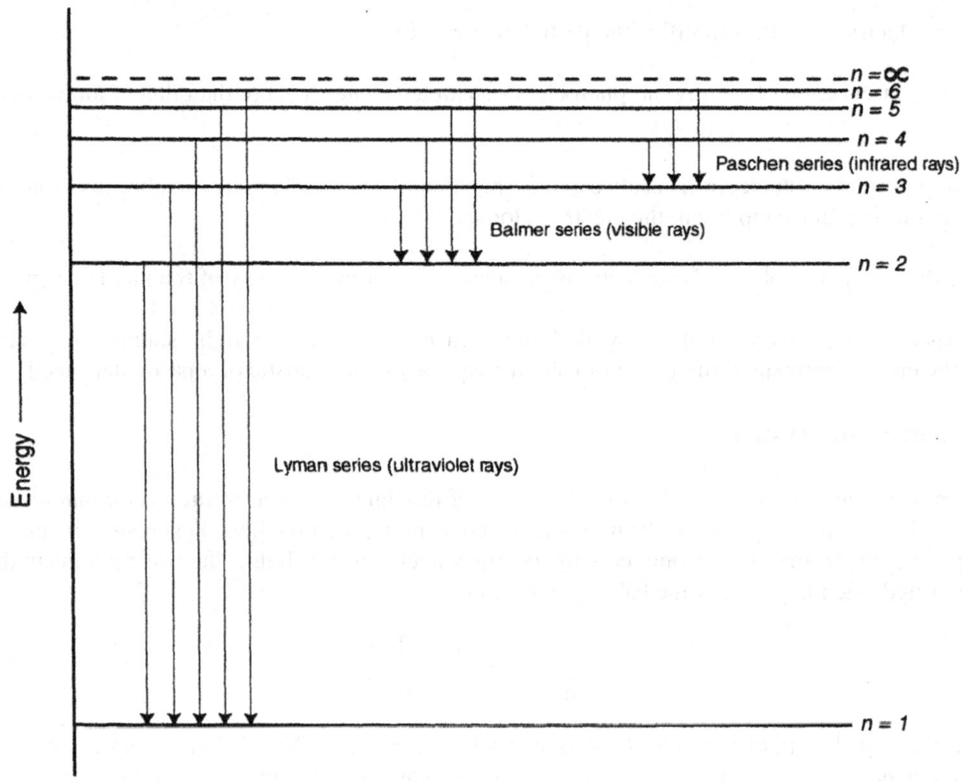


The energy of the emitted photon

$$hf = R\left(\frac{1}{n_{final}^{2}} - \frac{1}{n_{inital}^{2}}\right)$$

where $n_{final}$ and $n_{inital}$ are the principal quantum numbers of final and initial energy levels, and $R$ is the Rydberg constant (2.18 x $10^{-18}$ J). The figure given above shows the transitions that can result in the Lyman (ultraviolet region), Balmer (visible region), and Paschen series (infrared region) for $n_{final}$ values 1, 2, and 3, respectively.

> *A photon is emitted when an electron in an atom jumps from a higher to a lower energy level. The energy of the emitted photon is equal to the difference in energy between the two energy levels.*

## C. QUANTUM NUMBERS

All electrons present in an atom have specific addresses or attributes by which each electron can be referred to. The four quantum numbers are the ones with which we can describe each and every electron that is present in an atom. One of the quantum numbers describes the shape or the most probable area around the nucleus where we can find the particular electrons of interest. This wave function of an electron is called an *orbital*.

*Principal quantum number* ($n$). The principal quantum number denotes the energy level of electrons. The larger the principal quantum number is, the larger the energy. The smaller the principal quantum number is, the lower the energy. The shells are often named K, L, M, N, . . ., which correspond to the principal quantum numbers 1, 2, 3, 4, . . ., respectively.

| *Letter* | *K* | *L* | *M* | *N . . .* |
|---|---|---|---|---|
| $n$ | 1 | 2 | 3 | 4 . . . |

*Angular momentum quantum number* ($l$). Angular momentum quantum number (azimuthal quantum number) denotes the shape of the orbital. The values range from 0 to n – 1, where n stands for the principal quantum number. If an electron has a principal quantum number of 4, the values of angular momentum quantum numbers are 0, 1, 2, and 3. The angular momentum quantum numbers correspond to different subshells. An angular momentum quantum number 0 corresponds to *s* subshell, 1 to *p* subshell, 2 to *d* subshell, 3 to *f* subshell, and so on. For instance, 3*d* denotes a subshell with quantum numbers $n$ = 3 and $l$ = 2

| *Orbital* | *s* | *p* | *d* | *f . . .* |
|---|---|---|---|---|
| $l$ | 0 | 1 | 2 | 3 . . . |

*s* orbital

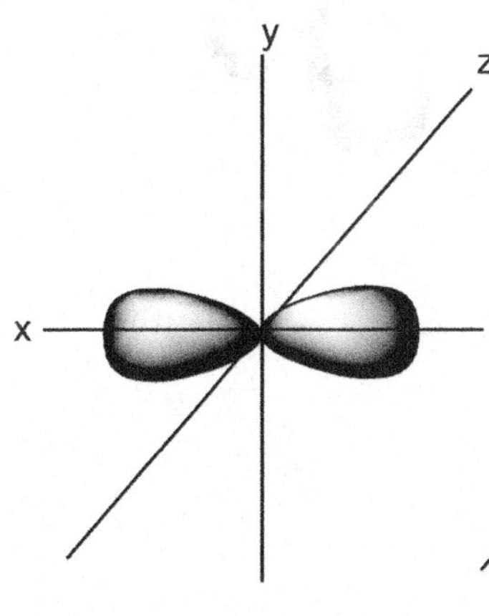


2 $p_x$ orbital

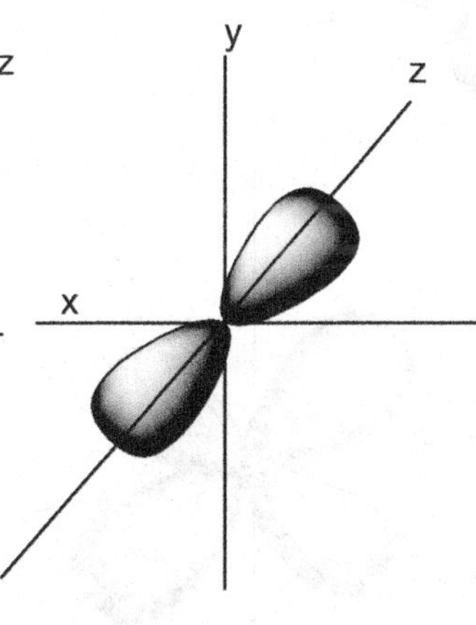


2 $p_y$ orbital

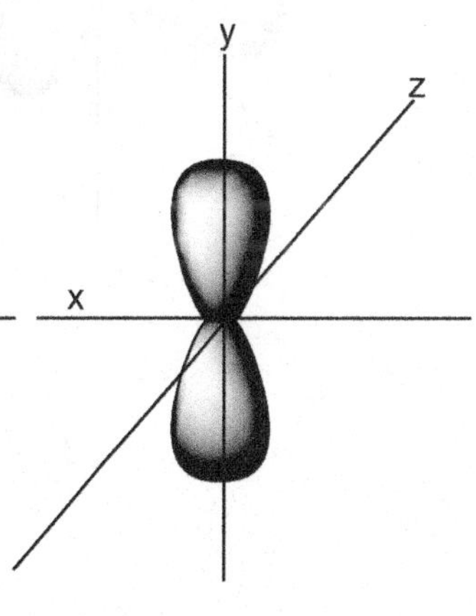


2 $p_z$ orbital

*Magnetic quantum number* ($m_l$). Magnetic quantum numbers define the different spatial orientations of the orbitals. The values from $-l$ to $+l$. For example, let's say the value of $l$ is 1. So the magnetic quantum numbers will be −1, 0, and +1. The $l$ value corresponds to $p$ sublevel and the three magnetic quantum numbers correspond to the three atomic orbitals in the $p$ subshell.

*Spin quantum number* ($m_s$). Spin quantum number has to do with the spin orientations of an electron. The two possible spins are denoted by the spin quantum numbers +1/2 and −1/2.

Spatial orientation of *d* orbitals

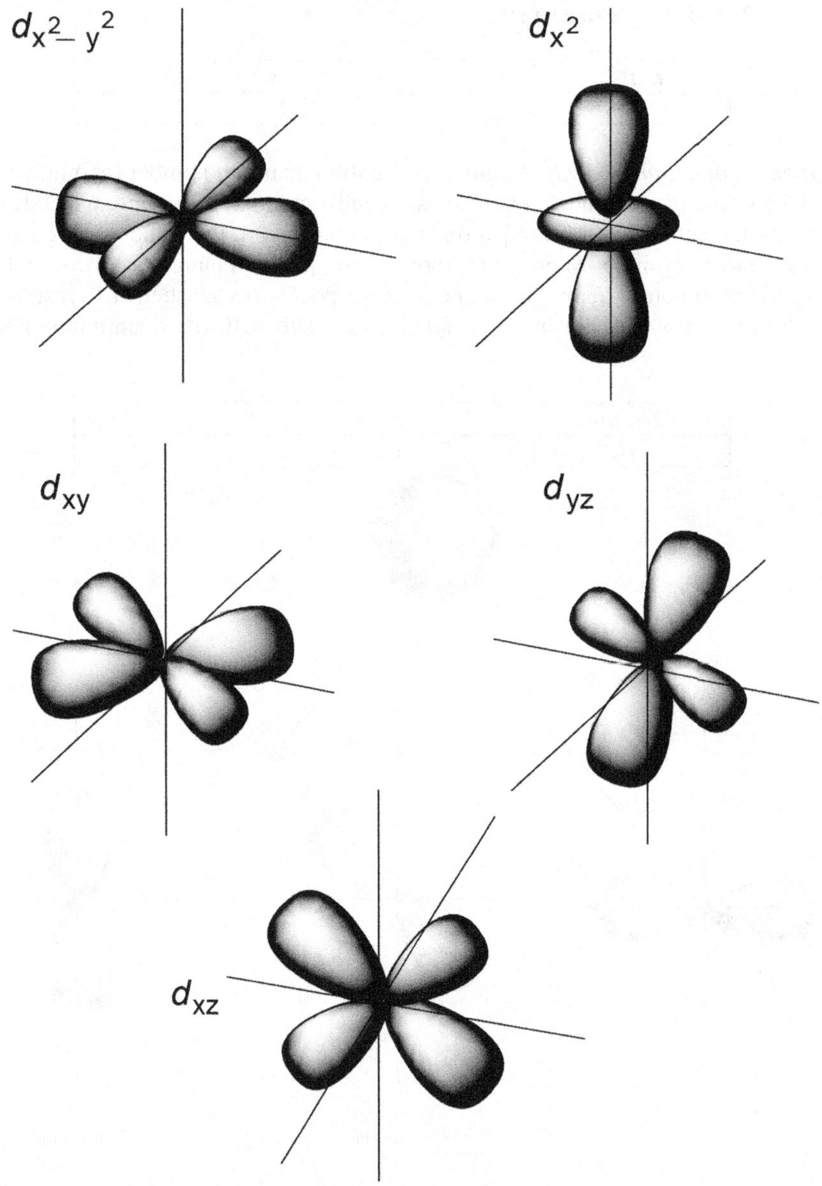

## D. ELECTRONIC CONFIGURATION

We talked about quantum numbers and atomic orbitals. In this section, we will focus our attention mainly on writing the electronic configuration of atoms and the rules associated with it. First, let's talk about the ground state electronic configuration. The ground state configuration means the electronic configuration at the lowest energy state.

There are certain rules that should be applied to the filling of orbitals. In order to do that properly, we need to know the Aufbau principle. According to the Aufbau principle, the filling order of electrons obeys a general pattern in which the electrons try to occupy the orbitals in such a way as to minimize the total energy; that is, they occupy the lowest energy orbitals first and then step-by-step go to the next available higher energy levels successively. Of course, there are some exceptions to these generalizations. Some students find it hard to remember the order of filling. The diagram given below may help you. So take a close look. Filling order can be depicted as follows:

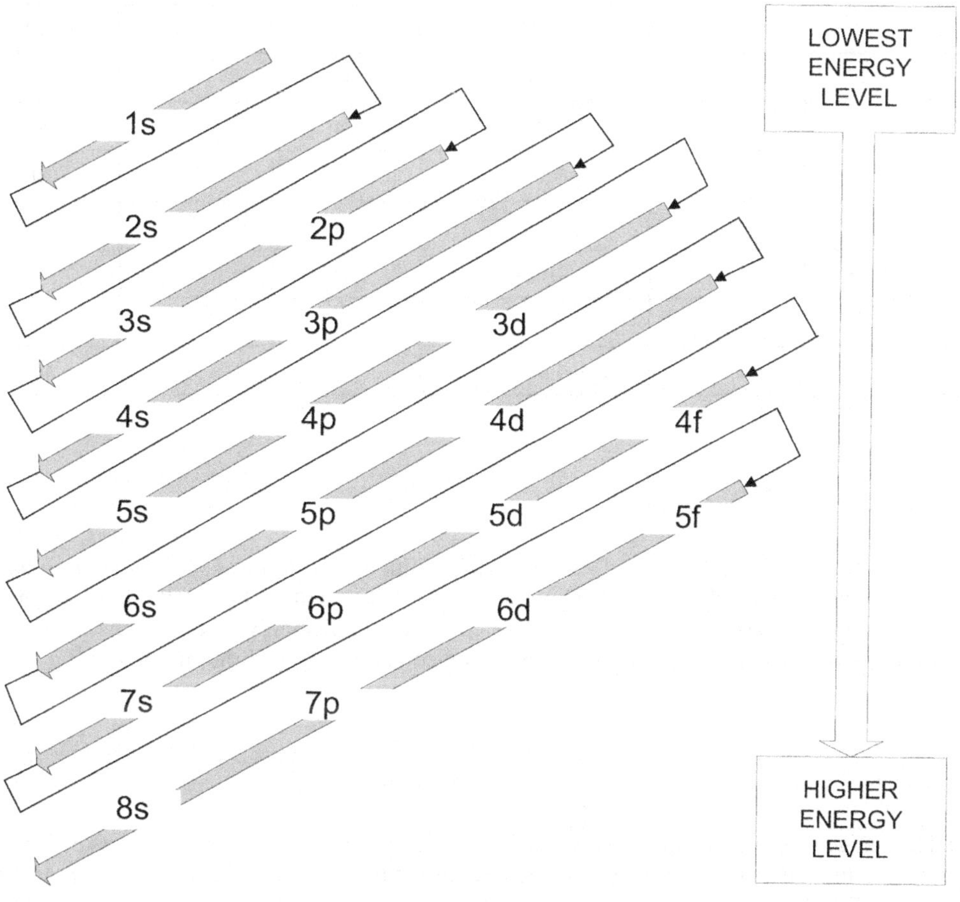


This is an easy way to remember the general order of electron-filling in the subshells. The order of filling is 1*s*, 2*s*, 2*p*, 3*s*, 3*p*, 4*s*, 4*p*, 5*s*, . . ., and so on.

An *s* orbital is spherical in shape. All *p* orbitals have dumbbell shape with two lobes aligned along an axis. All *d* orbitals have slightly complex shapes and are beyond the scope of our discussion. Each orbital can accommodate a maximum of two electrons. Hence, an *s* subshell, (only one orbital) can accommodate a maximum of two electrons. Similarly, *p* (three orbitals), *d* (five orbitals), and *f* (seven orbitals) subshells can have maximum of six, ten, and fourteen electrons, respectively.

**Table 3-1 Some possible combinations of quantum numbers for atomic orbitals**

| Principal quantum number ($n$) | Angular momentum quantum number ($l$) | Magnetic quantum number ($m_l$) | Subshells | Number of orbitals | Maximum number of electrons |
|---|---|---|---|---|---|
| 1 | 0 | 0 | 1*s* | 1 | 2 |
| 2 | 0 | 0 | 2*s* | 1 | 2 |
| 2 | 1 | -1,0,+1 | 2*p* | 3 | 6 |
| 3 | 0 | 0 | 3*s* | 1 | 2 |
| 3 | 1 | -1,0,+1 | 3*p* | 3 | 6 |
| 3 | 2 | -2,-1,0,+1,+2 | 3*d* | 5 | 10 |
| 4 | 0 | 0 | 4*s* | 1 | 2 |
| 4 | 1 | -1,0,+1 | 4*p* | 3 | 6 |
| 4 | 2 | -2,-1,0,+1,+2 | 4*d* | 5 | 10 |
| 4 | 3 | -3,-2,-1,0,+1,+2,+3 | 4*f* | 7 | 14 |

**Example 3-1**

Write the electronic configuration of lithium.

**Solution:** From the periodic table, we can get the atomic number of lithium. The lithium atom has 3 electrons. As we know, 1*s* subshell is the first one to be filled. The *s* orbital can hold a maximum of 2 electrons. We have one electron remaining. It will occupy the 2*s* subshell which is the next energy level. So the third electron will occupy the 2*s* subshell. Hence, the configuration of lithium is $1s^2 2s^1$.

**Example 3-2**

Write the electronic configuration of oxygen in its ground state form.

**Solution:** The oxygen atom contains 8 electrons. The first 2 electrons will go to the 1*s* level. The next 2 electrons will occupy the 2*s* level. We have 4 electrons remaining. What is the next subshell according to the filling order? It is 2*p*. The *p* subshell can hold a maximum of 6 electrons in its orbitals. So the remaining 4 electrons will occupy the 2*p* level. Hence, the configuration of oxygen is $1s^2\ 2s^2\ 2p^4$.

## Hund's Rule

We have learned the order of filling the subshells. Now let's take a closer look at the filling of electrons in an orbital level. Each orbital can be occupied by a maximum of 2 electrons, and these electrons will have opposite spins as dictated by the spin quantum number. Hund's rule describes the way the electrons fill up the orbitals. According to the Hund's rule, each electron starts filling up each orbital of a given subshell. After all the orbitals in a given subshell have been filled singly (half-filled), then the electrons start pairing. Let's look at some examples.

**Example 3-3**

Write the electronic configuration of sulfur and also show the filling of electrons with orbital notation.

**Solution:** Sulfur atom has 16 electrons. The electronic configuration is written as $1s^2\ 2s^2\ 2p^6\ 3s^2\ 3p^4$. To see the significance of the Hund's rule, look at the $3p$ subshell. In the $3p$ subshell, we have 3 orbitals.

Note that the electrons first occupy singly in the orbitals. Altogether there are 4 electrons in the $3p$ subshell. Instead of filling the orbitals in pairs, the first 3 electrons start filling the three orbitals singly, and then the remaining electron occupies the orbital with the other electron as paired electrons. (See the orbital notation of $3p$ subshell shown above.) If this were not the case, you would have seen two electron-paired orbitals followed by an empty orbital.

Another idea we want to touch on concerns paramagnetic and diamagnetic substances. Substances that have unpaired electrons are called paramagnetic. Substances that have only paired electrons are called diamagnetic.

# Chemistry Review 3 Problems

1. Which of the following shows the correct order of filling of subshells?

A. 3*s*, 3*p*, 4*s*, 3*d*, 4*p*
B. 3*s*, 3*p*, 3*d*, 4*s*, 4*p*
C. 3*s*, 3*d*, 3*p*, 4*s*, 4*p*
D. 3*s*, 3*d*, 3*p*, 4*p*, 4*s*

2. Which of the following represents the outer most shell configuration of an inert gas?

A. $4s^2 4p^2$
B. $4s^2 4p^3$
C. $4s^1 4p^6$
D. $4s^2 4p^6$

3. The electronic configuration shown is that of the element:

$$[\mathrm{Kr}]\ 5s^2\ 4d^{10}\ 5p^2$$

A. Sb.
B. Rb.
C. In.
D. Sn.

4. The maximum number of electrons that can occupy the energy levels is calculated using a formula, where *n* represents the number corresponding to the energy level. The formula is:

A. $n^2$
B. $2n^2$
C. $2n + 1$
D. $4n + 2$

5. Consider this statement: No two electrons of an atom can have the same sets of four quantum numbers. This is known as the:

A. Heisenberg's Uncertainty Principle.
B. Hund's Rule.
C. Pauli Exclusion Principle.
D. Aufbau Principle.

6. Which of the following is true with respect to subshells?

A. The 4*p* subshell has higher energy than the 5*s* subshell.
B. The 3*p* subshell can have a maximum of 3 electrons.
C. The 5*d* subshell has higher energy than the 6*s* subshell.
D. The 4*f* subshell has higher energy than the 5*d* subshell.

7. Which of the following is NOT isoelectronic with any of the noble gases?

A. $Ca^{2+}$
B. $Br^-$
C. $S^{2-}$
D. $Mg^+$

8. Which of the following is not a possible set of quantum number values for the nitrogen atom, in the order of principal quantum number, azimuthal quantum number, magnetic quantum number, and spin quantum number?

A. 1, 0, 0, –1/2
B. 2, 1, +1, +1/2
C. 2, 0, 0, –1/2
D. 2, 1, +2, +1/2

9. The electronic configuration [Ar] $3d^{10}\,4s^2\,4p^2$ is that of:

A. Ge.
B. Zn.
C. Se.
D. Ar.

10. The electronic configuration of a given element has a $4d$ subshell. This cannot be the electronic configuration of:

A. Os.
B. Cu.
C. Ag.
D. Ra.

# Solutions

1. The answer is A. The correct order of filling from 3*s* onward is 3*s*, 3*p*, 4*s*, 3*d*, and 4*p*.

2. The answer is D. Inert gases (noble gases) have the completed outermost *s* and *p* subshells. Helium is an exception to this.

3. The answer is D. The atomic number of Kr (krypton) is 36. After the representation of Kr, how many electrons do we have? There are 14 more electrons. That adds up to an atomic number of 50. The atomic number 50 corresponds to the element Sn.

4. The correct answer is B. This is a straightforward question. The maximum number of electrons that can occupy the principal energy level is calculated using the formula $2n^2$.

5. The answer is C. The statement "no two electrons of an atom can have the same set of four quantum numbers" is Pauli exclusion principle.

6. The answer is C. The only correct statement in the set of choices is "the 5*d* subshell has higher energy than the 6*s* subshell." All other choices are wrong.

7. The answer is D. The question asks us to pick the species that is not isoelectronic (same number of electrons) with the noble gases. In order to solve this problem, you have to find the number of electrons present in them, and if it matches with the number of electrons in any of the noble gases, you can eliminate that choice. All the ions match with one or the other noble gas in terms of the number of electrons, except $Mg^+$.

8. The answer is D. The nitrogen atom has a total of 7 electrons. The principal quantum number could be 1 or 2. With that alone, we cannot rule out any of the choices. The next quantum number given is the azimuthal quantum number, also known as the angular momentum quantum number. The angular momentum quantum number can have values from 0 to $n - 1$, where n represents the principal quantum number. Since 2 denotes the maximum $n$ value, we cannot rule out any of the choices yet. Look at the next quantum number, which is the magnetic quantum number. The magnetic quantum number can have values from $-l$ to $+l$, where $l$ represents angular momentum quantum number. So the magnetic quantum number cannot be +2 for nitrogen.

9. The answer is A. This question tests your ability to recognize a given electronic configuration. The fastest way to approach this type of question is to look at the root element given. The atomic number of argon is 18. We can count 14 more electrons in the given configuration to get a total of 32 electrons. The atomic number 32 is that of germanium (Ge).

10. The answer is B. Osmium, silver, and radium can have the 4*d* subshell. Copper belongs to the 4$^{th}$ period and therefore cannot have a 4*d* subshell. All the other elements listed belong to the periods 5, 6 and 7.

# PERIODIC TABLE OF ELEMENTS

s BLOCK | d BLOCK | p BLOCK

| IA | IIA | IIIB | IVB | VB | VIB | VIIB | VIII | VIII | VIII | IB | IIB | IIIA | IVA | VA | VIA | VIIA | O |
|---|---|---|---|---|---|---|---|---|---|---|---|---|---|---|---|---|---|
| 1<br>H<br>1.0 | | | | | | | | | | | | | | | | | 2<br>He<br>4.0 |
| 3<br>Li<br>6.9 | 4<br>Be<br>9.0 | | | | | | | | | | | 5<br>B<br>10.8 | 6<br>C<br>12.0 | 7<br>N<br>14.0 | 8<br>O<br>16.0 | 9<br>F<br>19.0 | 10<br>Ne<br>20.2 |
| 11<br>Na<br>23.0 | 12<br>Mg<br>24.3 | | | | | | | | | | | 13<br>Al<br>27.0 | 14<br>Si<br>28.1 | 15<br>P<br>31.0 | 16<br>S<br>32.1 | 17<br>Cl<br>35.5 | 18<br>Ar<br>39.9 |
| 19<br>K<br>39.1 | 20<br>Ca<br>40.1 | 21<br>Sc<br>45.0 | 22<br>Ti<br>47.9 | 23<br>V<br>50.9 | 24<br>Cr<br>52.0 | 25<br>Mn<br>54.9 | 26<br>Fe<br>55.8 | 27<br>Co<br>58.9 | 28<br>Ni<br>58.7 | 29<br>Cu<br>63.5 | 30<br>Zn<br>65.4 | 31<br>Ga<br>69.7 | 32<br>Ge<br>72.6 | 33<br>As<br>74.9 | 34<br>Se<br>79.0 | 35<br>Br<br>79.9 | 36<br>Kr<br>83.8 |
| 37<br>Rb<br>85.5 | 38<br>Sr<br>87.6 | 39<br>Y<br>88.9 | 40<br>Zr<br>91.2 | 41<br>Nb<br>92.9 | 42<br>Mo<br>95.9 | 43<br>Tc<br>(98) | 44<br>Ru<br>101.1 | 45<br>Rh<br>102.9 | 46<br>Pd<br>106.4 | 47<br>Ag<br>107.9 | 48<br>Cd<br>112.4 | 49<br>In<br>114.8 | 50<br>Sn<br>118.7 | 51<br>Sb<br>121.8 | 52<br>Te<br>127.6 | 53<br>I<br>126.9 | 54<br>Xe<br>131.3 |
| 55<br>Cs<br>132.9 | 56<br>Ba<br>137.3 | 57<br>La*<br>138.9 | 72<br>Hf<br>178.5 | 73<br>Ta<br>180.9 | 74<br>W<br>183.9 | 75<br>Re<br>186.2 | 76<br>Os<br>190.2 | 77<br>Ir<br>192.2 | 78<br>Pt<br>195.1 | 79<br>Au<br>197.0 | 80<br>Hg<br>200.6 | 81<br>Tl<br>204.4 | 82<br>Pb<br>207.2 | 83<br>Bi<br>209.0 | 84<br>Po<br>(209) | 85<br>At<br>(210) | 86<br>Rn<br>(222) |
| 87<br>Fr<br>(223) | 88<br>Ra<br>226.0 | 89<br>Ac**<br>227.0 | 104<br>Unq<br>(261) | 105<br>Unp<br>(262) | 106<br>Unh<br>(263) | 107<br>Uns<br>(262) | 108<br>Uno<br>(265) | 109<br>Une<br>(267) | | | | | | | | | |

f BLOCK

| | | | | | | | | | | | | | | |
|---|---|---|---|---|---|---|---|---|---|---|---|---|---|---|
| * | 58<br>Ce<br>140.1 | 59<br>Pr<br>140.9 | 60<br>Nd<br>144.2 | 61<br>Pm<br>(145) | 62<br>Sm<br>150.4 | 63<br>Eu<br>152.0 | 64<br>Gd<br>157.3 | 65<br>Tb<br>158.9 | 66<br>Dy<br>162.5 | 67<br>Ho<br>164.9 | 68<br>Er<br>167.3 | 69<br>Tm<br>168.9 | 70<br>Yb<br>173.0 | 71<br>Lu<br>175.0 |
| ** | 90<br>Th<br>232.0 | 91<br>Pa<br>(231) | 92<br>U<br>238.0 | 93<br>Np<br>(237) | 94<br>Pu<br>(244) | 95<br>Am<br>(243) | 96<br>Cm<br>(247) | 97<br>Bk<br>(247) | 98<br>Cf<br>(251) | 99<br>Es<br>(252) | 100<br>Fm<br>(257) | 101<br>Md<br>(258) | 102<br>No<br>(259) | 103<br>Lr<br>(260) |

# Chemistry Review 4
# Organic Chemistry

## A. INTRODUCTION

Organic Chemistry is the study of carbon compounds. Carbon can form a wide array of compounds, because of its size and ability to form covalent bonds with other carbon atoms. In addition, carbon can form bonds with many other elements. This property of carbon increases the facility of forming multitudes of different compounds. The particular electronegativity of carbon also plays a key role in its versatility. In this chapter, we will review some of the fundamental aspects of carbon atom and the main types of hybridizations involving carbon compounds.

## B. THE CARBON ATOM

Electrons are found in regions around the nucleus in an atom, and those regions are called *orbitals*. The orbitals can be defined and differentiated by size, shape, and orientation. *Valence electrons* are electrons that are found in the outermost shell. The carbon atom has four valence electrons. These valence electrons are involved in chemical reactions and bonding.

The electronic configuration:

| Carbon | | |
|---|---|---|
| | Atomic number – | 6 |
| | Electronic configuration – | $1s^2 2s^2 2p^2$ |

## C. BONDING

*Ionic bond* – Ionic bond is formed between an electropositive and electronegative atom (ion), or generally we can define it as an attractive force between a positive and a negative ion (e.g., KCl).

*Covalent bond* – Covalent bond is formed by the sharing of a pair of electrons between two atoms. Carbon compounds generally contain covalent bonds.

## D. HYBRIDIZATION OF ORBITALS

The six electrons of a carbon atom are distributed in the orbitals as follows:

| Electronic configuration – $1s^2 2s^2 2p^2$ |
|---|

Ground state carbon atom

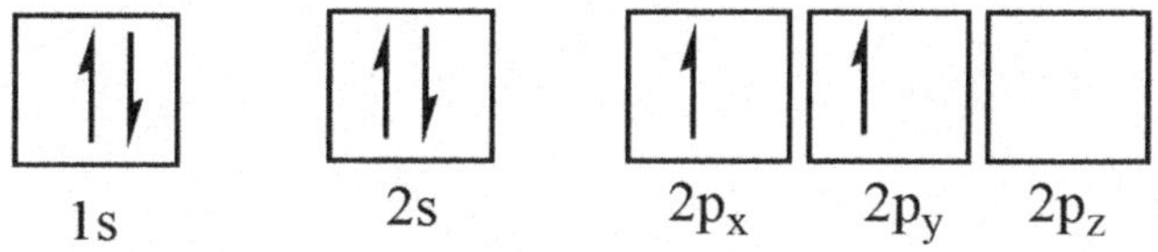

## $sp^3$ Hybridization

The carbon atoms of alkanes (e.g., methane, ethane) are $sp^3$ hybridized. In order to form the four bonds in methane, a carbon atom needs four half-filled orbitals. In order to have more free half-filled orbitals, the carbon atoms undergo hybridization.

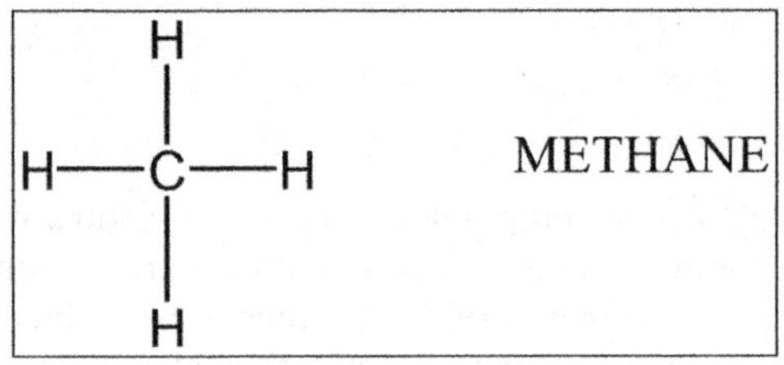


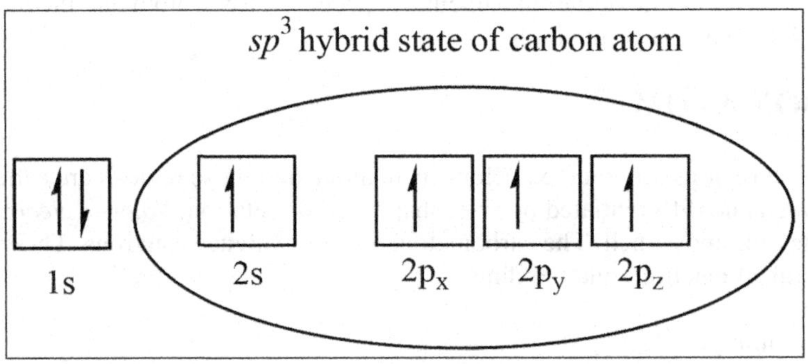


The hybridization results in one half-filled 2*s* orbital, and three half-filled 2*p* orbitals (a total of four half-filled orbitals). These unpaired electrons form the $sp^3$ hybridized carbon, which can form the four covalent bonds in the methane molecule. The four $sp^3$ hybrids are directed to the corners of a tetrahedron with bond angles of $109.5^0$.

## $sp^2$ Hybridization

In carbon-carbon double bonds, the carbons undergo another type of hybridization called the $sp^2$ hybridization. In this hybridization, only one 2*s*, and two 2*p* orbitals are involved. The C=C contains a sigma ($\sigma$) bond and a pi ($\pi$) bond. The pi bond is formed by the unhybridized 2*p* orbital overlap. The three equal hybrids lie in an xy-plane with bond angles of $120^0$.

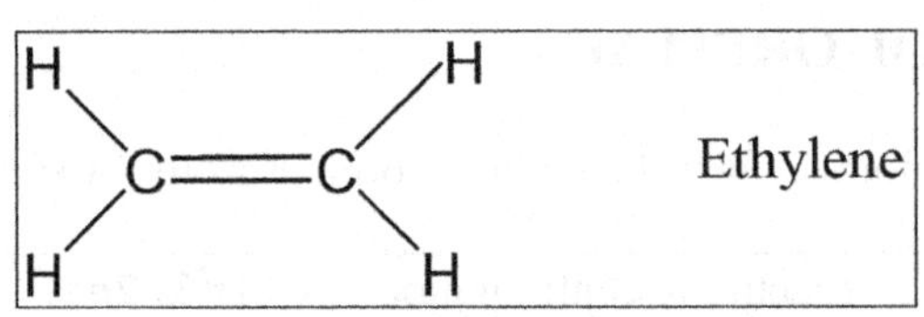

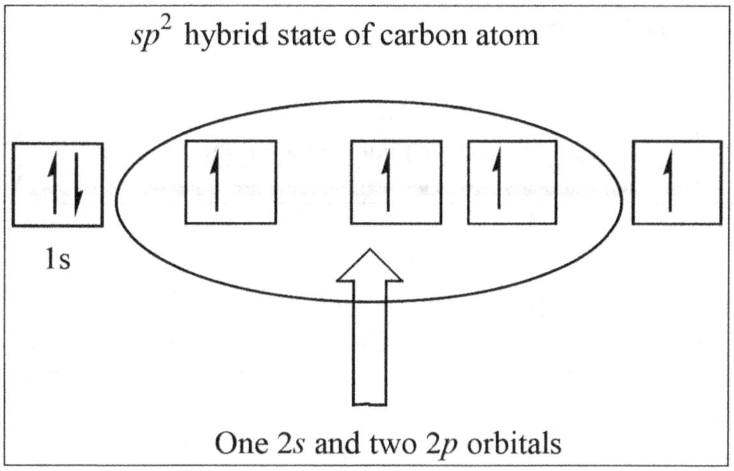


### *sp* Hybridization

Yet another hybridization called the *sp* hybridization exists in carbon-carbon triple bonds. An *sp* hybridized carbon atom is bonded only to two other atoms. In this type of hybridization, one 2*s* orbital and one 2*p* orbital are involved. A carbon-carbon triple bond contains one sigma bond and two pi bonds.

$HC \equiv CH$ Acetylene

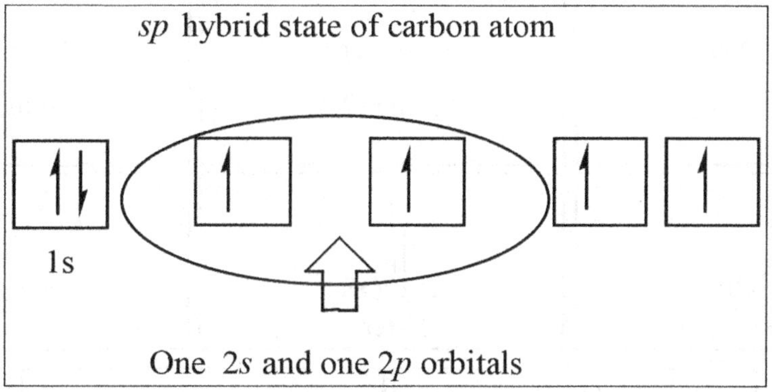


## E. RESONANCE

Resonance is an important aspect in Organic chemistry. Though we represent definite Lewis structures of molecules, in reality the electrons are not localized. They are shared and delocalized by the atoms in a molecule to have the most stable electron distribution. This is called *resonance*. Resonance promotes stability.

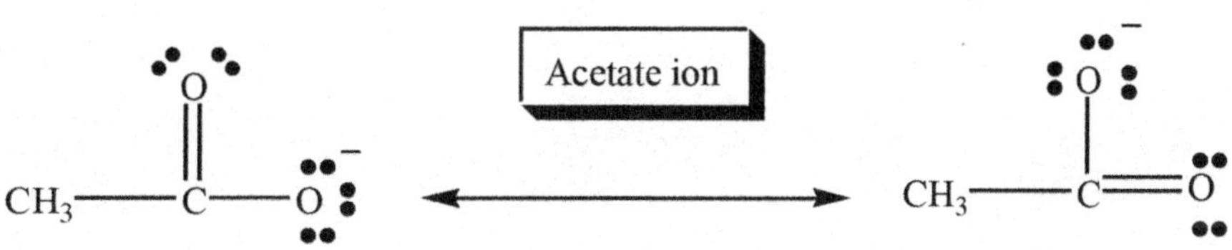

## F. FUNCTIONAL GROUPS

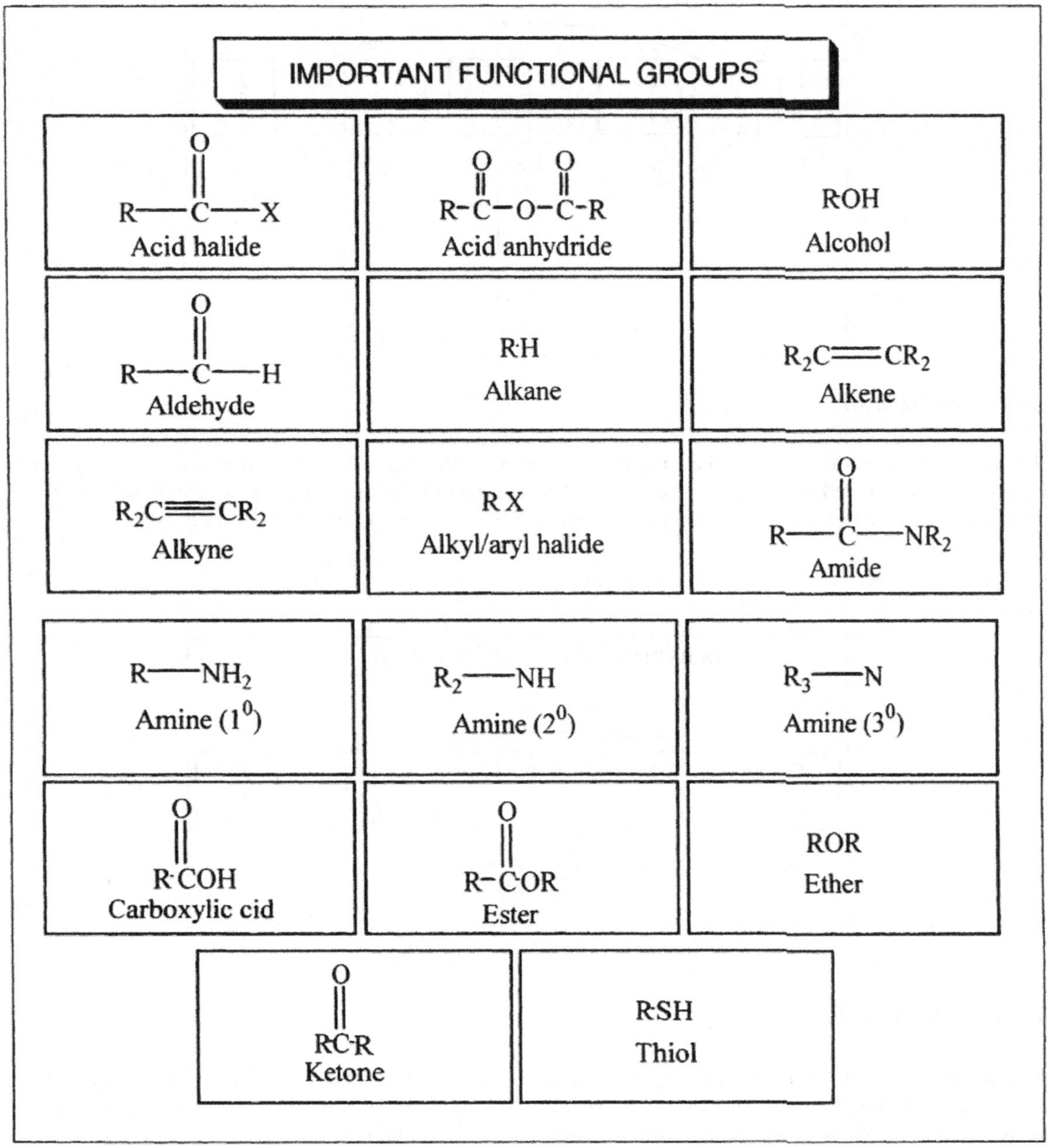

# Chemistry Review 4 Problems

1. The carbon indicated by the arrow has which of the following hybridizations?

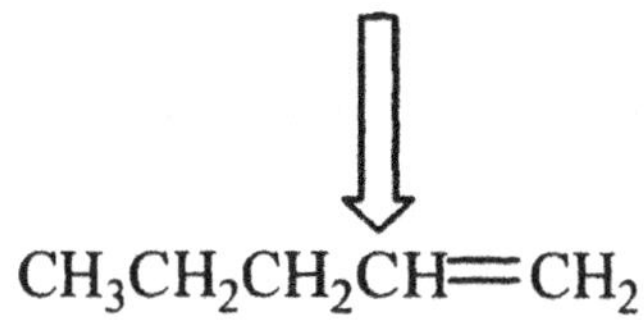


A. *sp* hybridization
B. $sp^2$ hybridization
C. $sp^3$ hybridization
D. $sp^3d^2$ hybridization

2. In acetylene, how many sigma bonds are present between the two carbons?

A. One
B. Two
C. Three
D. Four

3. The triple bond of 2-propyne contains:
A. one sigma bond and one pi bond.
B. one sigma bond and two pi bonds.
C. two sigma bonds and one pi bond.
D. three sigma bonds.

4. Which of the following represents the arrangement of the $sp^3$ hybrid orbitals in methane?

A. Planar
B. Tetragonal planar
C. Tetrahedral
D. Bipyramidal

5. How many electrons are actually involved in the carbon-carbon bond of acetylene?

A. 2
B. 3
C. 4
D. 6

6. Which of the following best represents the hydrogen-carbon-hydrogen bond angles in methane?

A. 90°
B. 109.5°
C. 120°
D. 180°

7. Which of the following bonds indicated by the arrows has the shortest bond length?

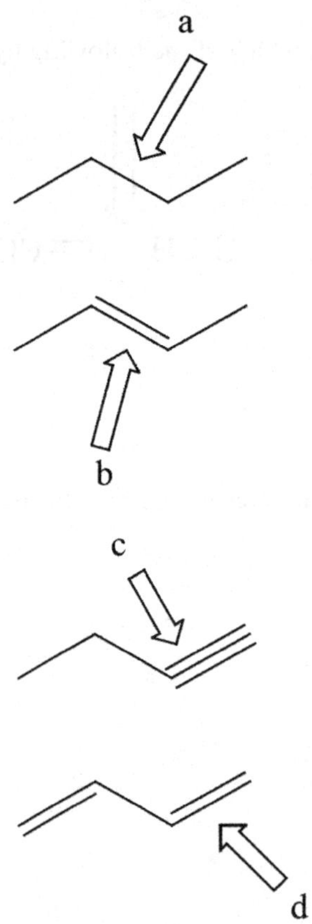


Figure for Q-7

A. a
B. b
C. c
D. d

8. The carbon-hydrogen bond in propane can be best described as:

A. an ionic bond.
B. a covalent bond.
C. a hydrogen bond.
D. a dipole-dipole bond.

# Solutions

1. The answer is B. The carbon indicated by the arrow is an $sp^2$ hybridized carbon atom. Notice that the indicated carbon has a double bond and two single bonds. The double bond connects this carbon to the end-carbon. A hydrogen atom and the 3rd carbon are attached to this carbon via single bonds.

2. The answer is A. Acetylene is the first member of the alkyne group. The carbon-carbon bond is a triple bond which contains only one sigma bond.

3. The answer is B. The triple bond of 2-propyne (an alkyne) contains one sigma bond and two pi bonds.

4. The answer is C. The $sp^3$ hybrid orbitals such as those seen in alkanes have a tetrahedral orientation.

5. The answer is D. This question also tests your basic knowledge of hybridization in carbon compounds. A triple bond means three bonds. Since there are three bonds, six electrons are involved in the bond.

6. The answer is B. Methane has a tetrahedral geometry. So the hydrogen-carbon-hydrogen bond angle is closest to $109.5^\circ$.

7. The answer is C. The triple bond is the shortest among the bonds indicated by the arrows. The only triple bond-bearing compound given is Choice C.

8. The answer is B. The carbon-hydrogen bond is a single covalent bond.

# Chemistry Review 5
# Alkanes

## A. INTRODUCTION

*Hydrocarbons* are compounds containing only carbon and hydrogen. They are classified into *aliphatic* and *aromatic hydrocarbons*. Aliphatic hydrocarbons can be divided into three major groups – *alkanes*, *alkenes*, and *alkynes*. Alkanes come under saturated hydrocarbons, because they have carbon-carbon single bonds. Alkenes and alkynes are unsaturated hydrocarbons, since they have carbon-carbon double and triple bonds, respectively.

## B. ALKANES

*Alkanes* have the general molecular formula $C_nH_{2n+2}$. Hence, if we know the number of carbons present in an alkane, we can calculate the number of hydrogens in it or vice versa. Methane is the first member of the alkane family. It has a molecular formula of $CH_4$. Natural gases which are found in petroleum deposits contain gases such as methane, ethane, and propane. They are the first three members of the alkane family.

**ALKANES WITH 1-12 CARBON ATOMS**

| | | | |
|---|---|---|---|
| Methane | $CH_4$ | Heptane | $C_7H_{16}$ |
| Ethane | $C_2H_6$ | Octane | $C_8H_{18}$ |
| Propane | $C_3H_8$ | Nonane | $C_9H_{20}$ |
| Butane | $C_4H_{10}$ | Decane | $C_{10}H_{22}$ |
| Pentane | $C_5H_{12}$ | Undecane | $C_{11}H_{24}$ |
| Hexane | $C_6H_{14}$ | Dodecane | $C_{12}H_{26}$ |

Methane is a colorless, odorless gas. Ethane, propane, and butane are also gases, with butane having the highest boiling point among these. What kind of trend can we see from these observations? As the number of carbons and hydrogens increases, the boiling point increases. Each carbon in an alkane is $sp^3$ hybridized.

## C. PROPERTIES OF ALKANES

At room temperature, the first four members of the alkane family are gases. The straight chain alkanes from pentane and up are liquids, and octadecane (18 carbons) and up in the alkane family are solids. As the number of carbons increases, the boiling point increases. Branched alkanes have lesser boiling points than their unbranched or less branched isomeric counterparts. The reason for this is that the unbranched molecules have more intermolecular interactions than the branched ones.

## D. STRAIGHT CHAIN AND BRANCHED ALKANES

Butane ($C_4H_{10}$) and the other alkanes above it can exhibit *constitutional isomerism*. If the alkane is unbranched and has a straight chain, it is called *n*-alkane. For example, the straight chain pentane is called *n*-Pentane.

*n*-Pentane $CH_3CH_2CH_2CH_2CH_3$

*n*-Butane $CH_3CH_2CH_2CH_3$

*Constitutional (structural) isomers are isomers with the same molecular formula, but are different in terms of the order in which the atoms are connected.*

Butane has two possible isomers: *n*-Butane and Isobutane

n-Butane Isobutane

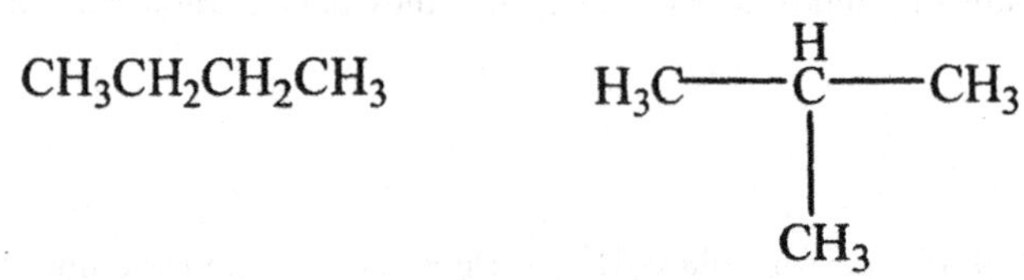

Pentane has three isomers: *n*-Pentane, Isopentane, and neopentane

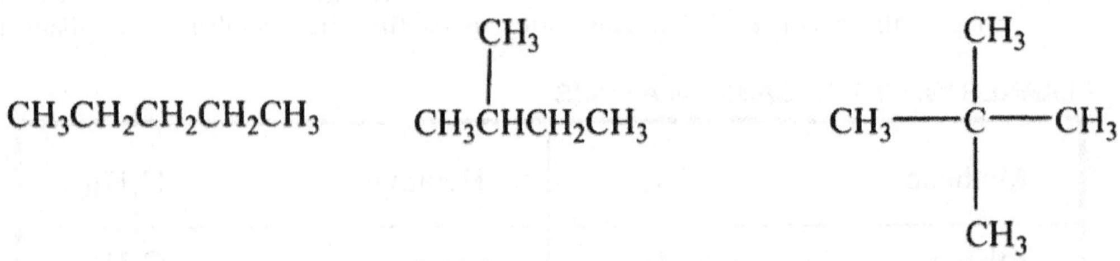

n-Pentane Isopentane Neopentane

The only way to find the number of possible isomers is by drawing out the structures sequentially and systematically, starting from the straight chain compound. There is no simple general formula to calculate the number of possible isomers of an alkane.

## E. ALKYL GROUPS

*Alkyl groups* are groups which lack one hydrogen atom compared to its parent alkane. For example, methyl group is $CH_3$—, which lacks one hydrogen atom with respect to its parent alkane, methane ($CH_4$). Similarly, ethyl group ($C_2H_5$—) lacks one hydrogen atom compared to ethane ($C_2H_6$).

***Some common alkyl groups***

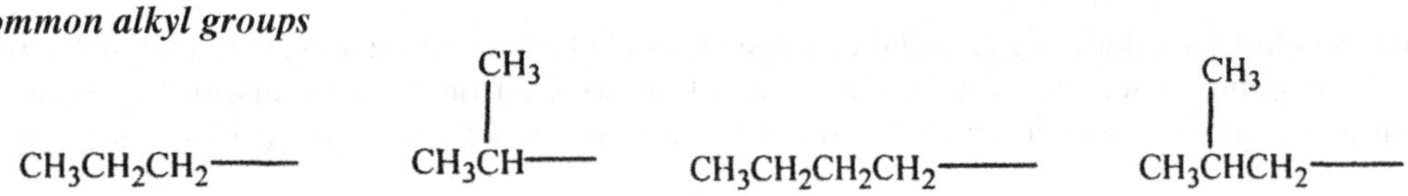

propyl (*n*-propyl) isopropyl butyl isobutyl

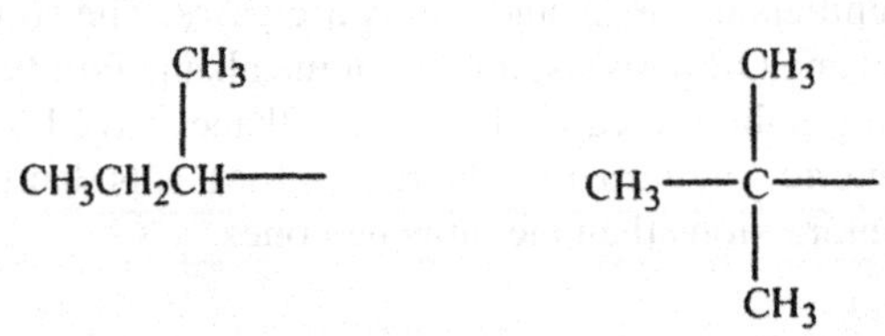

*sec*-butyl *tert*-butyl

# F. THE IUPAC NAMING OF ALKANES

**Main rules and strategies for the IUPAC naming of alkanes**

1. Write out the expanded structural formula, if it is not given in the expanded form.
2. Find the longest carbon chain.
3. Then identify the alkyl or other substituents that are connected to this long chain.
4. The numbering of carbons should start from the specific end of the long chain, so that the numbers assigned for the substituents are the lowest.

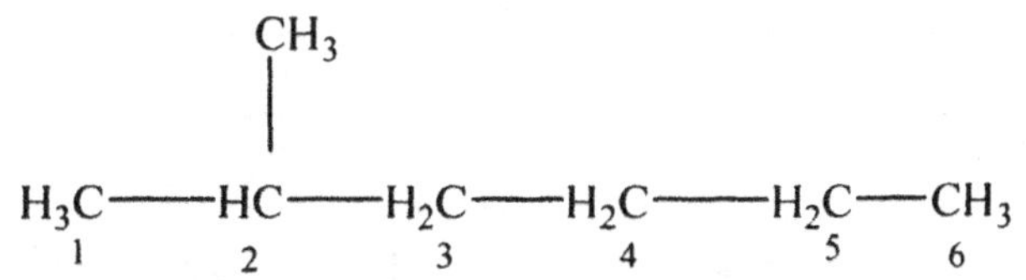

2-methyl hexane

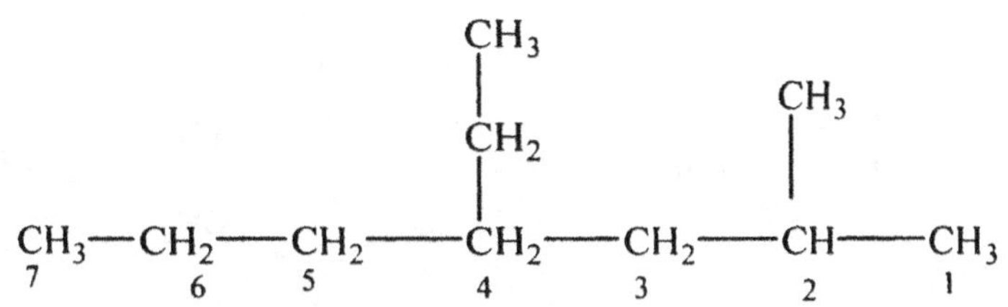

4-ethyl-2-methyl heptane

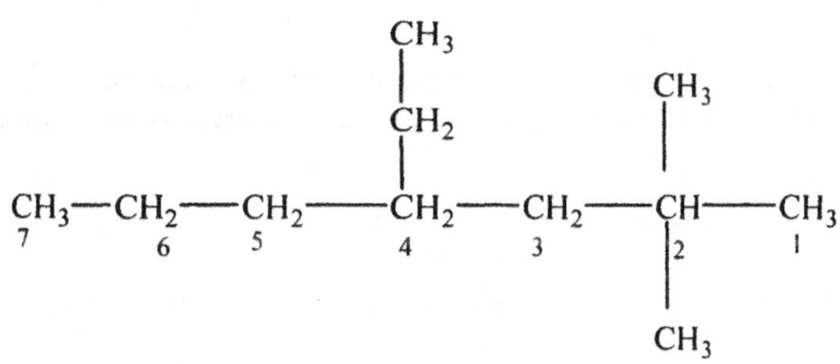

2,2-dimethyl-4-ethyl heptane

## G. CYCLOALKANES

*Cycloalkanes* are cyclic compounds with ring structures. The general molecular formula of a cycloalkane is $C_nH_{2n}$.

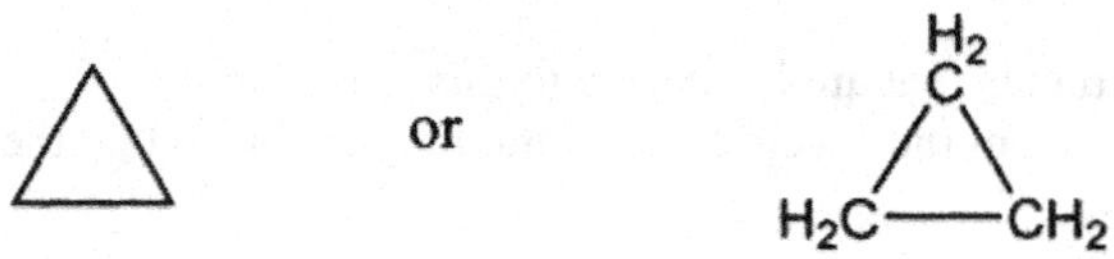


Cyclopropane

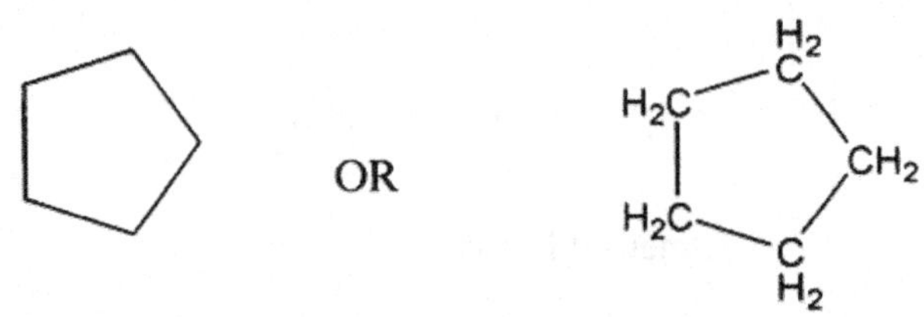


Cyclopentane

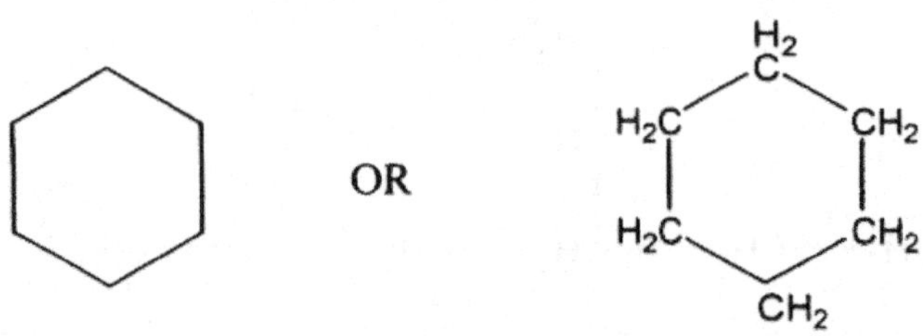


Cyclohexane

## H. REACTIONS OF ALKANES

### Combustion

Hydrocarbons undergo combustion reactions in the presence of oxygen to form carbon dioxide and water as products. Combustion reactions are very exothermic giving out energy, as they burn in the presence of oxygen.

**Sample reaction 15-1**

$$CH_4 + 2O_2 \longrightarrow CO_2 + 2\,H_2O \qquad \Delta H^0 = -890 \text{ kJ}$$

### Halogenation

The *halogenation reaction* can be generalized as follows:

$$\mathbf{RH} + \mathbf{X_2} \longrightarrow \mathbf{RX} + \mathbf{HX}$$

In this substitution reaction, the halogen (fluorine, chlorine, bromine, iodine) substitutes one hydrogen atom in the alkane, forming hydrogen halide and alkyl halide as the products. The reactivity of halogens in the halogenation reactions is as follows:

| Fluorine > Chlorine > Bromine > Iodine |
|---|

Fluorine is the most reactive among halogens in halogenation reactions, and iodine is the least reactive.

# I. MECHANISM OF FREE RADICAL SUBSTITUTION OF ALKANES

Halogenation reactions occur via a mechanism called *free radical substitution*. There are three main steps in the free radical mechanism. They are:

(1) initiation
(2) propagation
(3) termination.

The overall reaction of chlorination of methane.

$$CH_4 + Cl_2 \longrightarrow CH_3Cl + HCl$$

**(1) Initiation**

This step involves the dissociation of the halogen molecule (e.g., chlorine molecule) into two chlorine atoms. Even though the total reaction is exothermic, initially energy should be supplied for the reaction to proceed.

**(2) Propagation steps**

During the propagation step, the hydrogen atom is abstracted from methane by a chlorine atom. This is followed by the reaction between the methyl radical and the chlorine molecule.

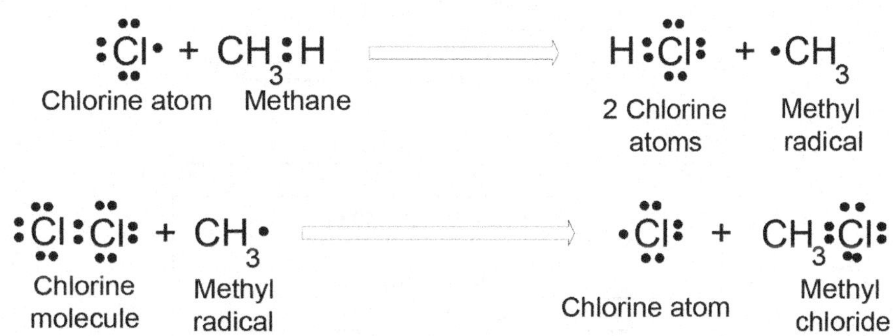


**(3) Termination**

The termination steps involve the combination of the radicals.

## J. REACTIVITY OF ALKANES

### Primary, Secondary, and Tertiary Carbons

A carbon which is attached directly to only one other carbon is called a primary ($1^0$) carbon. If it is attached directly to two other carbons, it is a secondary ($2^0$) carbon. A carbon is called a tertiary ($3^0$) carbon, if it is directly attached to three other carbons.

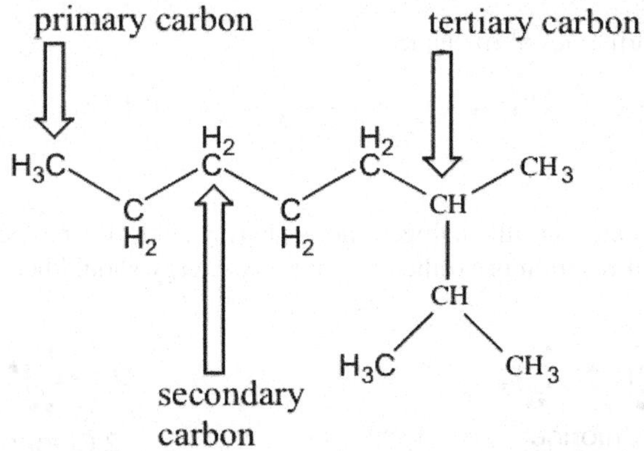


Though alkanes are not so reactive, they can undergo some reactions by forming intermediates. These intermediates can be alkyl radicals, carbocations, or carbanions. Alkyl radicals are intermediates of free radical reactions. Carbocations (carbonium ions) are species with a positive charge on one of the carbon atoms. A carbanion has a negative charge on one of its carbon atoms. Some major trends are given below:

| **Carbocation/Alkyl radical stability** |
| --- |
| **$3^0 > 2^0 > 1^0 >$ methyl** |

| **Carbocation/Akyl radical reactivity** |
| --- |
| **$3^0 < 2^0 < 1^0 <$ methyl** |

| **Carbanion stability** |
| --- |
| **$3^0 < 2^0 < 1^0 <$ methyl** |

| **Carbanion reactivity** |
| --- |
| **$3^0 > 2^0 > 1^0 >$ methyl** |

## K. CONFORMATION AND STABILITY OF ALKANES

### Conformations

Alkanes can have different conformations. By analyzing the structure of ethane, we can define certain aspects regarding its conformations. *Conformations* are different arrangements of the atoms in a molecule, as a result of rotation around a single bond.

**Figure 15-1**

STAGGERED CONFORMATION

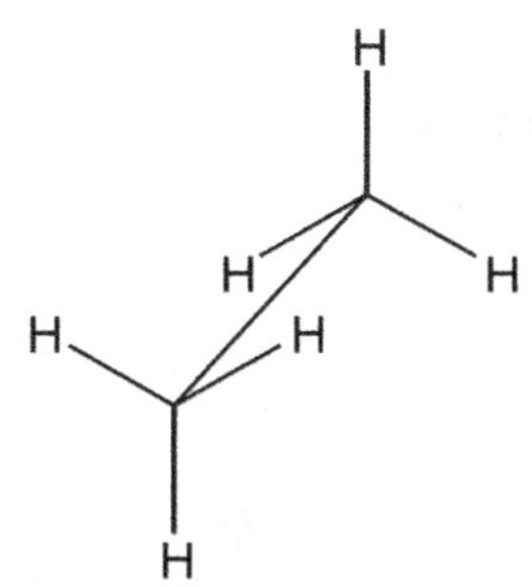

SAWHORSE REPRESENTATION

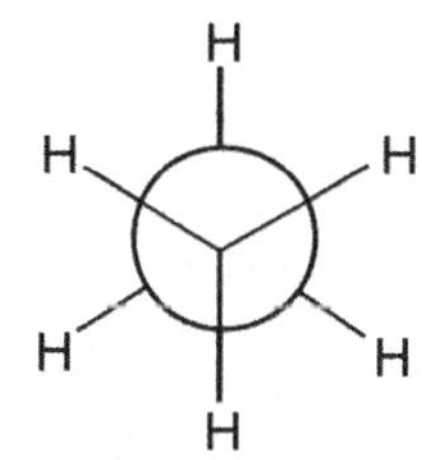

NEWMAN PROJECTION

In *staggered conformation*, the torsional angle is $60^0$. In *eclipsed conformation*, each carbon-hydrogen bond is aligned with the carbon-hydrogen bond of the next carbon.

**Figure 15-2**

ECLIPSED CONFORMATION

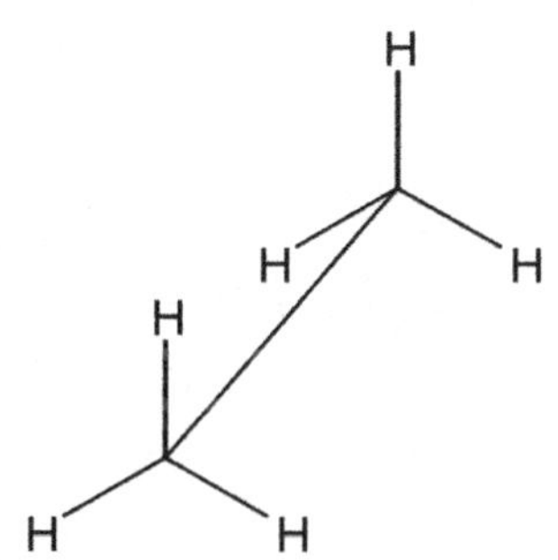

SAWHORSE REPRESENTATION

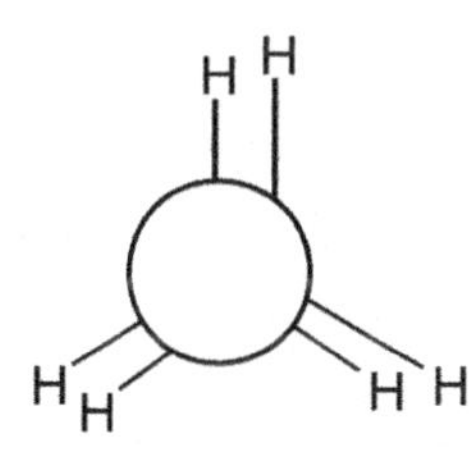

NEWMAN PROJECTION

In eclipsed conformations, the torsional angle is $0^o$. In staggered conformations, the torsional angles can either be $60^o$ (*gauche*) or $180^o$ (*anti*). The anti conformation is more stable than the gauche conformation. We should also consider the fact that in this analysis of staggered conformation, the ethane molecule looks the same in the Newman projections, whether it is gauche or anti. Reason: There are no substituents other than just hydrogens. To denote the positional significance, the hydrogens are indicated in bold in the diagrams shown in Figure 15-4.

Eclipsed

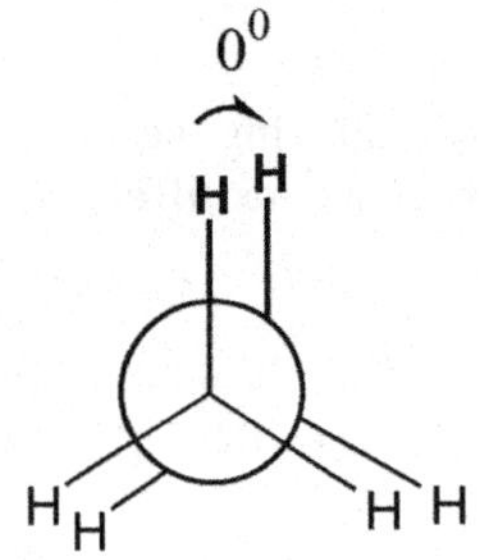


ANGLE OF TORSION IS $O^0$

**Figure 15-3**

Gauche | Anti

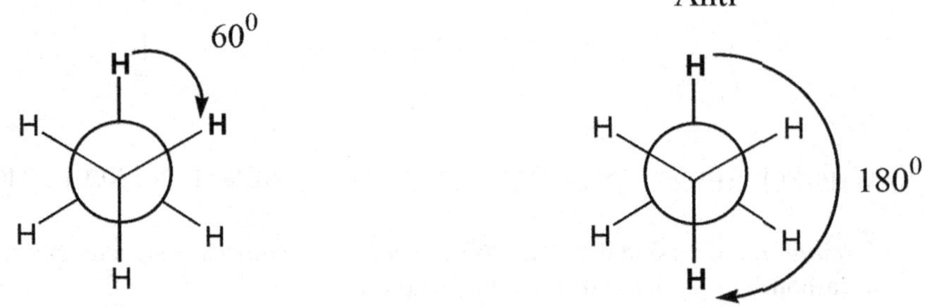


ANGLE OF TORSION IS $60^0$

ANGLE OF TORSION IS $180^0$

**Figure 15-4**

## L. CONFORMATION AND STABILITY OF CYCLOALKANES

According to Baeyer strain theory, the stability of a cycloalkane is based on how close its angles are to 109.5°. The closer the angle is to 109.5°, the more stable the cycloalkane. Deviation from this angle can cause angle strain. An increase in the angle strain means a decrease in the stability of the molecule.

### Conformations of Cyclohexane

Cyclohexane has a non-planar structure that makes it almost free from ring strain. The most important conformations that it can have include *chair conformation* and *boat conformation*. The chair conformation is more stable than the boat conformation. The boat conformation can sometimes be more stable than it usually is, by a slight rotation in the C-C bonds and is called the *skew boat conformation*. Nevertheless, the chair conformation is the most stable cyclohexane form.

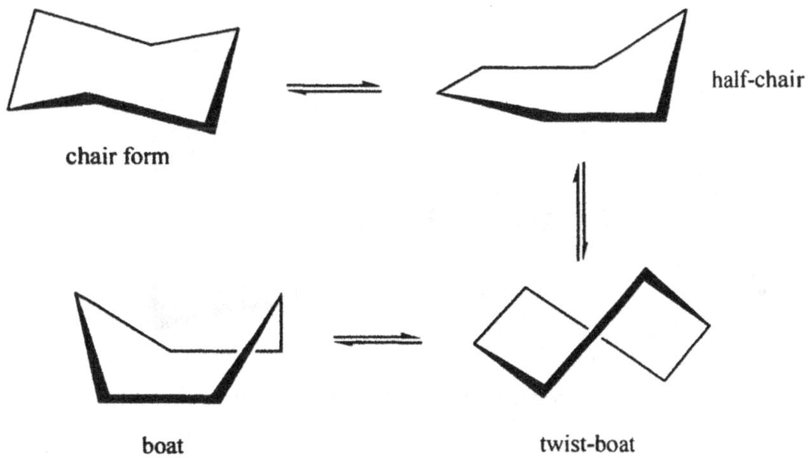


**Figure 15-5**

Relative energy levels of the conformations of cyclohexane

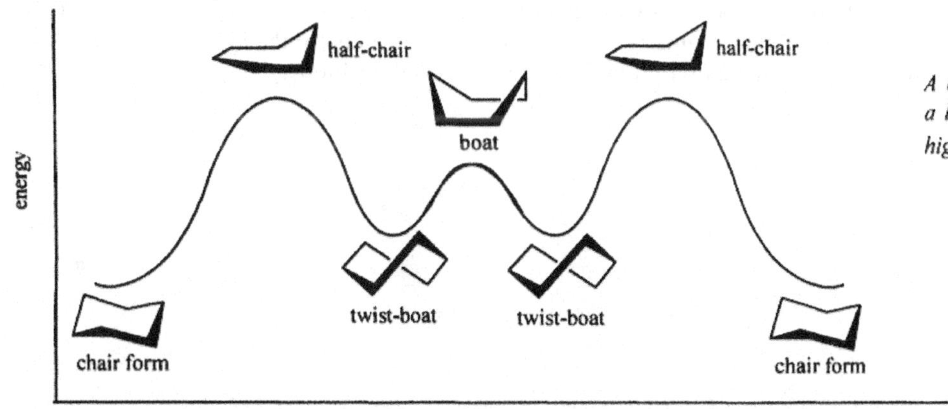


*A conformation having a lower energy implies higher relative stability.*

Based on the energy diagram of cyclohexane, the following inference can be made regarding the stability of the conformations:
chair conformation >twist boat conformation > boat conformation > half-chair conformation

⟸ increasing stability

In the chair form of cyclohexane, there are two different kinds of carbon-hydrogen bonds. Of the twelve carbon hydrogen bonds in a cyclohexane, six bonds are pointed up or down and are called axial bonds. The remaining six carbon-hydrogen bonds are pointed at an angle out of the ring and are called equatorial bonds. Each carbon atom in the ring is attached to one hydrogen atom by an axial bond, and to the other hydrogen atom by an equatorial bond.

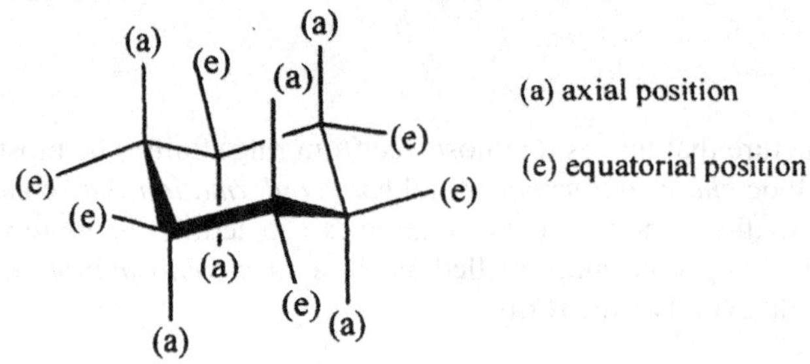


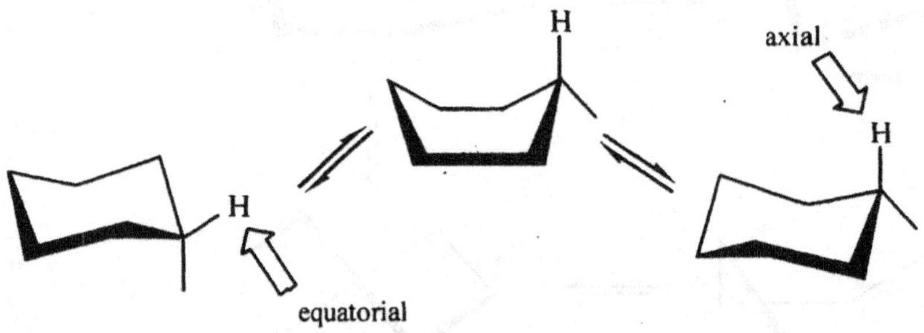


When ring flipping occurs between conformers, equatorial groups become axial, and axial groups become equatorial.

Experimental analysis has confirmed that among the two chair conformations of methylcyclohexane, 95% of the molecules have their methyl group in the equatorial position. In other words, the methyl group being in the equatorial position is more stable than methyl group in the axial position. The lower stability of the axial position can be attributed to the fact that there is an increased steric hindrance because of the proximity of the axial methyl group to the axial hydrogens that are attached to carbon atoms 3 and 5. This interaction called 1,3-diaxial interaction results in steric strain. This accounts for the increased relative stability of the conformation when the methyl group is in the equatorial position.

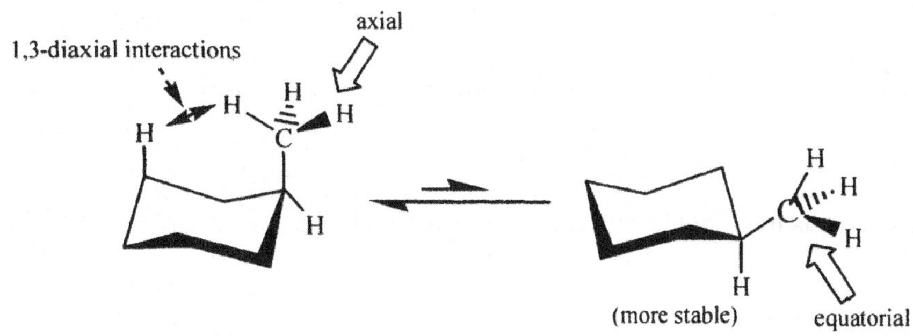


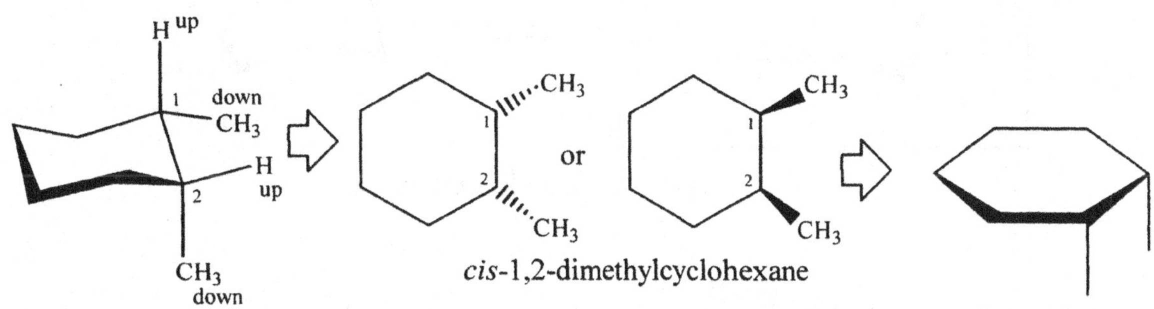


*cis*-1,2-dimethylcyclohexane

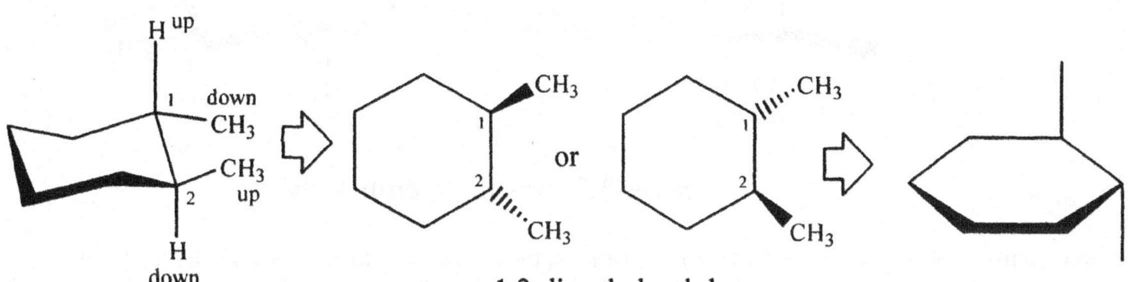


*trans*-1,2-dimethylcyclohexane

**Example 15-1**

The *cis*-geometric isomer of 1,3-dimethylclohexane is more stable than its *trans*-isomer. Why?

**Solution:**

To understand why the *cis*-geometric is more stable, let's draw the possible chair forms of 1,3-dimethylclohexane.

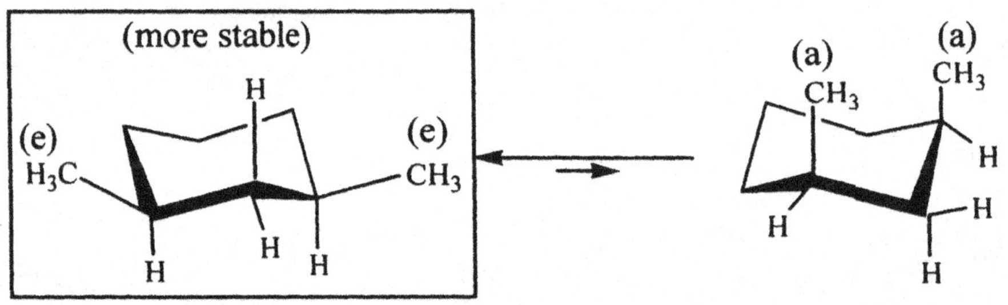


***cis*-1,3-dimethylcyclohexane**

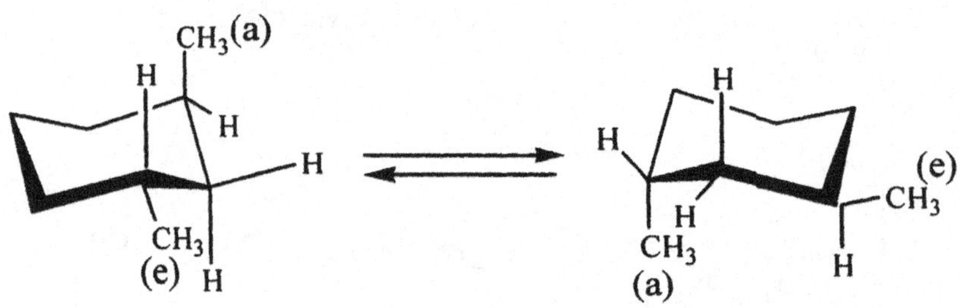


***trans*-1,3-dimethylcyclohexane**

When the two methyl substituents are in 1,3 positions, the *cis*-isomer can have both substituents in equatorial position. In the *trans*-isomer, one methyl group must always be axial.

# Chemistry Review 5 Problems

1. A student is assigned to identify an unknown compound in an organic chemistry class. She is sure that the compound contains only carbon and hydrogen atoms. In addition, the unknown compound is a saturated hydrocarbon. If there are fourteen hydrogens in a molecule of this unknown compound, and it does not have a ring structure, what is the most likely number of carbons in this compound?

A. 3
B. 6
C. 7
D. 14

2. Which of the following represents the general formula of an alkane?

A. $C_nH_{2n+2}$
B. $C_nH_{2n}$
C. $C_nH_{2n-2}$
D. $C_nH_n$

3. n-Butane and isobutane are best described as:

A. stereoisomers.
B. anomers.
C. diastereomers.
D. constitutional isomers.

4. Choose the correct name of the following compound from the choices given below.

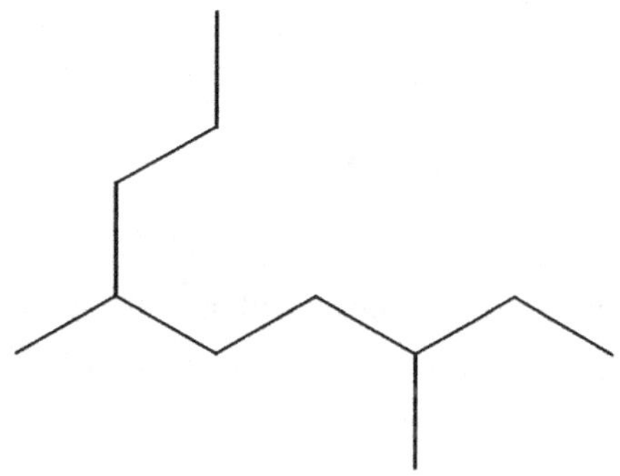

A. 2-propyl-5-methyl heptane
B. 3-methyl-6-propyl heptane
C. 3,6-dimethyl nonane
D. 4,7-dimethyl nonane

5. In the combustion reaction of butane, how many moles of carbon dioxide are formed, if one mole of butane undergoes complete combustion in a controlled environment in the presence of excess oxygen?

A. one
B. two
C. four
D. eight

6. Which of the following represents secondary carbons?

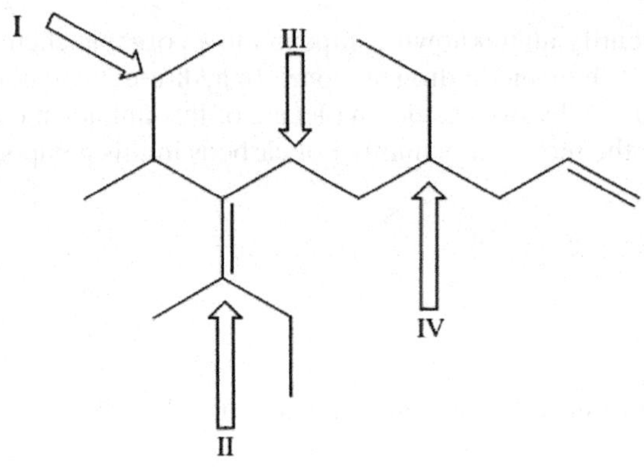


A. I & II only
B. I & III only
C. II & IV only
D. I, II, III & IV

7. Carbonium ions have:

A. a positive charge.
B. a negative charge.
C. no charge.
D. either a positive or a negative charge.

8. The gauche conformation is a form of:

A. eclipsed conformation.
B. anti-conformation.
C. staggered conformation.
D. none of the above conformations.

9. In cycloalkanes which of the following bond angles will have the least angle strain?

A. 90°
B. 110°
C. 125°
D. 180°

10. Which of the following alkanes has the highest boiling point?

I

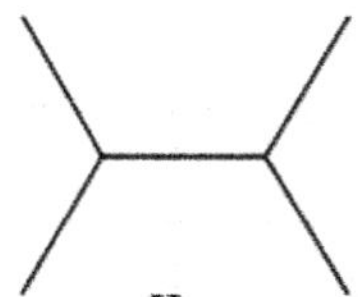

II

III

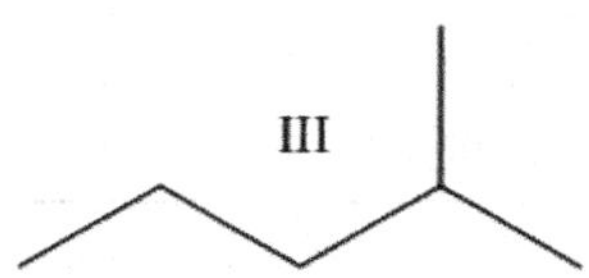

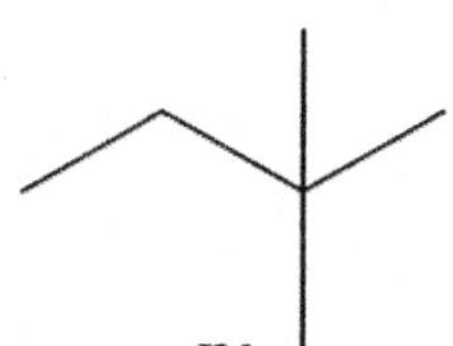

IV

A. I
B. II
C. III
D. IV

Questions 11-15 are based on the following passage.

**Passage**

Hydrocarbons are compounds composed of carbon and hydrogen atoms. Alkanes are hydrocarbons. As the number of carbons increases in straight-chain alkanes, there is a steady gradation of properties which can be easily compared and predicted. The properties of branched alkanes vary considerably and are hard to predict because of other intervening forces that come into play. The boiling points of a few alkanes are given below:

| | |
|---|---|
| methane | –161.6° C |
| propane | –88.6° C |
| pentane | 36.1° C |
| hexane | 68.7° C |
| octane | 125.6° C |
| 3-methylpentane | 63.3° C |

11. Which of the following intermolecular forces are important with respect to alkanes?

A. Hydrogen bonding
B. Dipole-dipole electrostatic forces
C. Ionic forces
D. Van der Waals forces

12. The melting point of butane is close to:

A. 37.5° C.
B. 55.1° C.
C. 24° C.
D. –138° C.

13. Alkenes can undergo free radical substitution reactions with halogens. Which of the following best represents a chain propagation step during the free radical chlorination of methane?

**A.** $Cl_2 \longrightarrow 2\ :\ddot{\underset{..}{Cl}}\bullet$

**B.** $\bullet CH_3 + :\ddot{\underset{..}{Cl}}\bullet \longrightarrow CH_3Cl$

**C.** $\bullet CH_3 + Cl_2 \longrightarrow CH_3Cl + :\ddot{\underset{..}{Cl}}\bullet$

**D.** $\bullet CH_3 + \bullet CH_3 \longrightarrow CH_3CH_3$

14. What is the most likely boiling point of 2,3-dimethylbutane?

A. 58° C
B. 63.3° C
C. 68.7° C
D. 75.8° C

15. The total number of possible structural isomers of heptane is:

A. three.
B. seven.
C. nine.
D. twelve.

# Solutions

1. The answer is B.

To answer this question, we can use the following clues from the question:

i) The compound is composed of only carbon and hydrogen atoms.
ii) It is a saturated hydrocarbon.
iii) It does not have a ring structure.

So the compound is an alkane. Alkanes have the general formula $C_nH_{2n+2}$. Now this becomes a simple algebra problem. Since there are 14 hydrogens, $2n + 2 = 14$.

So,

$$n = (14 - 2)/2 = 6$$

2. The answer is A.

3. The answer is D. Constitution isomers are compounds with the same molecular formula, but differ in the order with which the atoms are bonded. This definition matches the relationship between n-butane and isobutane.

4. The answer is C. First, we have to number the carbon chain to get the longest possible chain in the structure. The numbering should be done in such a way that the branches will have the smallest substituent number. The numbering is done as follows:

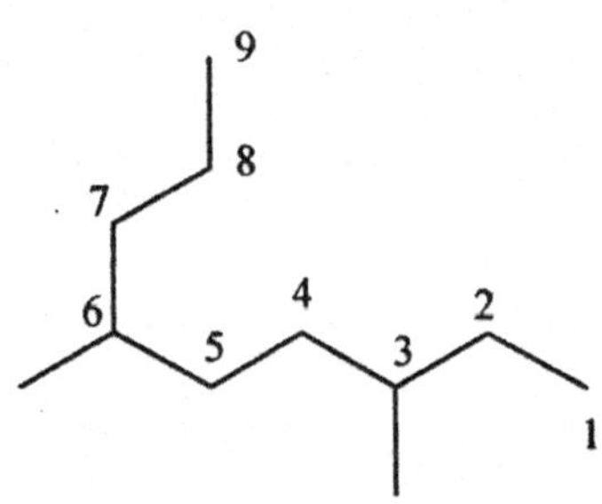


There are 9 carbons in the parent chain. So the parent chain is clear and the complete name ends with "nonane." This rules out Choices A and B. There are 2 methyl groups on carbons 3 and 6. So the compound is 3,6-dimethyl nonane.

5. The answer is C. Butane contains 4 carbon atoms and if one mole of butane is completely combusted, at least 4 moles of carbon dioxide will be formed provided that there is enough oxygen.

6. The answer is B. A secondary carbon atom should be attached to two other carbon atoms. The carbon atoms indicated by arrows I and III are secondary carbon atoms.

7. The answer is A. A carbonium ion is the same as a carbocation. A carbocation is a positively charged species.

8. The answer is C. Gauche conformation is a form of staggered conformation. The conformation is Gauche if the relative separation of two substituents on adjacent atoms is 60°. The Newman representation of a Gauche conformation is shown:

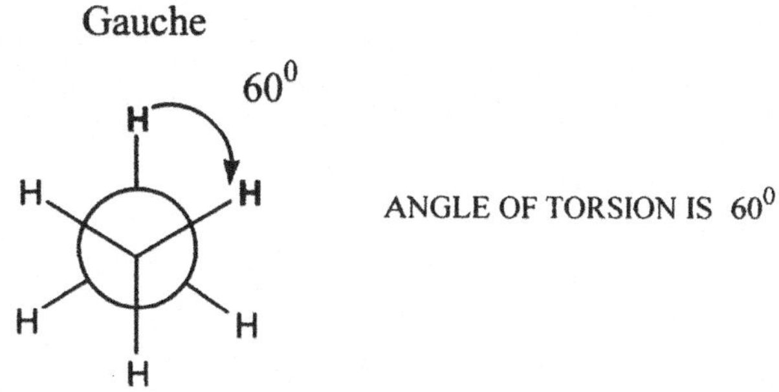


9. The answer is B. The angle closest to 109.5° is 110°. That is the most likely bond angle, having the least angle strain.

10. The answer is A. All the given choices are alkanes. Item I is an unbranched alkane. Items II, III, IV have increasing degrees of branching. Branching decreases the area of contact and thereby decreases the intermolecular interactions. So the unbranched alkane has the highest boiling point. Keep in mind that without actually measuring the properties, we can only compare compounds of comparable weights and number of carbons with reasonable accuracy.

11. The answer is D. The question asks for the type of force that is important with respect to alkanes. Alkanes cannot undergo hydrogen bonding. So Choice A is out. Choice B is generally not important in alkanes. Since there are no ionic bonds in an alkane, Choice C is wrong. The correct choice is Van der Waals forces.

12. The answer is D. To answer this question you have to do some intuitive thinking based on the information given in the passage. Butane is a four-carbon alkane with the formula $C_4H_{10}$. It is a bigger molecule than propane, but a smaller molecule than pentane. So the boiling point of butane should be between propane and pentane. We are talking in terms of the relative boiling points. But the question asks for the melting point. So let's translate this idea in terms of melting point. Obviously, the melting point of a substance should be below its boiling point. So the answer cannot be either 37.5°C or 55.1°C. This rules out Choices A and B. Choice C is not a sensible answer because butane is a gas at 25°C.

13. The answer is C. Choice A is an initiation step. Choice B is a chain termination step. Choice D is also a chain termination step. The only choice which represents a chain propagation step is Choice C.

14. The answer is A. This question tests your knowledge of intermolecular forces and other interactions that give rise to some of the chemical and physical properties of compounds. Let's take a look at the compound 2,3-dimethylbutane. We do not have a clue about the precise properties of the compound asked in this question. But what we do know is the boiling point of one of its relatives. In the table given in the passage, we are told that the boiling point of 3-methylpentane is 63.3° C. If we compare these two compounds, we see that they have the same number of carbons and hydrogens – both are six-carbon alkanes. If the boiling point of 3-methylpentane is 63.3° C, then 2,3-dimethylbutane will have a boiling point lower than that. The only choice with a lower value is Choice A.

15. The answer is C. The only way to answer this question is to draw the possible isomers by trial and error. There is no choice other than drawing out the possible structures, starting from the straight chain form of heptane.

Part Two

# THE TESTS

# Test 1

(35 Minutes—40 Questions)

**Passage 1**

Often, bacteria are grown on a medium that contains all of the components they need to survive. This includes an energy source, a carbon source, and sources of all the mineral salts the bacteria require to carry out their metabolic reactions. Sometimes, a mutant bacterium is isolated that needs more than the usual minimum requirements, because it cannot carry out the reactions necessary to produce a needed substance. These mutants are referred to as *auxotrophs*; if they are to grow, they must be supplemented with the substance they need but cannot manufacture.

*Experiment 1*

Bacteria were grown on media containing different radioactively labeled atoms to determine which types of biological molecules would be radioactively labeled. Molecules become labeled as they are synthesized by the bacteria using the raw materials provided by the medium. The bacteria were grown for a period of time that was sufficient to label all molecules that could be labeled, and the major types of biomolecules were isolated and tested for the presence of radioactivity. The following table shows the various strains of bacteria and the radioactive element present in the media in which they were grown.

| Bacterial strain | Radioactive element |
|---|---|
| strain 1 | carbon |
| strain 2 | phosphorous |
| strain 3 | nitrogen |
| strain 4 | oxygen |

*Experiment 2*

Auxotrophic bacteria were grown on media supplemented with the substance that they need but cannot manufacture. If they were not supplemented in this way, no growth would be observed. The following table shows two auxotrophic strains and the substance with which they were supplemented.

| Auxotrophic strain | Supplement |
|---|---|
| strain A | arginine, an amino acid |
| strain B | decanoic acid, a fatty acid |

*Experiment 3*

Auxotrophic strain A (from Experiment 2) was grown on a medium that contained supplemental arginine *and* radioactive phosphorus, and the major types of biomolecules were again isolated and tested for radioactivity.

1. Which of the following biomolecules isolated from Strain 2 would we expect to be radioactively labeled according to the table from experiment 1?
   A. RNA
   B. Proteins
   C. Glucose
   D. Amino acids

2. Which of the following biomolecules isolated from Strain 3 would we expect to be radioactively labeled according to the table from experiment 1?
   A. DNA, but not RNA
   B. DNA and RNA, but not proteins
   C. RNA and proteins
   D. DNA and fatty acids

3. Which of the following biomolecules isolated from Strain 1 would we expect to be radioactively labeled according to the table from experiment 1?
   A. Cholesterol
   B. Phospholipids
   C. Both A and B
   D. Neither A nor B

4. In experiment 2, the arginine requiring auxotrophic strain A was supplemented with arginine to allow growth. If the cells were not supplemented with arginine, the cells could not grow because arginine is directly required for the synthesis of:
   A. DNA.
   B. Proteins.
   C. Fatty acids.
   D. RNA.

5. In experiment 3, which of the following biomolecules would we expect to become labeled?
   A. All proteins, DNA, and RNA
   B. Only proteins containing arginine, but not DNA or RNA
   C. DNA, RNA, and only those proteins that contain arginine
   D. RNA and phospholipids

**Passage 2**

In a parallel-plate capacitor, two parallel metal plates are connected to a voltage source which maintains a potential $V$ across the plates. Positive charges collect on one side of the capacitor and negative charges on the other side, thus creating an electric field $E$ between the plates. The magnitude of the electric field is related to the potential and the separation between the plates according to

$$V = Ed$$

where $V$ is measured in volts, $E$ in J/m, and $d$ in m. A charged particle placed between the plates will experience a force given in magnitude by

$$F = qE$$

where $q$ is the charge of the particle in Coulombs, and $F$ is the force in N.

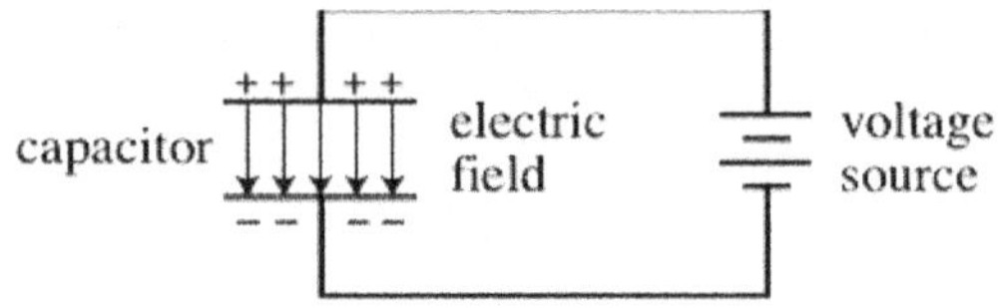


6. If a new battery is installed, so that the voltage between the plates is increased by a factor of 9, how is the electric field affected?
   A. It decreases by a factor of 9.
   B. It increases by a factor of 3.
   C. It increases by a factor of 9.
   D. It increases by a factor of 81.

7. If the voltage in a given experiment is held constant, but the distance between the plates is increased by a factor of 3, how is the electric field affected?
   A. It decreases by a factor of 9.
   B. It decreases by a factor of 3.
   C. It stays the same.
   D. It increases by a factor of 3.

8. In a given experiment, both a proton and a bare helium nucleus are between the plates. How does the force on the helium nucleus compare to the force on the proton?
   A. It is the same.
   B. It is twice as great.
   C. It is four times as great.
   D. There is no force on the helium.

9. In a given experiment, all other things being held constant, what happens to the force on a proton between the plates if the separation of the plates is increased by a factor of 2?
   A. It decreases by a factor of 4.
   B. It decreases by a factor of 2.
   C. It stays the same.
   D. It increases by a factor of 2.

10. Which graph best show the relationship between the potential $V$ and the electric field $E$?

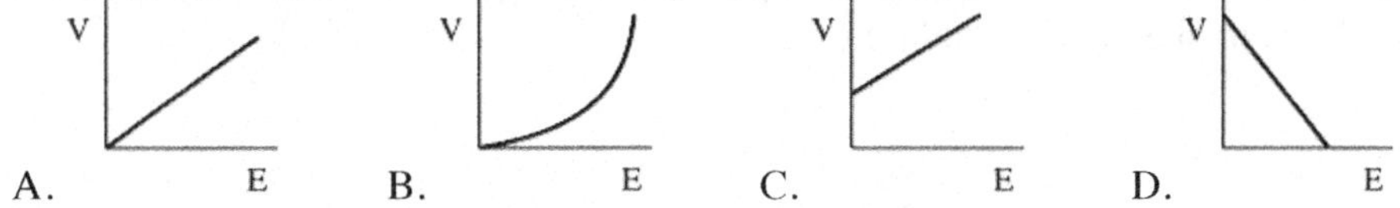


A. B. C. D.

**Passage 3**

The behavior of gases can be predicted to a large extent on the basis of various laws. Gases are expressed mainly in terms of pressure, volume, and temperature. Many gases obey the ideal gas laws and those gases are called ideal gases. If we are not doing precision experiments, we can normally ignore the slight deviations that occur under normal conditions. Nevertheless, we cannot completely ignore the deviation factors.

According to Boyle's law, the pressure is inversely proportional to the volume at a constant temperature. Charles' law states that the volume is directly proportional to the temperature at a constant pressure. Combining these laws and Avogadro's law gives the combined gas law.

$$PV = nRT$$

(The ideal gas law)

Not all gases behave ideally. There are often deviations from the ideal behavior. At low temperatures, gases often behave differently apart from what is described by kinetic-molecular theory of gases. Considering these correction factors, the modified gas equation is:

$$\left(P + \frac{n^2 a}{V^2}\right)\left(V - nb\right) = nRT$$

The correction constants or Vander Waal constants, *a* and *b*, of gases are experimentally found. The Van der Waal constants of some gases are given in Table 1.

**Table 1**

| Gas | a ($L^2$.atm/$mol^2$) | b (L/mol) |
|---|---|---|
| Carbon monoxide | 1.470 | 0.039 |
| Hydrogen | 0.245 | 0.026 |
| Chlorine | 6.340 | 0.054 |
| Fluorine | 1.170 | 0.029 |
| Hydrogen sulfide | 4.540 | 0.043 |
| Nitrogen | 1.380 | 0.039 |
| Carbon dioxide | 3.660 | 0.043 |
| Argon | 1.360 | 0.032 |
| Propane | 9.390 | 0.091 |

11. Under which of the following conditions will gases behave most ideally?
    A. Low temperature and high pressure
    B. High temperature and low pressure
    C. Low temperature and low pressure
    D. Gas behavior does not depend on temperature and pressure if the gas is an ideal gas, and such a gas will behave ideally regardless of the conditions

12. If the Vander Waal constant *a* is found to be zero for gas A, what are true regarding this gas?

I. Gas A does not behave ideally.
II. Gas A behaves ideally.
III. Gas A is Xenon.

A. I only
B. II only
C. II & III only
D. III only

13. Equal volumes of gases will contain the same number of molecules at the same temperature and pressure. This prediction is directly based on:
A. Charles law.
B. Boyle's law.
C. kinetic-molecular theory.
D. Avogadro's law.

14. Based on kinetic-molecular theory, which of the following are true?

I. At a given temperature, all gases have the same average kinetic energy.
II. At a given temperature, different gases have different average velocities.
III. The average kinetic energy is proportional to the absolute temperature.

A. I only
B. II only
C. I & III only
D. I, II & III

**Passage 4**

We perform an experiment which involves two masses *m* and *M* connected by a string which we will consider to be massless. Mass *m* hangs over the edge of a table. The string passes over a pulley at the edge of the table and mass *M* sits on the table, such that it moves along the table without friction. (See figure.) The tension in the string is the force that the string exerts where it is connected to another object or to more string. It is generally true that the tension anywhere along the string is the same as the tension anywhere else in the string.

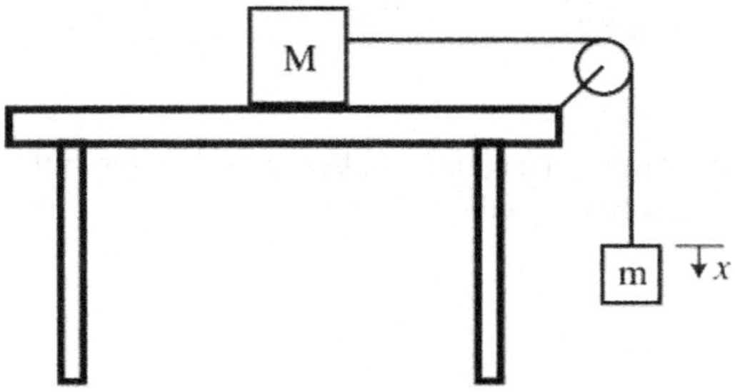


In this experiment the mass *m* is initially at rest and allowed to drop. Its position *x* is measured downward from its initial position. At various times, the position *x* and velocity *v* of mass *m* are measured and the results are recorded in the table which follows:

| *t* (s) | *x* (m) | *v* (m/s) |
|---|---|---|
| 0.0 | 0.00 | 0.0 |
| 0.2 | 0.01 | 0.1 |
| 0.4 | 0.04 | 0.2 |
| 0.6 | 0.09 | 0.3 |

15. Which of the following is evidence that the acceleration is uniform?
  A. The entries for *x* are nonnegative and increasing.
  B. The entries for *v* are nonnegative and increasing.
  C. The entries for *v* are always greater than *x*.
  D. Any interval $\Delta v$ is proportional to the interval $\Delta t$.

16. What are the forces on the mass *m*?
  A. The force of gravity.
  B. The force of gravity and the tension of the string.
  C. The force of gravity and the force due to mass *M*.
  D. The force of gravity, the tension of the string, and the force due to *M*.

17. What are the forces on the mass *M*?
  A. The force of gravity and the upward force of the table.
  B. The force of gravity, the upward force of the table, and the tension in the string.
  C. The force of gravity, the upward force of the table, and the force due to *m*.
  D. The force of gravity, the upward force of the table, the force due to *m*, and the tension in the string.

18. What is the average velocity $v_{avg}$ for the interval of time shown in the table?
  A. 0.0 m/s
  B. 0.1 m/s
  C. 0.15 m/s
  D. 0.3 m/s

19. After the experiment has run a while, the mass *m* hits the floor and the string goes slack. But mass *M* continues going forward until it hits the pulley. After the string goes slack but before *M* hits the pulley, what are the forces on mass *M*?
  A. There are no forces on *M*.
  B. The force of gravity.
  C. The force of gravity, and the upward force of the table.
  D. The force of gravity, the upward force of the table, and a forward force.

**Passage 5**

In general, enzymes are protein molecules which must be folded in a specific three-dimensional shape in order to function properly. Certain environmental parameters can affect enzyme activity, including pH and temperature. If an enzyme's shape changes significantly and it can no longer function, the enzyme is said to have become denatured.

The enzyme pancreatic amylase is manufactured and secreted by the pancreas into the duodenum (the large, beginning portion of the small intestine). Pancreatic amylase breaks down starch into maltose, a disaccharide. Pepsin is an enzyme that is released by the epithelium of the stomach, and functions in the stomach to break down proteins into smaller polypeptide units.

The following graphs show the activities of various enzymes under various environmental conditions.

*Graph 1*:

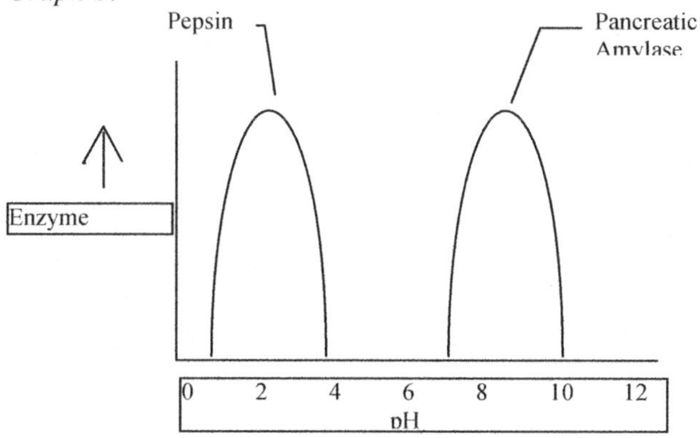


*Graph 2*:

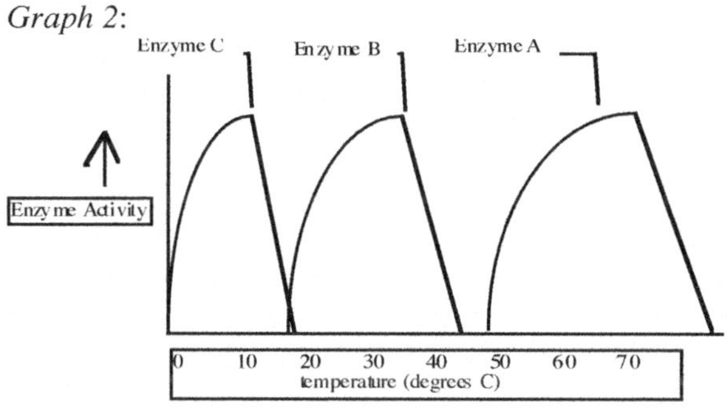


20. Which of the following statements is true with respect to graph 1?
    A. Pepsin and pancreatic amylase could never function together in the same part of the body at the same time.
    B. Pancreatic amylase could function in the stomach, but its activity would be low.
    C. The optimal pH for the functioning of pepsin is approximately 8.5.
    D. Normally, the small intestine must be slightly acidic.

21. Trypsin is a protein digesting enzyme that functions in the small intestine. Which of the following statements should be true about trypsin?
A. The optimum pH for the functioning of trypsin is approximately 2.
B. Both trypsin and pepsin would be expected to be found working together in the same part of the body.
C. The optimum pH for the functioning of trypsin is approximately 8.5.
D. Trypsin could function well in a solution containing 1 molar hydrochloric acid.

22. Graph 2 depicts the activities of three enzymes. Which curve illustrates the functioning of human DNA polymerase, which functions in the nucleus of cells?
A. Enzyme A
B. Enzyme B
C. Enzyme C
D. None of the above could represent the activity of human DNA polymerase.

23. Which curve illustrates the functioning of DNA polymerase from a shark?
A. Enzyme A
B. Enzyme B
C. Enzyme C
D. None of the above, since sharks, like all fish, do not contain DNA polymerase.

24. At what temperature would enzyme B be completely denatured?
A. 37° C
B. 15° C
C. 5° C
D. 50° C

**Passage 6**

A phase diagram illustrates the physical state of a substance with respect to the temperature and the pressure of the surroundings. The points plotted to draw the phase diagram are derived experimentally, and are usually plotted with temperature on the x-axis and pressure on the y-axis. A typical phase diagram is shown in Figure-1.

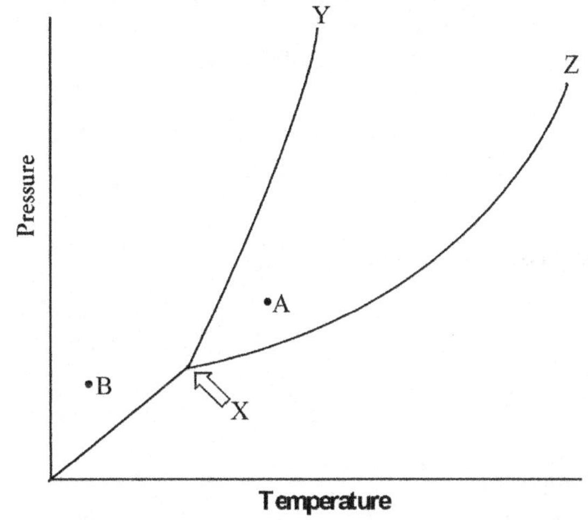


**Figure 1**

25. Which of the following represents the correct phase diagram of water?

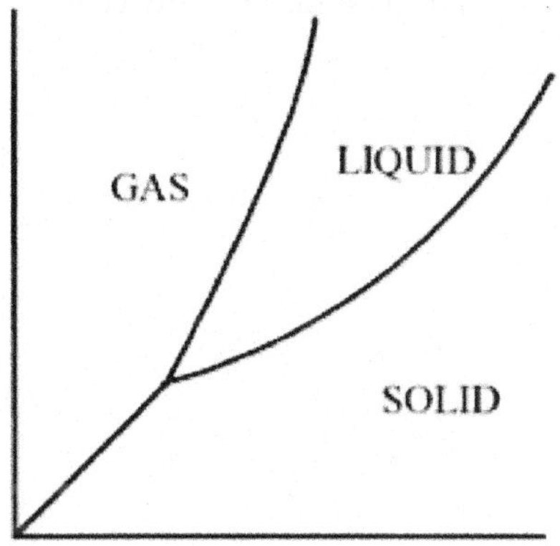


A.

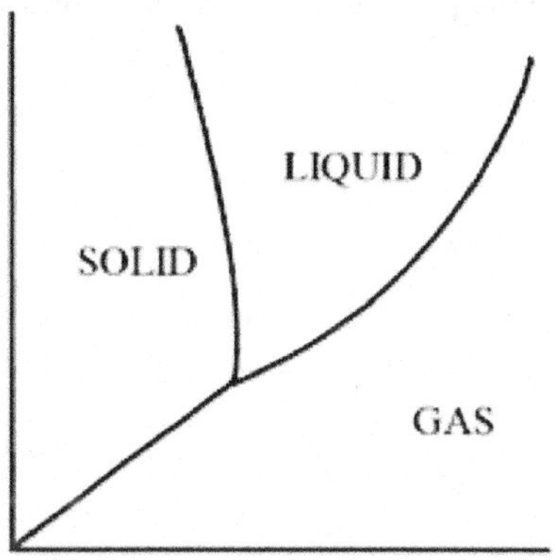


B.

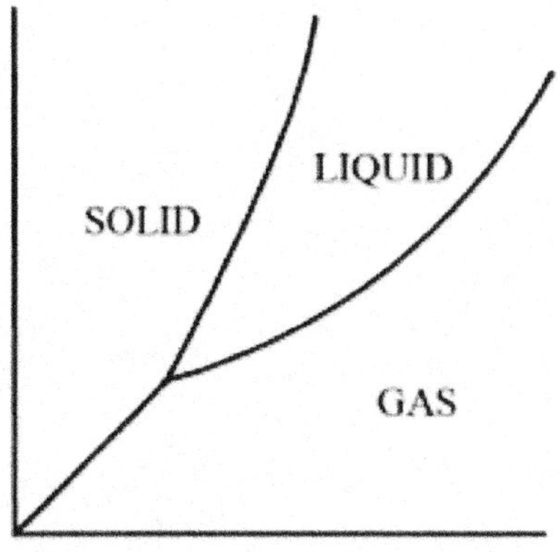


C.

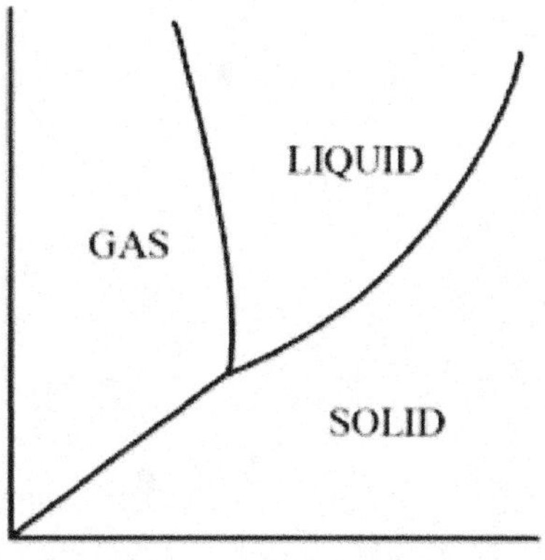


D.

26. The phase diagram given in the passage cannot be that of:
   A. $CO_2$
   B. $H_2O$
   C. CO
   D. $O_2$

27. From the phase diagram given in the passage, predict the likely phase change if the following changes happen. Consider point A as the current phase of substance M. What is the most likely effect on substance M, if the pressure was decreased without changing the temperature of the system?
   A. Gas to liquid
   B. Liquid to gas
   C. Liquid to solid
   D. Solid to liquid

28. The point marked 'X' in the phase diagram given in the passage is called:
   A. the critical point.
   B. the maximum point.
   C. the triple point.
   D. the critical pressure point.

29. The critical temperature of a substance is best defined as:
   A. the temperature below which a gas cannot be liquefied.
   B. the temperature above which a liquid cannot be vaporized.
   C. the temperature above which a substance exists only as gas.
   D. the temperature which is essential for a substance to obey ideal gas laws.

30. The segment XZ in the phase diagram given in the passage represents the equilibrium between:
   A. gas and liquid.
   B. liquid and solid.
   C. solid and gas.
   D. gas and plasma.

**Passage 7**

A sport at a nearby educational institute involves running along the roof of an apartment building, jumping off the edge, and falling into the pool below. This dangerous sport involves a combination of strength of spirit, braggadocio, and inebriation.

Let's say a student (50 kg) accelerates uniformly from rest at one side of the building to the jumping edge, a distance of 5 m. Just after his feet leave the building, he is traveling horizontally at a speed 5 m/s. The building is 7.2 m high, and the pool is 4.5 m from the edge of the building. Use $g = 10\ m/s^2$.

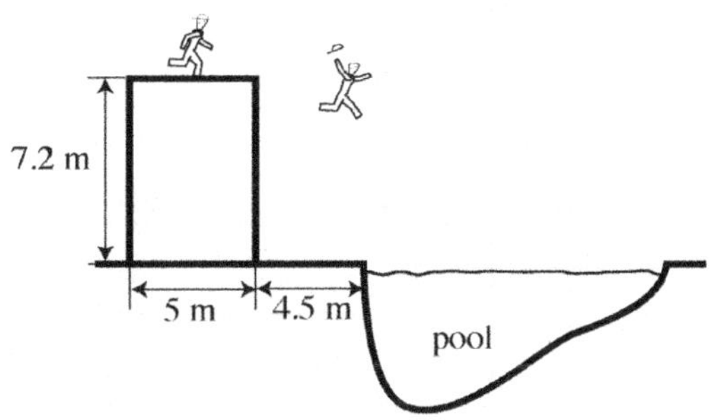


31. How much time does it take the student to accelerate as he is running along the roof?
    A. 1.00 s
    B. 1.20 s
    C. 1.44 s
    D. 2.00 s

32. Which diagram best represents the force diagram for the student while he is on the roof?

    A.  B. 

    C.  D. 

33. Which diagram best represents the force diagram for the student while he is in the air?

    A.  B. 

    C.  D. 

34. What is his horizontal velocity just before he lands?
    A. 0 m/s
    B. 5 m/s
    C. 12 m/s
    D. 13 m/s

**Passage 8**

Yeast are unicellular fungi that are considered to be facultatively anaerobic. This means that in the presence of oxygen they can and will undergo aerobic respiration, but at the same time they will likely be fermenting glucose. Furthermore, when no oxygen is present they can survive, potentially forever, by fermentation. Yeast use ethanol fermentation, and this produces carbon dioxide as one of its final products.

One way of quantifying the rate of fermentation in yeast is to measure the rate at which the volume of gas changes in a test tube connected to a solution of yeast being fed with glucose. Since carbon dioxide is the only gas involved in fermentation, the rate of gas production should mirror the rate of fermentation.

The following table shows the amount of gas evolved in such an experiment under different environmental conditions.

| pH | Gas evolved (in ml) |
|---|---|
| 5 | 5 |
| 6 | 10 |
| 7 | 20 |
| 8 | 30 |
| 9 | 25 |
| 10 | 10 |

35. At which pH is the most ethanol produced?
    A. 5
    B. 7
    C. 8
    D. 9

36. It is likely that oxygen is present in this system, so that aerobic respiration is going on at the same time as fermentation. Taking this into account, which of the following statements is true?
    A. Since oxygen will be used up, using the net change in overall gas production as an indicator of the fermentation rate will be inaccurate.
    B. It is irrelevant to our experiment whether or not aerobic respiration is taking place or not; using the net change in overall gas production as an indicator of the fermentation rate will be accurate.
    C. Under these conditions, no fermentation will occur, since according to the passage yeast only undergo fermentation when no oxygen is available.
    D. The experiment is further complicated by the fact that the yeast will use up some of the carbon dioxide during glycolysis, so using the net change in overall gas production as an indicator of the fermentation rate will be inaccurate.

**Passage 9**

Consider an object sitting on a scale at the surface of the Earth. The scale reading is the magnitude of the normal force which the scale exerts on the object. To a first approximation, there is force balance, and the magnitude of the scale's force is the magnitude of the gravitational force:

$$F_{\text{grav}} = \frac{GM_{\text{Earth}}m}{R^2_{\text{Earth}}} \qquad (1)$$

where $G$ is Newton's constant, $M_{\text{Earth}}$ is the mass of the Earth, and $R_{\text{Earth}}$ is the radius of the Earth. The simple result is that the force of gravity, and the reading of the scale, is proportional to the mass:

$$F_{\text{grav}} = mg \qquad (2)$$

where $g$ has the value $GM_{\text{Earth}}/R_{\text{Earth}}{}^2 = 9.8$ m/s$^2$. We have made several idealizations, however, and if we want to calculate the scale reading, we need to be more careful.

For example, we have ignored the rotation of the Earth. Consider a man standing on a scale at the equator. Because he is moving in a circle, there is a centripetal acceleration. The result is that the scale will not give a reading equal to the force of gravity (equation [1]).

We have also assumed that the Earth is a perfect sphere. Because it is rotating, the distance from the center of the Earth to the equator is greater than the distance from center to pole by about 0.1%.

A third effect we have ignored is that the Earth has local irregularities which make it necessary to measure $g$ in the local laboratory, if we need an exact value of the effective acceleration due to gravity.

37. For a man standing at the equator of a rotating Earth, which expression gives the best expression of his velocity? (Let $T_{\text{day}}$ be the time of one rotation, 1 day.)
A. $R_{\text{Earth}}/T_{\text{day}}$
B. $2\pi\, R_{\text{Earth}}/T_{\text{day}}$
C. $gT_{\text{day}}$
D. $2\pi\, gT_{\text{day}}$

38. Which is the best force diagram for a man standing at the equator of a rotating Earth?

A.
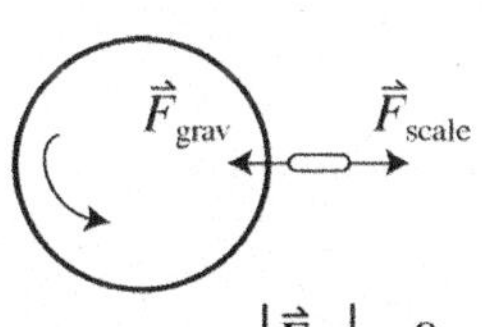


$|\vec{F}_{\text{net}}| = 0$

B.
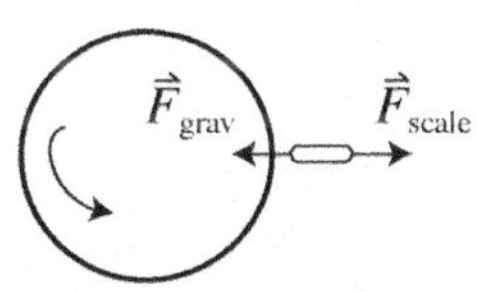


$\vec{F}_{\text{net}}$ points into the page.

C.
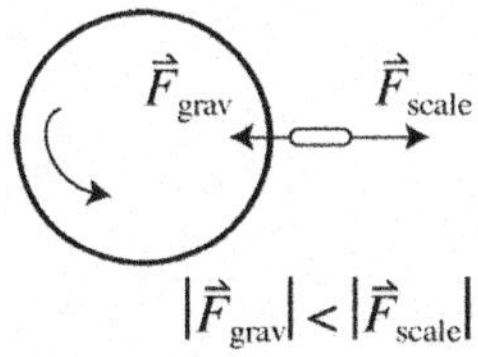


$|\vec{F}_{\text{grav}}| < |\vec{F}_{\text{scale}}|$

D.
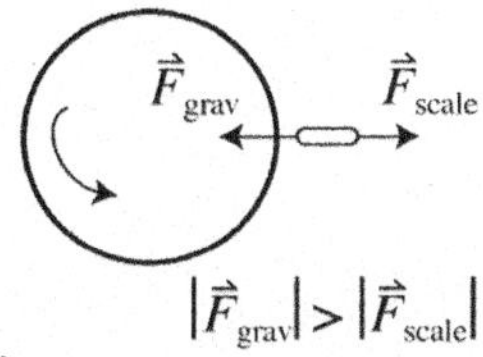


$|\vec{F}_{\text{grav}}| > |\vec{F}_{\text{scale}}|$

39. If the man at the equator stood on a scale, how would the scale read compared to the scale reading for an identical man standing at the equator of a nonrotating Earth?
    A. It would read less than on a nonrotating Earth.
    B. It would read the same as on a nonrotating Earth.
    C. It would read greater than on a nonrotating Earth.
    D. It would depend on where the man is.

40. If two identical men stood on scales at the south pole and at the equator of an Earth identical to this one but nonrotating, how would the reading of the polar scale compare to the equatorial one?
    A. It would be less.
    B. It would be the same.
    C. It would be greater.
    D. There is not enough information to answer this question.

# Test 1 Solutions

**Passage 1**

1. A.
RNA, a nucleic acid, is the only molecule listed which contains phosphorus (in the phosphate group portion of the nucleotide). Thus radioactive phosphorus will not show up in proteins, glucose (a carbohydrate), or amino acids (protein subunits).

2. C.
Both RNA and proteins contain nitrogen, RNA in the nitrogenous base portion of the nucleotide, and proteins in the amino group of every amino acid. A and B are not correct because they exclude RNA or proteins. D is incorrect because fatty acids, constituents of triglycerides, do not contain nitrogen.

3. C.
Both cholesterol and phospholipids contain carbon, as do all of the major biomolecules we discussed, as they are organic. Therefore, radioactive carbon would show up in all the biomolecules we mentioned.

4. B.
We are told that arginine is an amino acid, and amino acids are the building blocks of proteins. DNA and RNA are made up of nucleotides, while fatty acids are long chain hydrocarbons that are not considered polymers, and certainly are not formed from amino acids.

5. D.
This question combines aspects of both of the preceding experiments in an attempt to cause confusion. Arginine simply allows the cells to grow, as they could not make proteins without it. However, since neither arginine nor any other amino acids contain phosphorus, no proteins, whether they contain arginine or not, should be labeled. The only remaining choice recognizes that both RNA and phospholipids contain phosphorus, in the phosphate group that each contains as part of its structure.

**Passage 2**

6. C.
The separation of the plates is unchanged, so the electric field increases by a factor of 9.

7. B.
According to the first equation, the electric field and the plate separation are inversely related, so an increase in $d$ results in a decrease in $E$, so that B is the answer. If this is unclear, solve for $E$ in the first equation.

8. B.
For this question we need to remember that a helium nucleus has two protons (its atomic number on the periodic table), so the bare helium nucleus has twice the charge of a proton. Thus the force on it is twice as great.

9. B.
If the separation of the plates increases by a factor of 2, then the electric field decreases by a factor of 2. And the force on the proton decreases by a factor of 2.

10. A.
The voltage $V$ and electric field $E$ are proportional, so A is correct. For C to be correct, the equation would need to be $V = Ed + V_0$.

**Passage 3**

11. B.
Gases best behave the most ideally under conditions such as high temperature and low pressure. Choice D is wrong. The passage makes the behavior of gases clear: They depend on the prevailing temperature and pressure.

12. A.
A value of zero alone will not ensure ideal behavior. So gas A does not behave ideally. Hence item I is correct. Item II contradicts what we just said. We cannot say whether gas A is xenon. So item III is not a valid statement.

13. D.

14. D.
According to the kinetic-molecular theory, all the items are true.

**Passage 4**

15. D.
Choices A and B are irrelevant, and comparisons, such as those in choice C, are meaningless if the units do not match. The correct choice is D. From the table we can see that equal jumps of time (for example, 0.2 s to 0.4 s) result in equal jumps of velocity (0.1 m/s to 0.2 m/s). This is consistent with the statement that $\Delta v/\Delta t$ is a constant, or that acceleration is constant.

16. B.
Gravity certainly acts on *m*. The only thing touching mass *m* is the string, so B is correct. (See figure.)

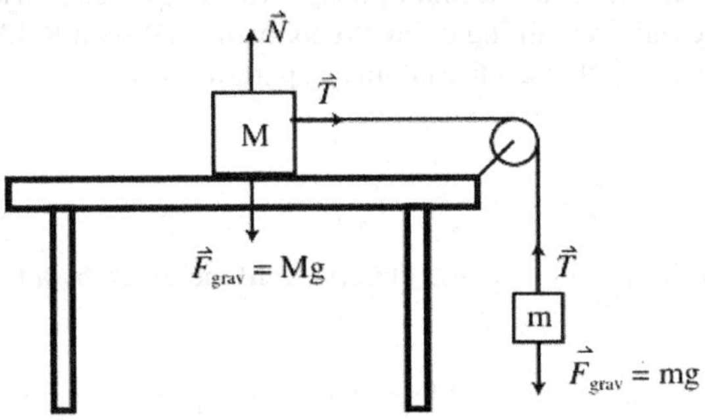


17. B.
The things touching mass *M* are the table and the string. It is true that the tension in the string is, in some sense, caused by mass *m*, but *M* does not know or care what the other end of the string is connected to. It only cares that there is a force due to a string which is directed to the right.

18. C.
By definition, we have $v_{avg} = \Delta x/\Delta t$ = (0.09 m – 0.0 m)/(0.06 s – 0.0 s) = 0.15 m/s.

19. C.
Only the table and the string are touching mass *M*, but the string has no tension in it, so the answer is C.

**Passage 5**

20. A.
We can tell that this is true because the ranges of activity depicted by the graph do not overlap. Therefore, no pH value exists at which both enzymes will be even slightly active. It is assumed that the general pH values are known for the stomach and small intestine, approximately 2 and 8 respectively. Therefore, the graph confirms that pancreatic amylase will have no activity at pH 2 (choice B), but that this is the optimum for pepsin (not 8.5, as is suggested in choice C.) The graph additionally shows that pancreatic amylase loses all activity at pH 7, so that it could not function in an acidic environment. Since the passage tells us that pancreatic amylase functions in the small intestine, choice D must be incorrect.

21. C.
The key here is to understand that trypsin, like pancreatic amylase, functions in the small intestine, and therefore must have a similar optimal pH as does pancreatic amylase. The graph shows that optimum as 8.5. For the same reasons stated above for pepsin and pancreatic amylase, pepsin and trypsin could never work under similar conditions (choice B). Additionally, both choices A and D must be incorrect because they demand that trypsin could function at pH 2 and 0 (the pH of 1M HCl, which you should at least know is acidic).

22. B.
Since humans are endotherms and maintain a constant body temperature of approximately $37^0$ C ($98.6^0$ F), all human enzymes should share this optimal temperature. Curves A and C depict very different optimal temperatures.

23. C.
Since sharks are fish, they are ectothermic and will assume the temperature of their environment. While there is no way for you to know exactly what this value is, it is certainly colder than $37^0$ C, and the optimal temperature of the enzyme shown in graph A, about $12^0$ C (about $50^0$ F), is reasonable for cool water, since you have to pick one choice. D cannot be true; everything that is alive must contain a DNA polymerase!

24. D.
Denaturation implies that the enzyme has lost all function because its shape has significantly changed. According to the graph, enzyme B has no activity below $17^0$ C or above $43^0$ C. Therefore, choice A is incorrect because this is the optimal temperature. Both choices B and C are out of the activity range, but cold temperatures do not cause denaturation, they simply cause a slowing of the reaction rate based on general kinetic principles until it is ultimately zero. Only high temperatures will cause denaturation, so choice D ($50^0$ C) must be correct.

**Passage 6**

25. B.
To answer this question, we need to recall some basic aspects regarding phase diagrams. Just by knowing those, we can eliminate some of the choices right away. In Choices A and D, the phases are wrongly labeled. The phases are correctly labeled only in Choices B and C. This rules out Choices A and D. Water has some anomalous properties. Such properties make water unique. In fact, the solid-liquid segment in the phase diagram of water has a negative slope.

26. B.
See the explanation for the previous question.

27. B.
The question describes a phase change scenario based on the diagram given in the passage. Substance M will be a liquid at point A. If the pressure is decreased, the most likely phase change will be from liquid (the phase before the change) to gas.

28. C.
This is a definition-type question. The point indicated by the X mark is called "triple point." At the triple point of a substance, the three phases coexist at equilibrium. Below this point, the solid to gas conversion or vice versa is direct without the intermediate liquid phase.

29. C.
The critical temperature is defined to be the temperature above which a gas cannot be changed into a liquid. The choice that best reflects this idea is C.

30. A.
In the phase diagram given in the passage, locate the segment XZ. This segment separates the liquid and the gas phases.

**Passage 7**

31. D.
For the student running along the roof, he starts from rest ($v_1 = 0$ m/s) and ends up running $v_2 = 5$ m/s. We have $\Delta x = 5$ m, and we want $\Delta t$. We use $\Delta x = 1/2\ (v_1 + v_2)\Delta t$ to obtain $\Delta t = 2$ s.

32. B.
We know gravity pulls down and the roof pushes up, and these forces add to zero. In addition, there must be a force accelerating the student forward (to the right). Surprisingly, it is the roof which exerts the force forward. His feet push backwards on the roof, and the roof (by the third law of motion) pushes forward on him.

33. A.
We know gravity pulls down. Since nothing else touches the student, there is no other force.

34. B.
All during the fall the student has the same horizontal velocity 5 m/s.

**Passage 8**

35. C.
Ethanol fermentation produces two products from the pyruvic acid made during glycolysis: ethanol and carbon dioxide. Logically, then, the pH at which the most carbon dioxide is produced will be the same pH at which the most ethanol is produced.

36. B.
If aerobic respiration is occurring, it should not affect the net amount of gas present because for every molecule of oxygen that is used up, a molecule of carbon dioxide will be produced. Thus the net production of gas will accurately reflect the carbon dioxide produced by fermentation. The passage clearly states that fermentation and aerobic respiration can occur simultaneously, so choice C must be incorrect. Choice D must be wrong because carbon dioxide is never used up or created by glycolysis.

**Passage 9**

37. B.
Each day the man travels a distance given by the circumference of the Earth $C = 2\pi R_{\text{Earth}}$.

38. D.
Since the man is traveling in a circle at constant speed, his acceleration vector points toward the center of rotation, and so does the net force vector. For this to be so, the magnitude of the gravitational force must be greater than the magnitude of the force of the ground.

39. A.
The gravitational force for the two men is the same. Since the scale reading on a rotating Earth is less than the gravitational force, the correct answer is A.

40. C.
According to Newton's law of gravitation, if the distance in the denominator is less, then the gravitational force is greater.

# Test 1 Conversion Chart

**Scale Score Conversion Chart**

| Raw Score | Scale Score | Percentile Ranking |
|---|---|---|
| 40 | 36 | 99 |
| 39 | 35 | 99 |
| 38 | 34 | 99 |
| 37 | 33 | 99 |
| 36 | 32 | 98 |
| — | 31 | 98 |
| 35 | 30 | 97 |
| 34 | 29 | 95 |
| 33 | 28 | 94 |
| 31–32 | 27 | 91 |
| 30 | 26 | 88 |
| 29 | 25 | 84 |
| 27–28 | 24 | 77 |
| 25–26 | 23 | 71 |
| 24 | 22 | 64 |
| 22–23 | 21 | 56 |
| 20–21 | 20 | 48 |
| 19 | 19 | 38 |
| 17–18 | 18 | 30 |
| 15–16 | 17 | 23 |
| 14 | 16 | 18 |
| 13 | 15 | 14 |
| 11–12 | 14 | 10 |
| 10 | 13 | 7 |
| 9 | 12 | 5 |
| 7–8 | 11 | 3 |
| 6 | 10 | 1 |
| 5 | 9 | 1 |
| 4 | 8 | 1 |
| 3 | 7 | 1 |
| — | 6 | 1 |
| 2 | 5 | 1 |
| — | 4 | 1 |
| 1 | 3 | 1 |
| — | 2 | 1 |
| 0 | 1 | 1 |

# Test 2

(35 Minutes—40 Questions)

**Passage 1**

The Drum of Discomfort is an amusement park ride which consists of a large vertical hollow cylinder which turns on its axis. A person of mass $M$ enters the drum (inside-radius $R$) while the drum is still and stands against the wall. The drum begins to turn, until it achieves uniform rotation with period $T$ and the rider feels as if some force is pushing him against the wall (see figure). Then the floor drops down, so there is nothing touching the bottoms of the rider's shoes.

Assume the coefficient of friction between the rider's clothes and the surface of the drum is $\mu$.

1. During uniform rotation, after the floor drops, what are the forces acting on a rider, besides gravity acting down and a force acting up?
   A. There is a force pointing inward.
   B. There is a force pointing inward and a force pointing in the same direction the rider is moving.
   C. There is a force pointing outward.
   D. There is a force pointing outward and a force pointing in the same direction the rider is moving.

2. After the floor drops, what force provides the centripetal force?
   A. The normal force.
   B. Friction.
   C. Gravitation.
   D. Tension.

3. After the floor drops, which direction does the acceleration vector point?
   A. Toward the center of rotation.
   B. Away from the center of rotation.
   C. In the direction of the rider is moving.
   D. The acceleration is zero.

4. Which gives an expression for the speed $v$ the rider is going?
   A. $R/2\pi T$
   B. $R/T$
   C. $2\pi R/T$
   D. $2\pi T/R$

5. What is the magnitude of the upward force on the rider?
   A. $Mg$
   B. $\mu Mg$
   C. $Mv^2/r$
   D. $\mu Mv^2/r$

**Passage 2**

Since only four types of nucleotides (adenine, guanine, cytosine and thymine) make up DNA, many scientists were skeptical that this was the hereditary molecule of life. Most believed proteins, with their building blocks of 20 amino acids, provided the complexity necessary to carry the genetic information.

In 1952, Alfred Hershey and Martha Chase designed an experiment to determine whether DNA or protein was the genetic material. They used a *bacteriophage* (a virus that infects bacteria) called T2. This bacteriophage was only composed of proteins and DNA. A short time after infecting a bacterial cell, the cell would *lyse* (break open) and release new T2 particles. T2 reprogrammed the cell to make more phage, but did it use DNA or proteins?

Hershey and Chase infected bacteria with T2 phage in growth medium containing radioactive sulfur (which labels proteins) and radioactive phosphate (which labels DNA). The resulting radioactive phages were then incubated with nonradioactive bacteria and allowed to infect these cells. After a short time, the mixture was placed in a blender to shake loose any phage particles remaining outside of or attached to the bacteria. The mixture was then centrifuged: bacterial cells would form a *pellet* at the bottom of the tube while any phage outside the cell would remain in the liquid portion (the *supernatant*). The radioactivity in each sample was measured.

6. Radioactive sulfur was found exclusively in the supernatant fraction, while the radioactive phosphate was found in the pellet fraction. This indicated:
   A. phage protein entered the cell, but the DNA did not.
   B. phage DNA entered the cell, but protein did not.
   C. both protein and DNA entered the cell.
   D. the phage did not infect the bacterial cell.

7. If the bacterial pellet were incubated in an appropriate growth medium, what would most likely occur?
   A. No phage would be produced as the protein coat was removed.
   B. Bacteria would survive as there is no proof they were infected with phage.
   C. Phage should be produced and kill the cells.
   D. The radioactive phosphate would kill the cells immediately.

8. A possible alternative to the blender technique used in this experiment would be:
   A. add a competitive inhibitor to the bacteria before adding phage.
   B. break apart the cell membranes with a strong detergent.
   C. incubate bacteria with a mutant phage that could not bind to the cells.
   D. shake the cell/phage mixture vigorously.

9. A modern experiment that confirms this experiment would be:
   A. injecting DNA into a cell which causes the cell to produce an enzyme it never had before.
   B. fertilizing an egg with a sperm in vitro.
   C. prescribing enzyme pills to help cystic fibrosis patients with digestion.
   D. consuming radioactive barium to view internal organs with X-rays.

**Passage 3**

The concentration of solutions are generally expressed in terms of molarity, molality, normality, and weight percent. The formation of the solutions itself has many ramifications. The solubility of solutes differ considerably from one other. Some of the factors that can influence the solubility include temperature and pressure. Solubility depends on other factors as well. Given below in Figure-1 is a graph which depicts solubility differences of some solutes.

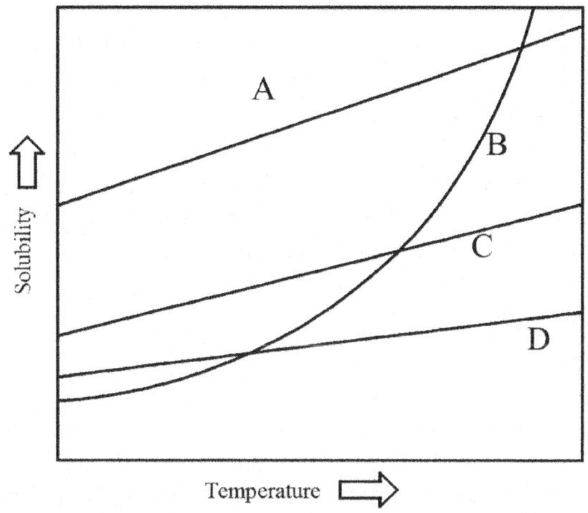


**Figure-1**

Quite often, the freezing and boiling point changes that are brought about by the dissolved solutes can be predicted reasonably. But this is not always the case.

The predictions and calculations are done for freezing point depression ($\Delta T_f$) based on the following formula:

$$\Delta T_f = K_f\, m,$$

where m is the molality and $K_f$ is the freezing point depression constant. ($K_f = 1.86\ ^{\circ}C/m$)

For boiling point elevation ($\Delta T_b$), the calculations are based on the following formula:

$$\Delta T_b = K_b\, m,$$

where m is the molality and $K_b$ is the boiling point elevation constant. ($K_b = 0.512\ ^{\circ}C/m$)

10. The graph in the passage shows the solubilities of some compounds (salts). Which of the following is most likely true regarding the Compounds A, B, C and D?
    A. The solubility process of Compound B is exothermic, and those of Compounds A, C and D are endothermic.
    B. The solubility process of Compound B is endothermic, and those of Compounds A, C and D are exothermic.
    C. The solubility processes of Compounds A, B, C and D are exothermic.
    D. The solubility processes of Compounds A, B, C and D are endothermic.

11. Two experiments were conducted for analyzing the solubility properties and their effects on colligative properties. In Experiment I, a 1.0 m solution of KBr was analyzed and in Experiment II, a 2.3 m solution of KBr was analyzed. The experimental values of freezing and boiling points were slightly different from the expected values based on the theoretical calculations discussed in the passage. Which of the following is most likely correct regarding the two experiments?
    A. The experimental values of the solution in Experiment I was more different from the theoretical calculations than the solution in Experiment II.
    B. The experimental values of the solution in Experiment II was more different from the theoretical calculations than the solution in Experiment I.
    C. Both experiments have the same extent of differences from the theoretical predictions.
    D. The discrepancy noted in the experiments is absolutely a result of instrumental error, because the differences in Experiments I & II cannot change colligative properties.

12. A solution was made by using 315 g of glucose in 750 g of water. The boiling point of this solution is 101.2°. If the amount of glucose used in this solution is doubled, while the amount of water stays the same, what must happen to the boiling point of the solution?
    A. It will double.
    B. It will quadruple.
    C. It will decrease slightly.
    D. It will increase slightly.

13. Which of the following statements is true regarding solutions?
    A. As the concentration of a solution increases, the vapor pressure increases.
    B. As the concentration of a solution decreases, the vapor pressure decreases.
    C. As the concentration of a solution increases, the vapor pressure decreases.
    D. None of the above

**Passage 4**

Engineers often make scale models of structures they plan to build in order to test function and stability. Sometimes, however, structures fail even when the models function, so engineers have developed extensive theory in order to determine how to build proper scale models and extrapolate reliable results from them.

At first we might assume that a model made of the same material as the intended final structure with each dimension scaled by a single factor will accurately reproduce the behavior of the final structure. That this is not so was known in antiquity by tragic observation, and it was first explained by Galileo around AD 1600. We will not present his detailed argument but will sketch some of the conclusions.

To summarize Galileo's conclusion on this point, when the linear dimensions of a structure are all increased by a factor, the load across any surface increases by the cube of that factor, whereas the strength, or the maximum force the structure can hold across any surface increases by the square of that factor. Therefore, as a structure gets larger, it tends to become unstable, more susceptible to failure.

To illustrate the point, let's consider a block of metal connected to a cylinder, which has much greater length than its diameter and is connected to the ceiling (part A in figure). The *stress* in the cylinder is the force per area across a cross section. Each material has a threshold stress, such that stress larger than the threshold causes the material to fail. If all the linear dimensions are increased by a factor (part B in figure), then the volume of the block increases by the cube of the factor, as well as the mass and weight of the block. The cross-sectional area of the cylinder increases by the square of the factor, so the stress increases as the factor itself.

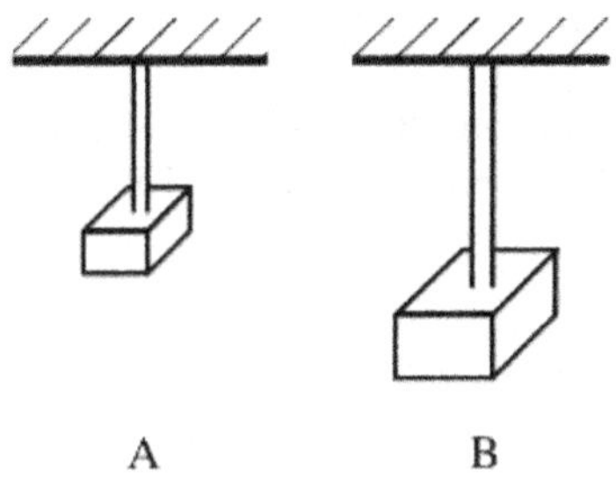


This is the simplest example of the subtlety involved in model building.

14. The figure below depicts a human biceps, which when flexed, has a cross-sectional area at its center of $5 \times 10^{-3}$ m$^2$ and cross-sectional area at the forearm of $5 \times 10^{-5}$ m$^2$. If the force exerted at the forearm is $F_0$, what is the force exerted at the shoulder?

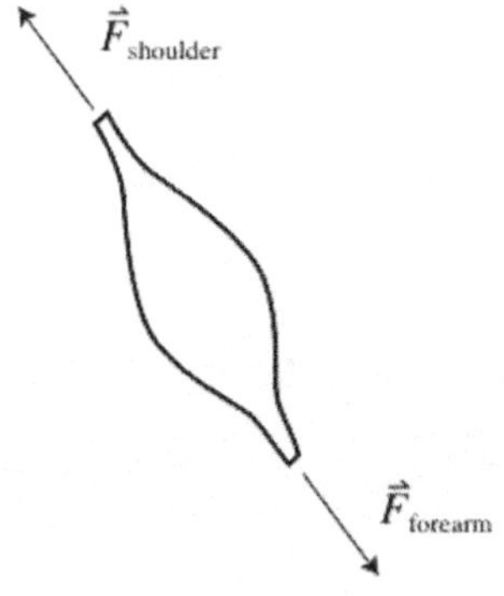


A. $F_0/100$
B. $F_0$
C. $100F_0$
D. $10^4 F_0$

15. Referring to the previous question, if the stress at the center of the biceps is $10^5$ Pa, what is the stress at the forearm? Note: Stress, force, and area are related by the equation

$$Stress = F/A$$

A. $10^3$ Pa
B. $10^5$ Pa
C. $10^7$ Pa
D. $10^9$ Pa

16. The figure below shows two elephant statues which are the same shape, both made of plaster of paris. Statue B has twice the linear dimension of statue A. If statue A weighs 40 N, how much does statue B weigh? Note: $mass = \rho V$, where $\rho$ is the density and $V$ is the volume. And $weight = mg$, where $m$ is the mass and $g$ is gravity.

A B

A. 48 N
B. 80 N
C. 160 N
D. 320 N

17. In the previous question, how does the pressure (stress) exerted by the right front foot of statue B compare with the pressure exerted by the corresponding foot of statue A?
A. It would be the same.
B. It would be twice as large.
C. It would be four times as large.
D. It would be eight times as large.

18. A lantern is hanging from a cable of negligible mass. The cable breaks when the weight of the lantern exceeds $W_{max}$. What would be the breaking weight for a similar cable (same material, same cross section) which was 10 times longer?
A. $W_{max}/100$
B. $W_{max}/10$
C. $W_{max}$
D. $10W_{max}$

**Passage 5**

The activity of bacteriophages is easily assayed. Bacteria are spread on a solid nutrient agar plate. Under appropriate growth conditions, the bacteria will form a cloudy layer completely covering the dish. In microbiology, this is called a *lawn*, as opposed to isolated spots, or *colonies*, of bacteria. If bacteriophages are introduced to the lawn of bacteria, the phages will infect single cells. Once the phages have reproduced, they will lyse the cells and the progeny phage will infect the surrounding bacterial cells. In a relatively short period of time, the phages will have lysed all the bacteria within a radius of the initial infection, causing a clear, circular spot to form in the lawn of bacteria. This is called a *plaque*.

A researcher working with bacteriophages set up the following experiment:

*Condition 1*:
Bacteria were spread on agar plates and incubated at 37°C for 24 hours. A lawn was clearly visible and no plaques formed.

*Condition 2*:
The same strain of bacteria was mixed with a preparation of lambda bacteriophage. The mixture was spread on agar plates and incubated for 24 hours at 37°C. An average of twenty plaques formed on the bacterial lawn on each plate.

*Condition 3*:
The same strain of bacteria was spread on agar plates and then exposed to UV light. After incubation at 37°C for 24 hours, lawns grew and approximately fifty plaques formed on each plate.

19. After incubating the plates from condition 2 for an additional 5 days, the plates became totally clear. What is the most likely explanation for this result?
    A. Newly produced phages infected all bacteria on the dish, thus killing everything.
    B. The bacteria died off due to lack of nutrients.
    C. The bacteria protected themselves from the phages by slowing their growth.
    D. The phage forced all the bacteria to go into the lysogenic cycle and stop growing.

20. What happened in condition 3?
    A. Phage contaminated the dishes causing infection and lysis of the bacteria.
    B. The nutrients in the agar were insufficient to support bacterial growth.
    C. The bacteria were probably a lysogenic strain induced by UV light.
    D. The bacteria were not healthy and died off when exposed to UV light.

21. If a fourth experiment was done, where bacteria were mixed with lambda and exposed to UV light, what would you predict?
    A. The results would be similar to condition 2, with approximately 20 plaques per plate.
    B. The results would be similar to condition 3, with approximately 50 plaques per plate.
    C. There would be approximately 35 plaques per plate.
    D. There would be approximately 70 plaques per plate.

22. Before the phage life cycle was understood, researchers called the results from condition 3 *autolysis*. They thought an enzyme in the bacteria caused its own destruction and the destruction of surrounding cells. Which of the following would lend support to the viral theory and help to disprove autolysis?
    A. Within a week, all the bacterial cells in condition 3 die.
    B. Viral particles can be purified from the plates in condition 3.
    C. Exposure of bacteria to X rays also causes plaques to form.
    D. With or without exposure to UV light, no bacteria grow at 4°C.

**Passage 6**

In nuclear reactions, significant changes occur in the composition of the nuclei of the atoms involved. These reactions usually release tremendous amounts of energy. One of the reasons for the nuclear changes can be attributed to the stability of the nucleus.

The formation of nucleus from the subatomic particles - neutrons and protons, results in the release of energy. The mass of these individual particles in the nucleus is greater than that of the actual nucleus that is formed. This loss of mass is due to the change of mass into energy. The energy-mass relation can be represented in terms of the equation:

$$E = mc^2,$$

where m represents the mass, and c represents the speed of light ($3 \times 10^8$ m/s).

If the nucleus of an atom is not stable, it can get transformed into another nucleus. A plot of the number of neutrons versus the number of protons is often used to assess the stability trends of elements. If the number of protons and neutrons are equal, the nucleus is considered to be reasonably stable. As the atomic number increases, the trend changes.

Isotopes of elements having atomic numbers above ≈83 are unstable atoms. These unstable atoms can undergo disintegrations. The half-lives of some radioactive elements are shown in Table 1.

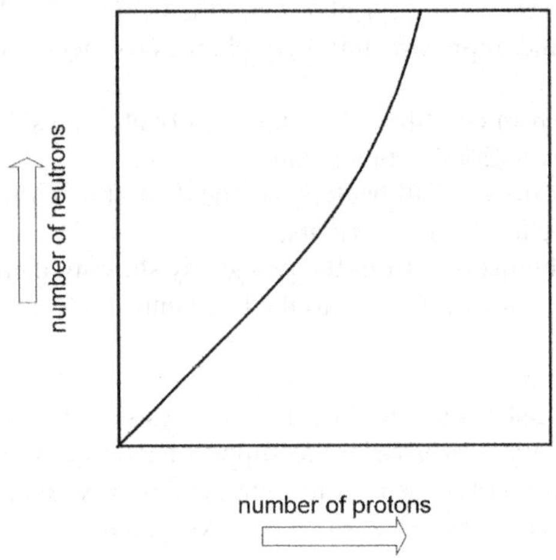


**TABLE 1**

| *atom* | *half-life* |
|---|---|
| Carbon 14 | $5.73 \times 10^3$ years |
| Uranium 238 | $4.47 \times 10^9$ years |
| Radon 222 | 3.82 days |
| Radium 226 | $1.6 \times 10^3$ years |
| Krypton 89 | 3.2 minutes |

23. A sample of $^{226}$Ra disintegrates to 3% of its original quantity. How many half-lives have this sample passed? Note: Half-life is the amount of time required for a radioactive substance to disintegrate to half its initial quantity.
   A. two
   B. three
   C. four
   D. five

24. Which of the following is true regarding radioactivity?
    A. As the atomic number increases, eventually the neutron-proton ratio values become $\leq 1$.
    B. As the atomic number increases, eventually the neutron-proton ratio values become $\geq 1$.
    C. As the atomic number increases, eventually the proton-neutron ratio values become $\geq 1$.
    D. None of the above

25. When a helium nucleus is formed, there is always some degree of loss of mass. If the loss of mass equals $3.1 \times 10^{-5}$ kg during the formation of one mol of it, what is the binding energy? Note: Mass loss can be calculated using the equation $E = mc^2$.
    A. $3.1 \times 10^{-5}$ J
    B. $1.8 \times 10^{19}$ J
    C. $9.3 \times 10^{3}$ J
    D. $2.8 \times 10^{12}$ J

**Passage 7**

A rocket engine operates on the principle that hot gas is expelled backwards through a nozzle in order to produce a thrust on the ship in the opposite direction. Since momentum is conserved in this operation, we can derive the result that the effective force on the ship is

$$F_{\text{effective}} = Mu$$

where $M$ is the mass expulsion rate, and $u$ is the exhaust velocity relative to the ship. Thus it is important for both $M$ and $u$ to be high. The exhaust velocity varies as the square root of the ratio of the temperature of the combustion chamber and the molecular mass of the exhaust.

In conventional rocket engine design, large fuel tanks carry liquid hydrogen and oxygen, and these react by chemical combustion to yield water vapor. The water vapor, due to its high temperature, shoots out of the nozzle, and the ship is thrust forward.

In an experimental engine design, nuclear fission of uranium is used to heat a supply of hydrogen to high temperatures (around 2200 K). The hydrogen is then expelled through a nozzle at $1.0 \times 10^4$ m/s, about twice the exhaust velocity as that for conventional rockets. One major engineering problem involves the heat exchange between the hydrogen gas and the site where the nuclear reaction takes place. Engineers are improving the design so that the hydrogen is heated at a faster rate than it is in current designs.

26. When a rocket ship expels gas in order to produce a thrust, this is an example of
    A. the first law of motion: If the forces on an object are balanced, then the object moves with constant velocity.
    B. the second law of motion: The force acting on an object is equal to the mass of that object times its acceleration ($F = ma$).
    C. the third law of motion: To every action there is an equal and opposite reaction.
    D. the law of universal gravitation: Any two bodies attract each other with a force that is directly proportional to the product of their masses and inversely proportional to the square of the distance between them ($F = Gm_1m_2/r^2$).

27. Could heavier neon gas work instead of hydrogen in the design of the nuclear engine?
    A. No, since neon is not a product of uranium fission.
    B. No, since neon is an inert gas.
    C. Yes, but the engine would not be as efficient because of exhaust velocity.
    D. Yes, it would work approximately as well.

28. What, according to the passage, is a disadvantage of nuclear engines compared to conventional engines?
    A. The molecular mass of the exhaust is too low.
    B. The molecular mass of the exhaust is too high.
    C. Some of the energy is lost as heat.
    D. The mass expulsion rate is too low.

29. A rocket ship is going forward at 2000 m/s and fires its engines in order to speed up but not turn. If the absolute velocity of the exhaust gases is 3000 m/s going backwards, what is the exhaust velocity $u$ relative to the ship?
    A. 1000 m/s
    B. 2000 m/s
    C. 3000 m/s
    D. 5000 m/s

30. We can model a rocket and its exhaust with two carts sitting on a level, frictionless one-dimensional track. The large cart (rocket) has a mass 10 kg, and the small cart (exhaust gas), 0.1 kg. There is a small explosive between them. At a certain time the explosive goes off and the two carts go flying apart. The less massive cart recoils with velocity 20 m/s. The explosion is over in about 0.2 s. What is the final velocity of the larger mass? Note: Use the principle of conservation of momentum: $P_{\text{before}} = P_{\text{after}}$.
A. 0.2 m/s
B. 1.0 m/s
C. 20 m/s
D. 2000 m/s

**Passage 8**

Interestingly, many, if not most, bacterial cells can communicate with other cells in the same species. The bacteria secrete proteins, and, once enough bacteria are present, the concentration of the proteins increases. A high concentration often triggers the bacteria to turn on other proteins so the colony can perform a function. For example, individuals who have impaired immune systems or who have the genetic disease *cystic fibrosis* (*CF*) are often infected by the respiratory bacterium, *Pseudomonas aeruginosa*. Only when enough bacteria have accumulated, *P. aeruginosa* will produce an enzyme that degrades lung tissue, which allows the bacteria to invade the blood stream.

Although many examples of this intraspecies communication exist, scientists wondered if different species can communicate with each other. To answer this question, researchers examined the relationship between *P. aeruginosa* and *Burkholderia cepacia*. *B. cepacia* causes fatal lung infections in CF patients, but only after these individuals have also been infected by *P. aeruginosa*.

*Experiment 1*:
*P. aeruginosa* were grown in an appropriate liquid medium in the laboratory. The culture was centrifuged to remove the bacterial cells. A culture of *B. cepacia* was then grown in the medium. These bacteria increased production of molecules necessary for survival.

*Experiment 2*:
Mutant *P. aeruginosa* were grown in the laboratory in liquid medium. After centrifugation, the medium was used to incubate cultures of *B. cepacia*. Very few "survival molecules" were produced in the *B. cepacia*.

31. The experiments indicate:
    A. the two bacterial strains can communicate with each other.
    B. the *P. aeruginosa* bacteria help turn on production of survival molecules in *B. cepacia*.
    C. *B. cepacia* bacteria help turn on production of survival molecules in *P. aeruginosa*.
    D. no communication exists between the two bacterial strains.

32. The key to communication between these two bacteria is probably due to:
    A. a soluble protein secreted by *P. aeruginosa*.
    B. direct interaction between the two bacteria, possibly via cells of each species binding to each other.
    C. a soluble protein secreted by *B. cepacia*.
    D. unable to determine based on the available data.

33. A similar communication system exists between:
    I. nerve cells
    II. hormones and receptors
    III. photoreceptors in the eye
    A. I only
    B. II only
    C. I and II only
    D. III only

34. One important control that should be included in this experiment is:
    A. grow *P. aeruginosa* in medium used to grow *B. cepacia* first.
    B. infect mice with both species of bacteria.
    C. grow *B. cepacia* in medium that had not been used to grow *P. aeruginosa*
    D. no other control is necessary for this experiment.

35. The mutant *P. aeruginosa* used in Experiment 2 were most likely deficient in:
    A. replication
    B. transcription
    C. translation
    D. secretion

36. If the doubling time of a strain of bacteria is 30 minutes, approximately how many cells would there be after 5 hours if the original culture had $1 \times 10^5$ cells?
    A. $5 \times 10^5$
    B. $3.2 \times 10^6$
    C. $1 \times 10^8$
    D. $1 \times 10^{25}$

**Passage 9**

In a certain experiment, a piston chamber is used as part of a primitive engine. The apparatus consists of a pipe closed at one end with a piston at the other end. A valve in the cylinder allows fuel gases to be introduced or waste gases to be expelled.

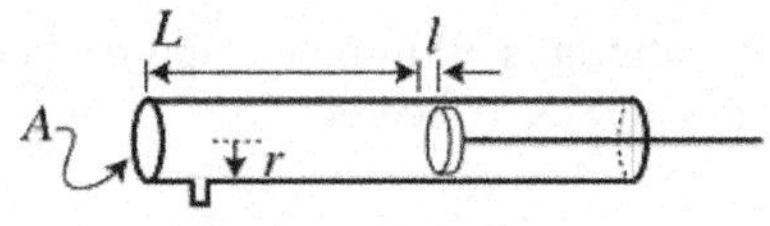


In the operation of this engine, hydrogen and oxygen are introduced in a 2:1 ratio (in order to ensure complete combustion) at ambient temperature $T_{amb}$ and atmospheric pressure $P_{atm}$. The following reaction is ignited

$$2H_{2(g)} + O_{2(g)} \rightarrow 2H_2O_{(g)}$$

with a heat of reaction

$$\Delta H_{rea} = -4.8 \text{ x } 10^5 \text{ J/mol}$$

The pressure rises to $P_{burn}$.

Next the piston slowly moves back a distance $l$, from which the engine derives useful work. The distance $l$ is short enough that the pressure and temperature inside the chamber remain roughly constant.

The waste gases are then expelled, and the piston is restored to its original position.

The radius of the cylinder is $r$, and the cross-sectional area is $A$. The length of the cylinder before the piston moves back is $L$, which is much larger than $l$. The number of moles of oxygen introduced is $n$.

37. After the combustion occurs, why does the pressure go up?
    A. There are more gas particles on the left side of the reaction.
    B. The temperature rises considerably.
    C. The reaction is spontaneous.
    D. The heat of reaction is negative.

38. What would happen if the ratio in the second paragraph were not 2:1?
    A. The heat of reaction would be less than 4.8 x $10^5$ J/mol.
    B. The combustion would not ignite.
    C. Some of the waste gas would be intermediate products of incomplete combustion.
    D. Some of the waste gas would be oxygen or hydrogen.

39. If the reaction shown were performed in a closed chamber isothermally at 500°C , what would happen to the pressure?
    A. The pressure would decrease.
    B. The pressure would stay the same.
    C. The pressure would increase.
    D. There is not enough information to solve this problem.

40. What would be the consequence of making $l$ larger? During the piston movement,
    A. pressure in the chamber would decrease, and temperature would decrease.
    B. pressure in the chamber would decrease, and temperature would increase.
    C. pressure in the chamber would increase, and temperature would decrease.
    D. pressure in the chamber would increase, and temperature would increase.

# Test 2 Solutions

**Passage 1**

1. A.
We draw a force diagram including the force of gravity.

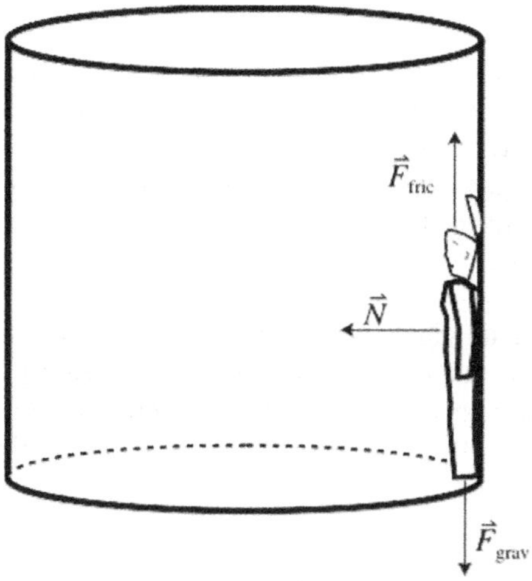


The only thing touching the rider once the floor drops is the side of the drum, which exerts a normal force that ends up being toward the axis of rotation. It also exerts a frictional force, which is up (balancing gravity). For uniform circular motion, we know the net force must be toward the center of rotation, so we can see that this force diagram is complete.

2. A.
From the diagram we can see that the normal force provides the centripetal force.

3. A.
Because the motion is uniform circular rotation, the acceleration vector points toward the center of rotation.

4. C.
The rider traverses the circumference of the circle ($2\pi R$) during each period of time $T$. Thus his speed is $2\pi R/T$.

5. A.
The upward force must balance the gravitational force, so the magnitude of the upward force must be $Mg$ as well.

**Passage 2**

6. B.
Detection of radioactive phosphate in the pellet indicates the bacteria now contain the DNA of the bacteriophage. Radioactive sulfur in the supernatant suggests the protein of the phage did not enter the cell. Since the bacteria took up something from the phage, answer D is unlikely, as it does appear the phage infected the cell.

7. C.
The experiment verified that the DNA of the phage entered the cell. Therefore, the bacteria were indeed infected. It is not likely the protein coats are necessary for further functioning of the phage as they never entered the cell. It is also unlikely the radioactive phosphate would immediately kill the cells (after all, phage were propagated in bacteria grown in medium containing the isotope). The most likely answer is that the phage would reproduce as normal.

8. D.
Answer D is the only choice that makes any sense. All the other answers defeat the purpose of the experiment: a competitive inhibitor would prevent the phage from entering the cells at all; breaking apart the cell membrane would cause the contents of the cell to be released, where they would be found in the supernatant, along with the components of the phage that did not infect the cell; a mutant phage that could not bind to the cell would probably not infect the cell.

9. A.
Bacteriophages work by making the target cell do something it could not do before. Fertilization results in a new type of cell and is the beginning of a developmental cycle, not the best analogy to the phage experiment. Consuming enzymes would not fundamentally change an organism. Although the T2 experiment used radioactivity, not all experiments involving radioactive isotopes examine the same phenomenon.

**Passage 3**

10. D.
All the solubilities shown in the graph are endothermic. In fact most compounds (solids) dissolve endothermically. The graph shows that as the temperature increases, the solubility of the salts increases. In other words, as the temperature increases, more and more of the salt present is dissociated into ions. This is exactly how an endothermic process works.

11. B.
The observed changes in colligative properties are less than what was theoretically predicted. This is because the ionization of salts in water is not as complete as we expect. Some of the oppositely charged ions stick together and function as a unit. So the number of particles (units) is lower than what we expect. As the concentration increases, the possibility of the oppositely charged ions sticking together is greater than that in dilute solutions.

12. D.
This question is a bit tricky. You may have selected Choice A. But you have to understand what exactly is asked in the question. The question is not asking about the change in boiling point ($\Delta T$). Instead, it asks what happens to the boiling point. It will not double; it will only increase slightly.

13. C.
Choice C is the only true statement given. Looking closely at Choices A and B, you should realize that they represent the same relationship. Since you cannot have two correct answers, this rules out Choices A and B. Thus, such logical elimination-techniques can be used to solve many problems on the test and save valuable time.

**Passage 4**

14. B.
Don't let the information in the problem distract you from the fact that there must be a force balance on the muscle, since the muscle is not accelerating. Thus the magnitudes of $F_{forearm}$ and $F_{shoulder}$ are the same.

15. C.
The force exerted at the center of the biceps is the same as the force exerted at the forearm (force balance, see previous question), and the area is 100 times smaller. Thus the stress at the forearm is 100 times larger. (You may have done this problem by an intermediate step by calculating that force to be 500 N.)

16. D.
Statue B has 8 times the volume of statue A. Since $mass = \rho V$, statue B has 8 times the mass. Since $weight = mg$, statue B has 8 times the weight as well.

17. B.
Stress is force/area. The force (weight) increases by a factor of 8, and the area increases by a factor of 4, so the stress increases by a factor of 8/4 = 2. Also, see the last sentence of the fourth paragraph of the passage.

18. C.
The passage indicates that the breaking point is related to the threshold stress for the material in the cable. Stress is force per area. Increasing the length does not change the force of the load (the lantern weight) or the cross-sectional area, so the breaking weight would be the same.

**Passage 5**

19. A.
All the bacteria were infected by the phage and lysed. The other conditions, whether true or not, deal with growth or death of bacteria and would not cause lysis of the bacteria. Therefore, the plate would remain cloudy. Only lysis causes the plates to appear clear.

20. C.
The control in this experiment (condition 1) rules out answers A and B. If UV light killed the bacteria, as suggested in D (and as can happen, depending on conditions), then all the bacteria would die. (Although it is possible that only a small percentage would die, it would be highly unlikely that this lethality would show a plaque pattern.) C is the only plausible answer, and we know that UV light can induce lysogenic strains to become lytic.

21. D.
Under both conditions, infection and induction, lysis would occur. Therefore, on average for these experimental conditions, infection would produce 20 plaques per plate and induction would produce 50 plaques per plate, for a total of 70 plaques per plate.

22. B.
Once induction occurs, viral particles are produced that can go on to infect other bacterial cells. Therefore, the isolation of bacteriophage from cultures would not support the theory of autolysis. Answer A is possible in the realm of either theory, C would not distinguish between the two, and D has nothing to do with either theory, as bacteria that grow well at 37°C will not grow at 4°C.

**Passage 6**

23. D.
The sample described in the question has disintegrated to 3% of its original quantity. The disintegration can be analyzed as follows:

100% ⇨ 50% ⇨ 25% ⇨ 12.5% ⇨ 6.25% ⇨ 3.125%

This means, at least 5 half-lives must have elapsed.

24. B.
According to the figure given in the passage, as the number of protons increases the neutron-proton ratio increases and becomes greater than or equal to one. (Clue: Look at the slope)

25. D.
The mass loss can be equated with the binding energy. In essence, mass loss can be calculated using the equation $E = mc^2$.

$$\text{Energy} = (3.1 \times 10^{-5}\ \text{kg})(3 \times 10^{8}\ \text{m/s})^2 = 2.8 \times 10^{12}\ \text{kg. m/s}^2 = 2.8 \times 10^{12}\ \text{J}.$$

**Passage 7**

26. C.
Both the first and second laws of motion concern unbalanced forces. But the third law of motion states that, if the ship exerts a force on the gas, then the gas exerts an equal and oppositely directed force on the ship, so C is correct.

27. C.
Choice A is a true statement: Neon is not a product of uranium fission, but neither is hydrogen. The passage says that hydrogen is heated and then expelled. This is different from conventional rockets in which the products of the reaction themselves are expelled. Choice B is a true statement but also irrelevant, because the hydrogen does not react chemically in this process either. Concerning choice C, let's think of exhaust velocity. It is related to temperature and molecular mass of the exhaust gas. Since neon is more massive, the exhaust velocity will be less, and the thrust will be less. Choice C is correct.

28. D.
According to the passage, the hydrogen cannot be heated fast enough. This would result in a low mass expulsion rate.

29. D.
It helps to visualize this problem if we draw a diagram (see figure).

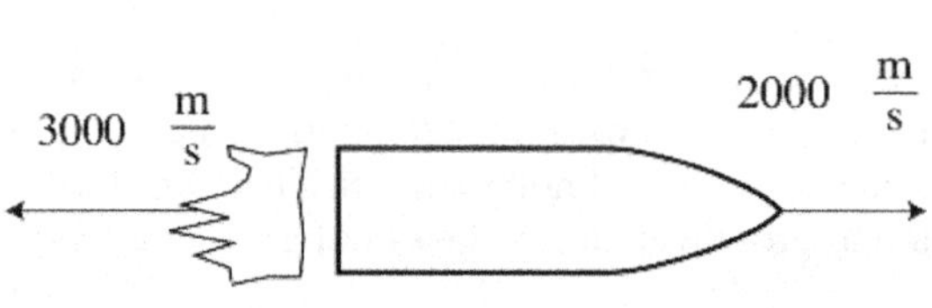


From the figure we can see that, relative to the ship, the gases are going 5000 m/s.

30. A.
Let's draw a diagram of the system.

before:

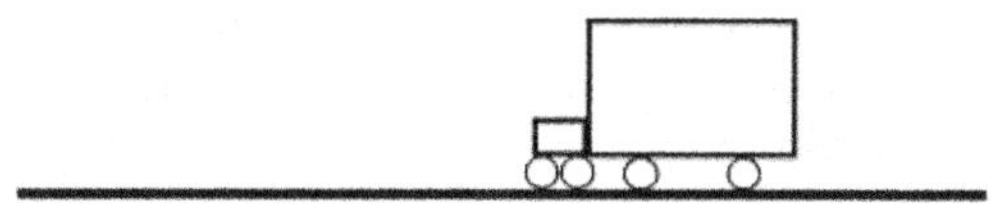

after:

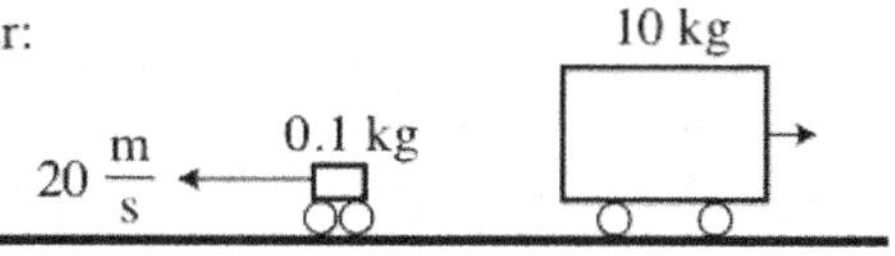


We can do this by momentum conservation:

$$P_{\text{before}} = P_{\text{after}}$$

$$0 = (0.1 \text{ kg})(-20 \text{ m/s}) + (10 \text{ kg})v_t$$

$$v_t = 0.2 \text{ m/s}$$

**Passage 8**

31. B.
The experiment shows that a soluble factor released by *P. aeruginosa* is probably present in the supernatant (the medium which was then used to grow *B. cepacia*). This factor induced the production of survival molecules in *B. cepacia*. Therefore, we know that *P. aeruginosa* can communicate with *B. cepacia*. We are given no experimental evidence to indicate if *B. cepacia* is able to communicate with *P. aeruginosa* .

32. A.
Since *B. cepacia* were grown in the liquid medium after the *P. aeruginosa* had been removed by centrifugation, a soluble factor, probably a protein, must have been present. No direct cell-to-cell contact could have taken place. In addition, we have no direct evidence that *B. cepacia* secreted any proteins.

33. C.
Nerve cells communicate chemically by releasing factors into the synapse between each other. Hormones are released from one gland and travel to a target tissue or organ where they are detected by receptors. Both systems use communication systems similar to the release of protein from one bacterial cell and detection by another.

34. C.
It is important to control this experiment. This can be accomplished by incubating *B. cepacia* in medium that had not been used to grow another strain of bacteria. In this manner, we could directly see the effects of *P. aeruginosa* on *B. cepacia*.

35. D.
The most likely answer to this question is that the *P. aeruginosa* were defective in secretion. If they had mutations in any other system, they would be unlikely to survive and reproduce.

36. C.
After five hours, there would be ten doublings since the doubling time is 30 minutes. Initially, there are $10^5$ cells. So, after 30 minutes, there are $2 \times 10^5$ cells. After 1 hour, there are $2 \times 2 \times 10^5$ cells. After 10 iterations, we get

$$2 \times 2 \times \cdots \times 2 \times 10^5 = 2^{10} \times 10^5 = 1024 \times 10^5 \approx 10^3 \times 10^5 = 10^8$$

**Passage 9**

37. B.
It is true that there are more particles on the left side of the reaction, but that would tend to make the pressure go down, not up, so A is false. The pressure goes up because the temperature goes up, so B seems a good choice. Spontaneous reactions can have either an increase or a decrease in pressure, so C is incorrect. D is incorrect for the same reason.

38. D.
One of the reactants would be in excess. The heat of reaction is unchanged, so A is false. The combustion would still ignite, so B is out. The answer is C or D. If any of the intermediates in the reaction were stable compounds, then some of those compounds could end up in the waste gas. But the reaction of hydrogen with water is clean, and the waste gas would contain only leftover hydrogen or oxygen.

39. A.
If the reaction were performed isothermally, then the number of gas particles would be less after the reaction than before the reaction. Since pressure at a given volume and temperature is proportional to the number of gas particles by the ideal gas law, the pressure would decrease.

40. A.
We have assumed that the piston movement is small and the pressure stays about the same. If the piston movement were larger, the increase in volume would decrease the pressure. The gas in the chamber is doing work against the piston, so the internal energy of the gas must decrease. The temperature must decrease as well.

# Test 2 Conversion Chart

**Scale Score Conversion Chart**

| Raw Score | Scale Score | Percentile Ranking |
|---|---|---|
| 40 | 36 | 99 |
| 39 | 35 | 99 |
| 38 | 34 | 99 |
| 37 | 33 | 99 |
| 36 | 32 | 98 |
| — | 31 | 98 |
| 35 | 30 | 97 |
| 34 | 29 | 95 |
| 33 | 28 | 94 |
| 31–32 | 27 | 91 |
| 30 | 26 | 88 |
| 29 | 25 | 84 |
| 27–28 | 24 | 77 |
| 25–26 | 23 | 71 |
| 24 | 22 | 64 |
| 22–23 | 21 | 56 |
| 20–21 | 20 | 48 |
| 19 | 19 | 38 |
| 17–18 | 18 | 30 |
| 15–16 | 17 | 23 |
| 14 | 16 | 18 |
| 13 | 15 | 14 |
| 11–12 | 14 | 10 |
| 10 | 13 | 7 |
| 9 | 12 | 5 |
| 7–8 | 11 | 3 |
| 6 | 10 | 1 |
| 5 | 9 | 1 |
| 4 | 8 | 1 |
| 3 | 7 | 1 |
| — | 6 | 1 |
| 2 | 5 | 1 |
| — | 4 | 1 |
| 1 | 3 | 1 |
| — | 2 | 1 |
| 0 | 1 | 1 |

# Test 3

(35 Minutes—40 Questions)

**Passage 1**

When we place a straw (a hollow cylindrical tube) in water, the water inside the straw rises above the surface level outside and a meniscus (curvature of the surface) forms. Consider the column of water in the straw from the height of the water outside the straw to the top of the column. (See figure, where the column is shown shaded.) Let $P_1$ be a point in the straw at the bottom of the column and let $P_2$ be a point inside the water column at the top.

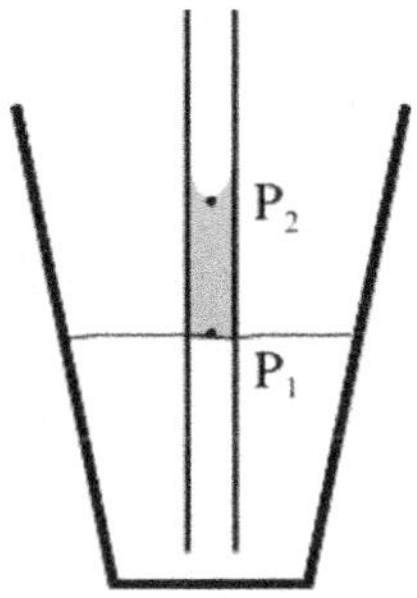


The force due to surface tension for a maximally stretched surface is given by $F_{surf} = \gamma L$, where $\gamma$ is the coefficient of surface tension (which depends only on the substance) and $L$ is the length of the line of contact between an object and the fluid. The pressures at two points separated vertically by height $h$ are related by $P_2 = P_1 + \rho gh$. For this problem use the following:

$\rho$ = density of water
$\gamma$ = surface tension of water
$r$ = radius of the straw
$h$ = height of the column
$P_{atm}$ = atmospheric pressure
$g$ = acceleration due to gravity

1. Which pressure is greater, the pressure at $P_1$ or at $P_2$?
   A. Pressure at $P_1$ is greater than at $P_2$ by the term $\rho gh$.
   B. Pressure at $P_1$ and at $P_2$ are both $P_{atm}$.
   C. Pressure at $P_1$ and at $P_2$ are the same but not $P_{atm}$.
   D. Pressure at $P_2$ is greater than at $P_1$ by the term $\rho gh$.

2. What are the forces acting on the column?
   A. Gravity, down.
   B. Gravity, down; force due to pressure on top surface; force due to pressure on bottom surface.
   C. Gravity, down; force due to pressure on top surface; force due to pressure on bottom surface; and surface tension, up.
   D. Gravity, down; force due to pressure on top surface; force due to pressure on bottom surface; and surface tension, down.

3. What expression gives the magnitude of the force due to pressure on the top of the column? Note: Force equals pressure times area: $F = PA$.
   A. $F = \pi r^2(P_{atm} - \rho gh)$
   B. $F = 2\pi r^2(P_{atm} - \rho gh)$
   C. $F = \pi r^2 P_{atm}$
   D. $F = 2\pi r^2 P_{atm}$

4. What expression gives the force due to gravity on the column? Note: $F_{grav} = mg = \rho Vg$, where $V$ stands for volume.
   A. $F_{grav} = r^2hg$
   B. $F_{grav} = 2r^2hg$
   C. $F_{grav} = \pi r^2h\rho g$
   D. $F_{grav} = 2\pi r^2h\rho g$

5. What expression approximates the force due to surface tension?
   A. $2r\gamma$
   B. $\pi r\gamma$
   C. $2\pi r\gamma$
   D. $2\pi r^2\gamma/h$

**Passage 2**

Lichens are difficult to classify. They are an interaction between a fungus and a photosynthetic species, usually green or blue green algae. The species form a symbiosis, but there is a debate as to what the association between the two species really is. To answer this question, researchers designed a series of experiments. They took advantage of the photosynthetic abilities of an algae to examine the role of association with fungi in the production of sugar, a product of photosynthesis. The carbon in the sugar molecules comes from carbon dioxide ($CO_2$) in the air. Researchers exposed the algae to $CO_2$ molecules containing radioactive carbon. The following experiments were done and the data is graphed (note: all graphs use the same scale).

*Experiment 1*:

Algae were grown in culture in the lab and exposed to radioactive $CO_2$. The growth rates and production of sugar containing radioactive carbon are shown in the graphs below.

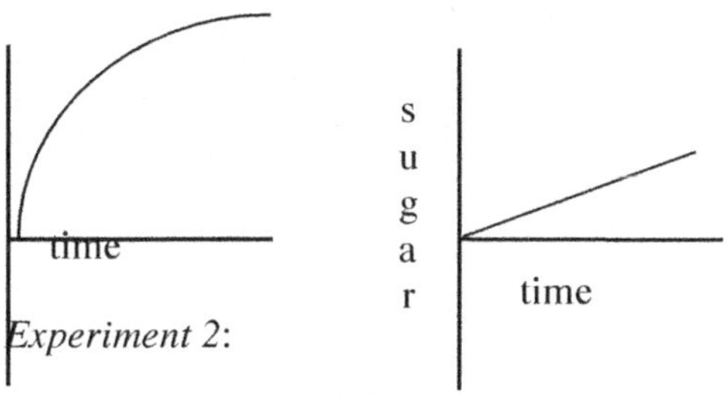


*Experiment 2*:

Algae were grown in the presence of fungi (therefore, this is lichen grown in the laboratory). The lichen was exposed to radioactive $CO_2$. The growth rate of the algae species and the production of sugar containing radioactive carbon are shown in the graphs below.

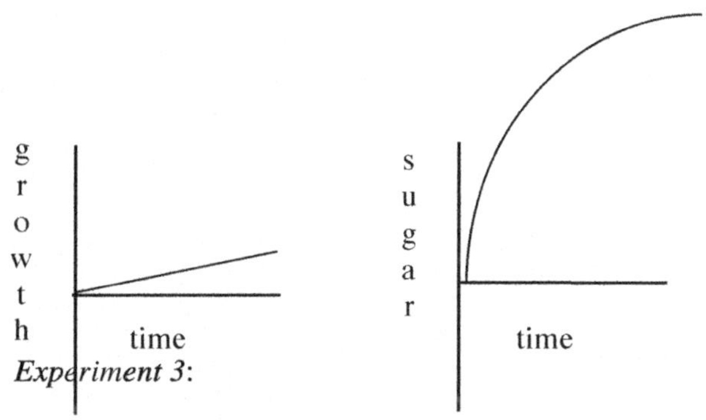


*Experiment 3*:

Lichen were grown under dehydration conditions while exposed to radioactive $CO_2$. The growth rate of the algae and the production of sugar containing radioactive carbon are shown in the graphs below.

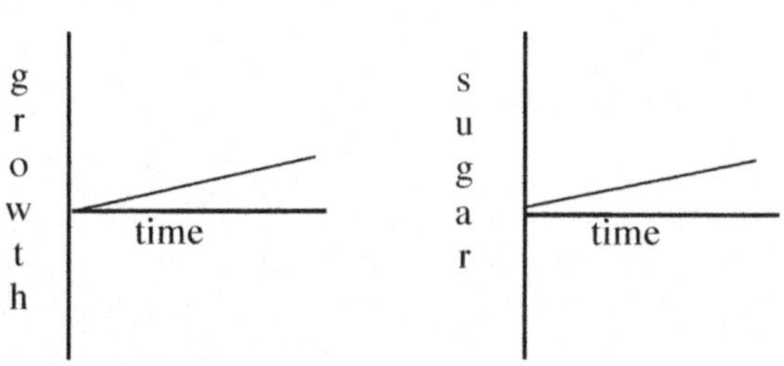


6. The conclusion that can be drawn from experiments 1 and 2 is
   A. the fungal species causes a decrease in sugar production in the algae.
   B. the fungal species causes an increase in sugar production in the algae.
   C. the algae cannot use the radioactive $CO_2$ efficiently to produce sugar.
   D. the algae grows better in the presence of the fungus.

7. The symbiotic relationship between the fungus and the algae is most likely a form of:
   A. mutualism
   B. parasitism
   C. commensualism
   D. cannot be determined

8. Sugar production in the lichen in dehydration conditions (experiment 3) indicates that
   A. there is reduced photosynthesis in the algae.
   B. there is reduced photosynthesis in the fungus.
   C. the fungi have been killed.
   D. there is a decomposition of the algae by the fungus.

9. This lichen would probably not be found in
   A. the tropics
   B. old growth forests
   C. suburban settings
   D. desert

**Passage 3**

When a guitar string is plucked, there are many frequencies of sound which are emitted. The lowest frequency is the note we associate with the string, while the mix of other frequencies gives the sound its *timbre*, or sound quality. The lowest frequency is the *fundamental*, while the higher frequencies make up the *harmonic series*. The next lowest frequency is the second harmonic; the next to lowest, the third harmonic, and so on. The timbre depends on the material of the string (steel or plastic or catgut), on the way it is plucked (middle or at the end), and on the sounding board.

Sometimes some of the frequencies may be suppressed, for example, by lightly holding a finger at a point along the string to force a node there. This is not the same as fretting the string, which involves holding the string all the way down to the neck in order to effectively change the length of the string.

The wave velocity is given by

$$v = \sqrt{T/\mu}$$

where $T$ is the tension in the string, and $\mu$ is the linear mass density, which is the product of material density and cross-sectional area.

For the following questions, consider an E string (frequency 660 Hz) which is made of steel. It has a mass of 0.66 grams for each meter of wire and has a circular cross section of diameter 0.33 mm. The string length when strung on a guitar is 0.65 m.

Also note that the D string has a wave velocity of 382 m/s.

Often waves have a characteristic frequency $f$ and a wavelength $\lambda$, and these are connected by the wave velocity $v = f\lambda$.

10. What is the velocity of a wave on the E string mentioned in paragraph 4?
    A. 214.5 m/s
    B. 429 m/s
    C. 858 m/s
    D. 1716 m/s

11. What is the frequency of the fundamental of the D string?
    A. 294 Hz
    B. 588 Hz
    C. 882 Hz
    D. 1175 Hz

12. If the guitarist places her left finger lightly on the D string one fourth way from the neck end to the base, what is the lowest frequency that will be heard?
    A. 588 Hz
    B. 882 Hz
    C. 1175 Hz
    D. 2350 Hz

13. What is the wavelength of the sixth harmonic of the E string?
    A. 0.22 m
    B. 0.325 m
    C. 3.25 m
    D. 20.8 m

**Passage 4**

Cells grown in culture are often used to investigate the role of organelles. As culture conditions can be easily manipulated, cells can be subjected to different chemicals and their responses studied and analyzed. In one such experiment, mouse cells were incubated in the presence of high concentrations of sucrose. The cells were examined under the microscope. Crystals of sucrose were found in the lysosomes of all the cells. When sucrose was depleted from the culture medium, the sucrose crystals in the lysosomes persisted for many days, but did not affect the growth of the cells. Eventually, the sucrose crystals disappeared. The rate of disappearance is shown in the graph below (note: all graphs are the same scale).

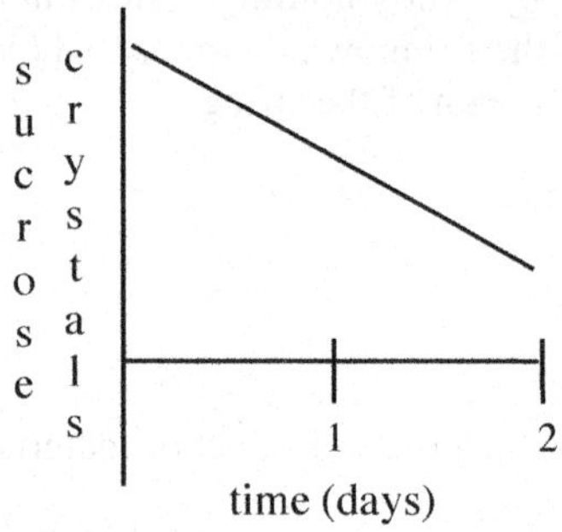


Researchers performed two additional experiments:

*Experiment 1*:

Immediately after sucrose was removed from the culture medium, the enzyme *invertase* was added. Invertase catalyzes the cleavage of sucrose into monosaccharides. The cells were then examined for the presence of sucrose crystals. The results are shown below.

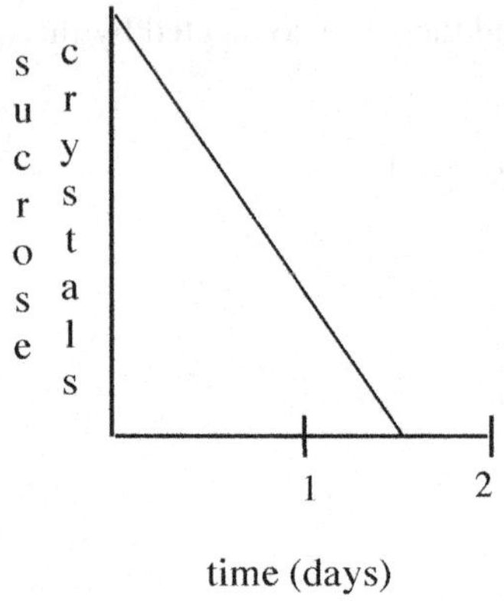

*Experiment 2*:

Mouse cells grown in the presence of invertase (cell type A) were mixed with cells grown in the presence of sucrose (cell type B). A reagent that fuses cells was also added to the culture. After fusion, the cells (cell type AB) were grown in the absence of sucrose. The cells were then examined for the presence of sucrose crystals. The results are shown below.

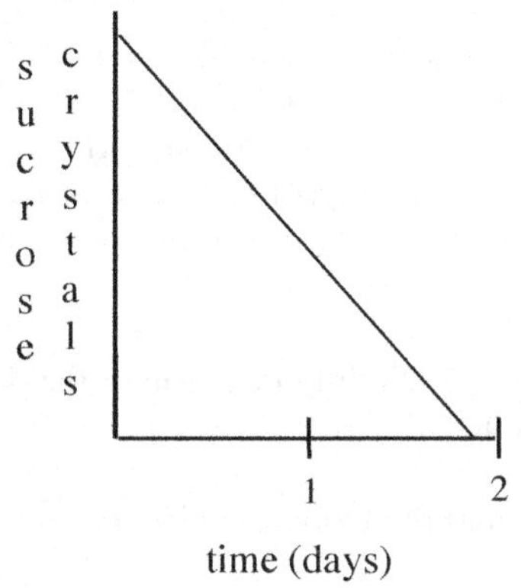


14. From the results in Experiment 1, what can you conclude about invertase?
    A. It irreversibly inhibits sucrose production.
    B. It catalyzed the break down of sucrose in the medium.
    C. It was internalized by the lysosomes.
    D. It changes the pH of the cytoplasm to catalyze sucrose.

15. Which of the following statements is true for mouse cells grown in culture?
    A. They do not normally have the ability to break down sucrose.
    B. They make invertase but only in low levels.
    C. Fusion with other cells reduces the ability of cells to produce invertase.
    D. High concentrations of sucrose are lethal.

16. Based on the results in Experiment 2, which of the following could be true?
    I. Invertase diffused from the lysosomes of cell type A into the lysosomes of cell type B.
    II. The lysosomes of cell type A fused with the lysosomes of cell type B.
    III. Invertase destroyed cell type B and only allowed cell type A to survive.
    A. I only
    B. II only
    C. I and II only
    D. III only

17. Another sugar, mannose, was added to mouse cells, and mannose crystals formed inside lysosomes. Similar to sucrose crystals, the mannose crystals disappeared in a few days after the sugar was removed from the medium. If invertase were added to the culture immediately after mannose was removed, what would you predict would happen?
    A. The mannose crystals would disappear in less than two days.
    B. The mannose crystals would disappear within a few days.
    C. Sucrose crystals would form.
    D. Mannose crystals would appear in the cytoplasm.

**Passage 5**

Bats are mammals which have acquired the ability of flight and of echolocation. Echolocation involves using vibrating membranes to direct a high frequency sound, with frequencies ranging from 12 kHz to 150 kHz. If the sound encounters a flying insect or obstacle which is larger than the wavelength of the sound, then a portion of the sound wave is reflected, and the bat detects it.

Beyond this basic framework, different species of bats use different strategies in echolocation. Some species emit a series of pulses, determining the distance to an object by the delay in return of the signal. Some emit a constant frequency, using the frequency of the returned sound to determine information about the velocity of the insect. Others use a sweep of frequencies, presumably to determine size information or directional information. Some emit a sound with a high harmonic content. Many use some combination of these strategies.

Several adaptations provide for better processing of the returned signal, including isolation of the detection apparatus from the emitting apparatus and specializations in the middle ear.

For the questions, use the following: The speed of sound is 343 m/s, and the Doppler-shifted frequency for a detector and emitter moving relative to each other is

$$f_{det} = \frac{v_s \pm v_{det}}{v_s \pm v_{em}} f_{em}$$

Note: The frequency, $f$, of sound is given by the formula $f = v/\lambda$, $v$ is the speed to the sound and $\lambda$ is the wavelength.

18. What frequency would a bat use to locate an insect 1 cm wide which is 10 m away?
   A. Any frequency less than 34 kHz.
   B. Any frequency greater than 34 kHz.
   C. Any frequency less than 34 Hz.
   D. Any frequency greater than 34 Hz.

19. If an insect is 3 m away, and a bat emits a pulse signal, how long is the delay in the return signal? Remember, distance equals rate times time, or $d = rt$.
   A. 0.009 s
   B. 0.017 s
   C. 0.09 s
   D. 0.17 s

20. A bat is in pursuit of an insect. The bat is flying 10 m/s to the east, and the insect is flying 10 m/s to the east. If the bat emits a constant frequency sound of 30 kHz, what frequency will he detect?
   A. 29 kHz
   B. 30 kHz
   C. 31 kHz
   D. 32 KHz

21. What is a possible reason for using a sound with high harmonic content?
   A. The harmonic frequency can determine the distance to an insect.
   B. The harmonic frequency can stun the insect.
   C. The harmonic frequency might be reflected even if the fundamental is not.
   D. The harmonic frequency might be Doppler shifted even if the fundamental is not.

22. A bat is traveling west at 15 m/s, emitting a constant–frequency sound of 50 kHz. If it encounters an obstacle, such as a tree, what frequency sound does it detect?
    A. 2.2 kHz
    B. 48 kHz
    C. 50 kHz
    D. 55 kHz

23. Which is an adaptation which might aid an insect?
    A. An ability to emit a sound with frequency much lower than that of a bat.
    B. An ability to emit a sound with frequency about the same as that of a bat.
    C. An ability to emit a sound with frequency much higher than that of a bat.
    D. A secretion of an obnoxious tasting chemical.

**Passage 6**

Cells will often delay entry into mitosis when the DNA has not completed replication, or when the cell is not yet big enough to divide. In yeast, many proteins and enzymes are known which control the cell's ability to detect unreplicated DNA and to determine exactly how big the cell is. This allows for control of the cell cycle.

You wish to study these cell cycle control proteins in detail. You grow yeast cells and then expose them to a chemical mutagen, know to cause mutations in the DNA. You then examine the cells to see if they exhibit any abnormal growth patterns, indicating cell cycle control proteins may have been mutated. Each mutant cell you obtain has only one mutation in one protein.

*Experiment 1*:
You expose your cells to a drug which speeds up the cell cycle, but you grow the cells in a nutrient poor medium. This medium normally prevents the cells from growing quickly, although DNA replication still occurs at a normal pace. Most cells take a long time to enter into mitosis. However, you isolate several cells that do not delay entry into mitosis. You call these mutants 1, 2, 3, etc.

*Experiment 2*:
You add a drug to your yeast cells which does not allow replication to occur. Most of the yeast cells arrest and do not grow. However, several do enter mitosis. You call these mutants A, B, C, etc.

*Experiment 3:*
Upon careful examination of your mutants, you find a few that allow cells to enter mitosis when either the cell is small or when the DNA has not completed replication. These mutants you call by both previous names (e.g., 5C, 9G, etc.).

Using this data, you try to determine where in the cell cycle these mutations are exerting their influence, and hence, where their normal counterparts act.

24. At what stage in the cell cycle is mutant 8 most likely defective?
    A. $G_1$
    B. S
    C. $G_2$
    D. M

25. At what stage in the cell cycle is mutant D most likely defective?
    A. $G_1$
    B. S
    C. $G_2$
    D. M

26. At what stage in the cell cycle is mutant 9G most likely defective?
    A. $G_1$
    B. S
    C. $G_2$
    D. must be at more than one stage

27. Under your experimental conditions, you find that your mutants quickly die out after a few generations. However, when you culture them without drugs they typically do well. Which of the following mutants would probably die off more quickly than the others under your experimental conditions?
    A. 1
    B. A
    C. 5C
    D. All will die off at the same rate.

**Passage 7**

Electromagnetic radiation from an incandescent source, such as a light bulb, is unpolarized, which means that the electric field of the wave points in random directions perpendicular to wave travel. One way to produce polarized radiation involves applying an alternating voltage to a straight piece of wire to form an antenna (See figure below). Radiation is emitted from the antenna perpendicular to the wire with a polarization which is parallel to the wire.

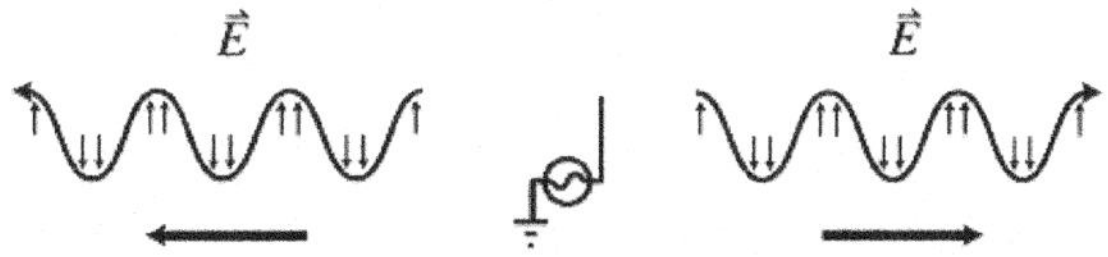


In this diagram the small arrows show the direction of the electric field, and the large arrows show the direction of the wave.

Another way to obtain polarized radiation involves allowing unpolarized radiation to be incident on a film or material which transmits radiation of one polarization but absorbs radiation of the perpendicular polarization. Such a film is called a *polarizer*. If unpolarized light of intensity $I_0$ is incident on a vertical polarizer, the radiation that passes through is vertically polarized with intensity 1/2 $I_0$. The figure below shows this schematically.

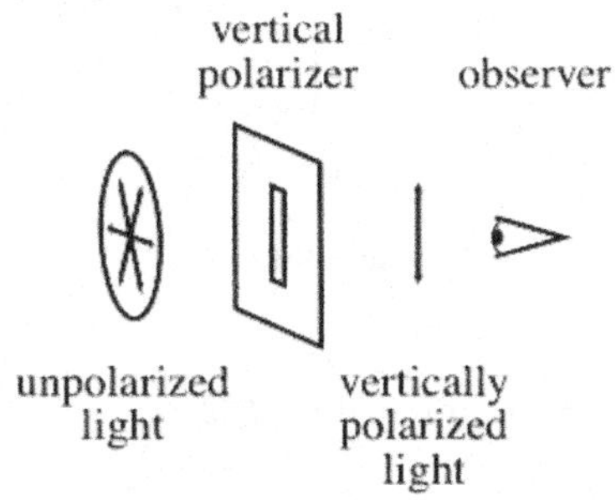


If polarized radiation is incident on a polarizer, the amount of energy that is transmitted depends on the relative angle of the radiation polarization and the polarizer axis. If they are aligned, then all the radiation is transmitted. If the angles differ by $\theta$, then the intensity of the transmitted radiation is $\cos^2\theta$ of the original intensity.

All of the foregoing refers to ideal polarizers. All manufactured polarizers have less than ideal efficiency which comes from reflection off the two surfaces and absorption of the parallel component.

In questions 1–4, an unpolarized radiation source is incident on a series of polarizers. Without the polarizers the intensity of the source is $I_0$. Assume the polarizers are ideal.

28. In the figure below, unpolarized light is incident on polarizers A and B in series. Polarizers A and B are both oriented vertically. What is the intensity of the resultant beam?

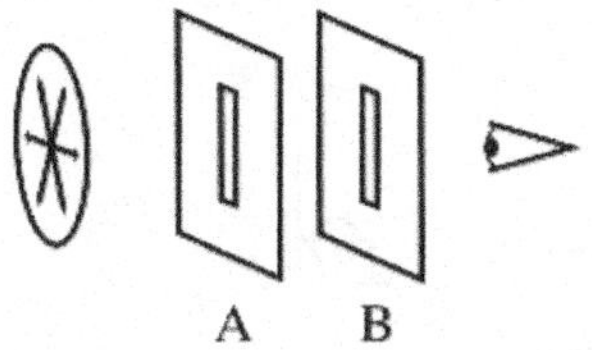


A. $I_0$
B. $I_0/2$
C. less than $I_0/2$ but greater than zero intensity
D. 0

29. In the figure below, unpolarized light is incident on polarizers A and B in series. Polarizer A is oriented vertically, while polarizer B is oriented horizontally. What is the intensity of the resultant beam?

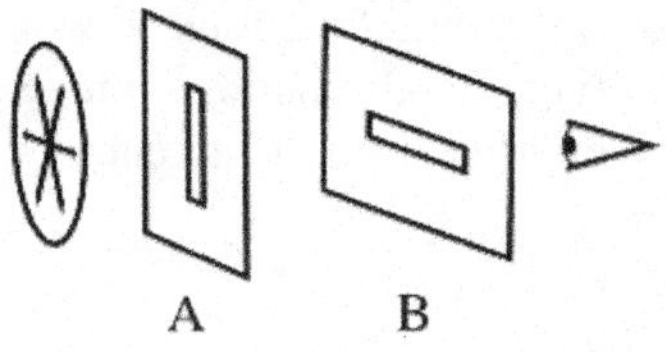


A. $I_0$
B. $I_0/2$
C. less than $I_0/2$ but greater than zero intensity
D. 0

30. An optically active substance is a substance which rotates the plane of polarization of a beam. The figure shows a modification of the figure in Problem 2, with an optically active substance between the polarizers. What is the intensity of the resultant beam?

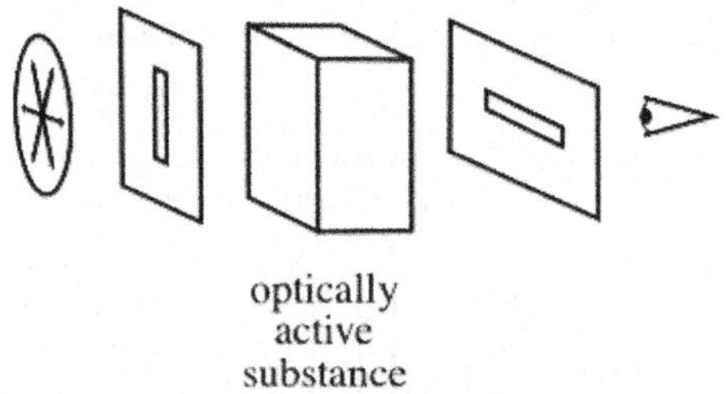


A. $I_0$
B. $I_0/2$
C. It could have any intensity less than (or equal to) $I_0/2$
D. 0

31. Where does the energy of the original beam go which is not in the resultant beam?
A. chemical energy
B. potential energy
C. heat
D. nuclear energy

32. A horizontal antenna is aligned along a north-south axis. This antenna has an alternating voltage applied to it, so it is emitting electromagnetic radiation. An observer is due north of the antenna. (See the figure.) What polarization does he detect from the antenna?

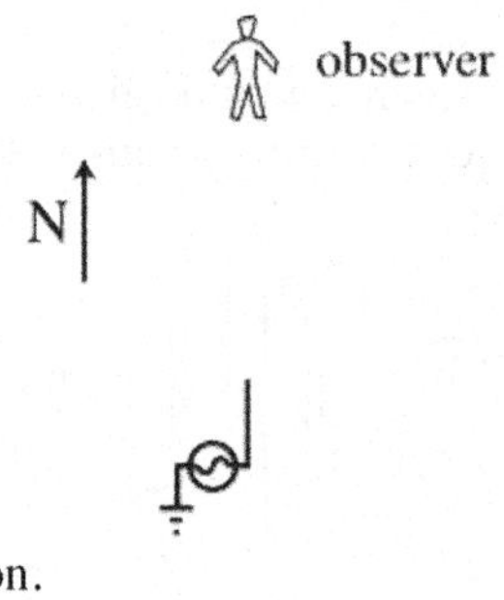


A. He observes horizontal polarization.
B. He observes vertical polarization.
C. He observes an unpolarized beam.
D. He observes no radiation.

**Passage 8**

Cancer is a disease in which cells begin to divide when they shouldn't, producing masses of cells called tumors. Cancerous tumors are called "malignant" because of their ability to grow aggressively, invade surrounding healthy tissue, and "metastasize", or spread to different locations in the body as tumor cells break off, travel through the blood or lymph, and establish new tumors.

Cancer is generally caused by environmental agents known as "carcinogens", which include chemicals (such as those found in cigarette smoke) and radiation (like the UV rays of sunlight) that mutate cells, causing them to become cancerous.

Tumors that originate from epithelial cells are generally known as carcinomas, and it is estimated that about 90% of all human cancers are carcinomas. Tumors that originate from muscle tissue are called "sarcomas", and tumors can also originate from nervous and connective tissues. Epithelial tissue exists in various locations in the body, including the skin, the mucous membranes lining the lungs and gastrointestinal tract, and the linings surrounding major organs.

33. A person who develops cancer from continually smoking cigarettes would be most likely to develop a:
    A. Carcinoma of the skin.
    B. Carcinoma of a mucous membrane.
    C. Sarcoma of the diaphragm.
    D. Carcinoma of the epithelium surrounding the heart.

34. The observation that 90% of human cancers are carcinomas implies that:
    A. Most carcinogens, like UV light, penetrate very deeply into underlying tissues.
    B. Epithelial cells by their very nature are more susceptible to mutation than cells of other tissue types.
    C. Cigarette smoking causes more tumors in humans than does UV light.
    D. Most carcinogens do not penetrate tissues very deeply.

35. A lab worker in Alaska develops a carcinoma of the skin. This particular worker is very dedicated, and almost never leaves the lab. The tumor is most likely the result of:
    A. Excessive cigarette smoking by coworkers.
    B. Excessive ingestion of toxic chemicals.
    C. Exposure to radiation from phosphorus-32, a radioactive isotope the worker routinely uses.
    D. Exposure to ultraviolet rays of sunlight.

36. Which of the following statements is correct?
    A. A tumor located in the liver must be a sarcoma.
    B. A tumor located in a muscle must be a sarcoma.
    C. A tumor composed of epithelial cells must be located in an epithelial tissue.
    D. A tumor in the breast could theoretically consist of cells of any tissue type.

37. Cancer cells can often secrete a substance that has the ability to dissolve basement membranes. This could account for the ability of the cells of a malignant tumor to:
    A. Invade adjacent tissue.
    B. Divide rapidly.
    C. Stop dividing.
    D. Become carcinogenic.

**Passage 9**

The radio waves, which carry information in a standard broadcast, are an example of electromagnetic radiation. These waves are a disturbance, not of a material medium, but of electric and magnetic fields. When the wave is linearly polarized, the electric field points in a direction perpendicular to the propagation of the wave, although its magnitude varies, of course, in space and time. The magnetic field points in a direction perpendicular to the wave propagation and to the electric field, and the two fields propagate in phase.

The electromagnetic radiation is generated by an antenna, which is a wire or metal rod which points perpendicular to the direction of the intended wave propagation. An alternating current is generated in the antenna, whose frequency is the same as that of the radiation to be produced. The electric field of the resulting electromagnetic radiation points along the same axis as the current.

The electric field of the electromagnetic radiation encounters electrons on the receiving antenna, which is another wire or metal rod. The electric field creates a current along the receiving antenna. One way to have good transmission and reception is to have the length of the antenna be one quarter of the wavelength of the electromagnetic wave.

For the following questions, consider a situation in which a transmitting antenna points vertically, and the receiving antenna is directly to the north. The speed of light is $3 \times 10^8$ m/s.

38. For a point between the two antennas, what is the direction of the electric field vector for the radiation?
    A. North/south.
    B. East/west.
    C. North/south or east/west.
    D. Up/down.

39. For a point between the two antennas, what is the direction of the magnetic field vector for the radiation?
    A. North/south.
    B. East/west.
    C. North/south or east/west.
    D. Up/down.

40. Which of the following best describes the energy flow?
    A. Electrical to electromagnetic to electrical.
    B. Electromagnetic to electrical to electromagnetic.
    C. Mechanical to electromagnetic to mechanical.
    D. Kinetic to electromagnetic to kinetic.

# Test 3 Solutions

**Passage 1**

1. A.

Well, points $P_1$ and $P_2$ are in the water, and $P_2$ is above $P_1$, so the pressure at $P_1$ should be greater by $\rho gh$. It is confusing, though, since the pressure at $P_1$ must be $P_{atm}$. The air just above $P_2$ must have pressure $P_{atm}$. Usually the pressure on one side of a boundary between two substances is the same as the pressure on the other side, but this is not true if the boundary is curved, as when a meniscus forms.

2. C.

Certainly there are pressure forces, but these balance (both are $F = P_{atm}\pi r^2$). The surface tension pulls the column up, balancing the force of gravity.

3. C.

The pressure at the top of the column is given by $F = PA = P_{atm}\pi r^2$. We use the air pressure since it is the air that exerts the downward force on the column.

4. C.

The force due to gravity is $F_{grav} = mg = \rho Vg = \rho\pi r^2 hg$.

5. C.

The length of the line of contact is the circumference of the straw, so $L = 2\pi r$. Hence, $F_{surf} = \gamma L = \gamma(2\pi r) = 2\pi r\gamma$.

**Passage 2**

6. B.

The graphs clearly show that, when the fungal species is present, the production of sugar is greatly increased and growth of the algae is decreased. Since sugar is made, the algae has no problem using the radioactive isotope of carbon.

7. C.

It appears that the presence of the fungus decreases the growth of the algae, and forces it to produce more sugar. However, the algae does not necessarily appear to be harmed by this interaction.

8. A.

Since there is less sugar production, but not less growth of the algae, it appears that dehydration causes photosynthesis to decrease.

9. D.

It appears that the lichen need abundant supplies of water in order to thrive. Desert conditions would probably not supply enough water.

**Passage 3**

10. C.

Our first thought is to use the information about linear density, but there is no way to find the tension in the string. What other formula do we have for wave velocity? Well, $v = \lambda f$, and we know $f = 660$ Hz. We can get the wavelength just by knowing that the note 660 Hz refers to the fundamental. The fundamental has a wavelength $\lambda = 2\ (0.65\text{ m}) = 1.3\text{ m}$.

11. A.

For the D string we can use $\lambda = 1.3$ m, so that $f = v/\lambda = 294$ Hz. We do not have to do the calculation. If we look at the choices, we see only one choice less than 382.

12. C.
The first node occurs one fourth of the way from the neck end. Since the neck is a node as well, there must be a node at the midpoint and three fourths of the way down. This corresponds to a wavelength 0.325 m, as in problem 3 (above).

13. A.

The sixth harmonic has five nodes (not including the ends). This is shown in the figure. The wavelength is $\lambda = 1/3\ (0.65\ \text{m}) = 0.22\ \text{m}$.

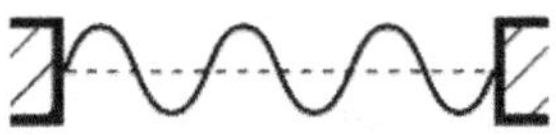

**Passage 4**

14. C.
Since the sucrose crystals appeared in the lysosomes of the cell, the invertase had to have been internalized in order to catalyze the reaction of the crystals into monosaccharides. Given the circumstances, none of the other answers are appropriate.

15. B.
Since we are told that the sucrose crystals are broken down within several days of sucrose being eliminated from the medium, it appears that the cells have the ability to produce invertase, but probably in low levels.

16. C.
Either conditions I or II would explain the results seen in Experiment 2. There is no indication that invertase would destroy one cell type, so III is not a possibility.

17. B.
Recall that enzymes are specific for their substrates. Invertase would probably not catalyze the break down of mannose. Therefore, even in the presence of invertase, the mannose crystals would take several days to be eliminated from the cells.

**Passage 5**

18. B.
According to paragraph 1, the wavelength must be shorter than the target insect. Thus the frequency must be greater than $f = v/l_{insect} = 34300\ \text{Hz} = 34.3\ \text{kHz}$.

19. B.
The sound pulse must travel from the bat to the insect, be reflected, and travel back again, for a total of 6 m. The time required is $t = d/r = 0.017$ s.

20. B.
Since the insect and the bat are moving in the same direction at the same speed, there is no Doppler shift, and the detected frequency is simply 30 kHz.

21. C.
Higher harmonic content refers to higher frequencies being present in addition to the fundamental. These higher harmonic frequencies thus have shorter wavelength. This would not really aid in distance measurement, so choice A is incorrect. Stunning the insect? Choice B is also incorrect. There is a sentence in paragraph 1 which indicates that C is plausible, since shorter wavelengths are more likely to be reflected if the wavelength of the fundamental is too large. If the harmonics are Doppler shifted, then so is the fundamental, so choice D is incorrect.

22. D.
If the tree could detect a frequency, it would detect

$$f_{\text{tree}} = \frac{343}{343-15} 50 \text{ kHz}$$

If it re-emitted this frequency, then the bat would detect a frequency given by

$$\begin{aligned} f_{\text{det}} &= \frac{343+15}{343} f_{\text{tree}} \\ &= \frac{358}{343}\frac{343}{328}(50 \text{ kHz}) \\ &= 55 \text{ kHz} \end{aligned}$$

There are two Doppler shifts.

You do not need to do the calculation to obtain the answer. Given the fact that the bat and the tree approach each other, the frequency must be increased, so D is the only possible answer.

23. B.
The bat does not hear frequencies which are too far from those it sends out. Choice D is definitely not right, but it is interesting to think why. Why is such an adaptation completely useless?

**Passage 6**

24. C.
Inability to detect whether the cell was large enough to divide would occur at $G_2$.

25. B.
The cell should not be allowed into $G_2$ if all the DNA has not yet been replicated. Therefore, the mutation probably occurred in S phase.

26. D.
It appears that these mutants exert their controls at two different stages in the cell cycle, perhaps at both S and $G_2$.

27. C.
Each mutant would eventually die out under the experimental conditions. However, the mutants that are defective at both checkpoints would probably die out most quickly.

**Passage 7**

28. B.
The light emerging from the first (left) polarizer has half the intensity of the original source, since the horizontal component has been taken out of it. All of the resulting light passes through the second (right) polarizer, since it is already vertically polarized. Thus the final beam has half the intensity of the original.

29. D.
The light emerging from the first polarizer has half the intensity of the original source, and it is vertically polarized. The second polarizer cuts out the vertical component and passes the horizontal component, so the resulting intensity is zero.

30. C.
Again, the light emerging from the first polarizer is vertically polarized and has intensity $1/2\ I_0$. After it goes through the optically active substance, it may have *any* orientation, since that information is not given in the problem. If the beam ends up horizontally polarized, then the resulting intensity is $1/2\ I_0$. If it ends up vertically polarized, then the intensity is zero. If the more probable situation arises that the beam is somewhere between these two extremes, still polarized but at some oblique angle, then the resultant intensity is between 0 and $1/2\ I_0$.

31. C.
Since the light is absorbed, the energy turns to heat.

32. D.
The passage states (paragraph 1) that radiation is emitted perpendicular to the wire. For the antenna in question 5, most of the radiation is emitted east, west, up, down, and so on, but none is emitted north and south.

**Passage 8**

33. B.
Since cigarette smoke is inhaled, it is most likely to contact the mucous membrane of the lungs, whose outer layer is composed of epithelium.

34. D.
Since carcinogens, whether chemicals or radiation, are contacting the body from the environment, and we know that epithelium is always the tissue that separates the body from the environment, we would expect the carcinogens to initially contact epithelium. If they penetrated tissue deeply, all types of cancers would be equally prevalent; the fact that carcinomas make up 90% of all cancers implies that after the carcinogen contacts the epithelium, it generally does not penetrate much deeper. There is no reason to believe that choice B is correct, and choice C is irrelevant to the question.

35. C.
The cause of a carcinoma of the skin would most likely be radiation exposure from the environment, or contact with the skin by carcinogenic chemicals. Since the worker lives in Alaska and rarely goes outside, it is unlikely that excessive exposure to sunlight is the culprit, and radiation from the isotope is a more likely causative agent. Cigarette smoke or ingestion of chemicals would likely affect mucous membranes or internal body parts.

36. D.
Due to the possibility of metastasis, a tumor that originated in one part of the body from a particular type of cell may "spread" to virtually any body location, producing a new tumor there. The tumor, however, will still be made up of the type of cells from which it originally grew.

37. A.
Since the basement membrane holds epithelium to underlying tissue, it would have to be penetrated by any cells wishing to invade another tissue. It has nothing to do with cell division (leaving out choices B and C) and choice D doesn't make sense -- the cells themselves cannot be carcinogens. The carcinogen is the environmental agent that caused the mutation in the first place.

**Passage 9**

38. D.
According to the passage, the electric field "points along the same axis as the current" that produces it. Since the antenna is vertical, the current it carries is vertical, so the electric field is vertical as well.

39. B.
The magnetic field is perpendicular to the direction of propagation (north/south) and to the electric field (up/down). So the magnetic field must point east/west.

40. A.
The energy starts as electrical.

# Test 3 Conversion Chart

**Scale Score Conversion Chart**

| Raw Score | Scale Score | Percentile Ranking |
|---|---|---|
| 40 | 36 | 99 |
| 39 | 35 | 99 |
| 38 | 34 | 99 |
| 37 | 33 | 99 |
| 36 | 32 | 98 |
| — | 31 | 98 |
| 35 | 30 | 97 |
| 34 | 29 | 95 |
| 33 | 28 | 94 |
| 31–32 | 27 | 91 |
| 30 | 26 | 88 |
| 29 | 25 | 84 |
| 27–28 | 24 | 77 |
| 25–26 | 23 | 71 |
| 24 | 22 | 64 |
| 22–23 | 21 | 56 |
| 20–21 | 20 | 48 |
| 19 | 19 | 38 |
| 17–18 | 18 | 30 |
| 15–16 | 17 | 23 |
| 14 | 16 | 18 |
| 13 | 15 | 14 |
| 11–12 | 14 | 10 |
| 10 | 13 | 7 |
| 9 | 12 | 5 |
| 7–8 | 11 | 3 |
| 6 | 10 | 1 |
| 5 | 9 | 1 |
| 4 | 8 | 1 |
| 3 | 7 | 1 |
| — | 6 | 1 |
| 2 | 5 | 1 |
| — | 4 | 1 |
| 1 | 3 | 1 |
| — | 2 | 1 |
| 0 | 1 | 1 |

# Test 4

(35 Minutes—40 Questions)

**Passage 1**

In a certain experiment, we place two metal plates of area $A$ parallel to each other and separated by a distance $d$ to form a capacitor. The copper disks are mounted on nonconducting stands in a dry room which does not allow the conduction of charge through the air. The two plates are connected with wires to the opposite ends of a DC cell. The capacitance of such a device is given by

$$C = \varepsilon_0 \frac{A}{d}$$

where

$$\varepsilon_0 = 8.85 \times 10^{-12} \frac{\text{C}^2}{\text{Nm}^2}$$

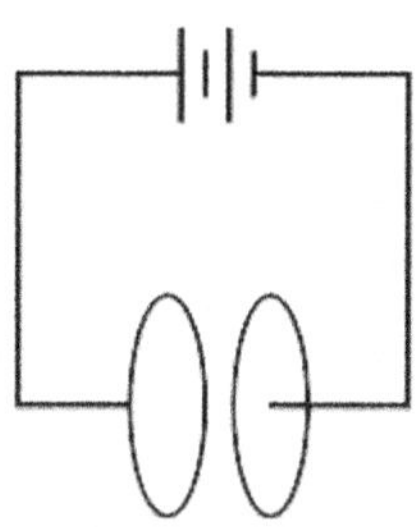

In Experiment 1 we place two copper circular disks of radius $R_1$ a distance $d_1$ apart and connect them to a battery which produces a potential $\Delta V_{bat}$. This produces a positive charge $Q_1$ on one of the plates and an electric field $E_1$ between the plates.

In Experiment 2 we use two copper circular disks of the same radius ($R_1$) and place them three times as far apart ($3d_1$) as in Experiment 1. Again, the two disks are connected to the opposite ends of the same battery.

In Experiment 3 we reproduce the setup in Experiment 1. Then the wires are removed from the copper plates. The copper plates are carefully separated to a new distance $2d_1$.

1. An alpha particle is a bare helium-4 nucleus, which has twice the charge of a proton. If an alpha particle is placed between the plates in Experiment 1, how would the electric force on it compare to the electric force on a proton? Note: the force on a charged particle is given by $F_{elec} = QE_1$.
   A. Both experience zero force.
   B. The force on the alpha particle would be the same as that on the proton.
   C. The force on the alpha particle would be double that on the proton.
   D. The force on the alpha particle would be four times that on the proton.

2. What charge is on the positive plate in Experiment 2?
   A. $Q_1/3$
   B. $Q_1$
   C. $3Q_1$
   D. $9Q_1$

3. What is the magnitude of the electric field between the plates in Experiment 2?
   A. $E_1/3$
   B. $E_1$
   C. $3E_1$
   D. $9 E_1$

4. What charge is on the positive plate at the end of Experiment 3?
   A. $Q_1/2$
   B. $Q_1$
   C. $2 Q_1$
   D. $4 Q_1$

5. What is the magnitude of the potential difference between the plates at the end of Experiment 3? Note: $\Delta V = Q/C$.
   A. $\Delta V_{bat}/2$
   B. $\Delta V_{bat}$
   C. $2\Delta V_{bat}$
   D. $4\Delta V_{bat}$

**Passage 2**

Both the nervous and endocrine systems are involved with communication among body parts, but they exert their actions by different methods. The nervous system uses electrochemical impulses to convey messages through nervous tissue at high speeds, while the endocrine system uses chemical hormones that are released into the blood and directly interact with their target cells. There is some evidence that the nervous and endocrine systems are evolutionarily related.

Hormones can be of two major types chemically: protein (including amines and peptides) or steroid. Each type of hormone normally exerts its effects differently. Protein hormones, such as insulin, adrenalin, ADH, and gonadotropin-releasing hormone (GnRH) interact with a membrane bound extracellular receptor on their target cells, which causes changes inside of the cell. This usually leads to the activation or inhibition of enzymes through a second-messenger system utilizing cyclic AMP (cAMP) as the second messenger. Steroid hormones such as testosterone, estrogen, and cortisol can diffuse through cell membranes, and usually act by affecting gene expression, but only after interacting with an intracellular receptor.

McCune-Albright syndrome is a genetic disease in which cAMP is synthesized in large quantities even when a signaling hormone is not present; it results in the over-secretion of many hormones, and increased effects of many hormones on their target cells.

6. Target cells of someone who suffers from McCune-Albright syndrome are likely to overreact to the presence of a hormone or respond to a hormone message even when the hormone is not present. Which of the following hormones would not cause overreaction or false message reception in an affected individual?
   A. Insulin
   B. ADH
   C. Testosterone
   D. Adrenalin

7. Some evidence that the nervous and endocrine systems are evolutionarily related comes from the fact that certain molecules function similarly to hormones during transmission of nerve impulses. Which of the following types of molecules fits this description?
   A. The sodium potassium pump
   B. Gated potassium channels
   C. Neurotransmitters
   D. Neurotransmitter receptors

8. The major connection between the nervous and endocrine systems that allows them to coordinate their activities is represented by:
   A. The cerebral cortex directly stimulating the thyroid gland.
   B. The brainstem controlling heart and breathing rate, which is also controlled by adrenalin.
   C. The ability of the cerebellum to coordinate the actions of several endocrine glands.
   D. The intimate connection between the hypothalamus and pituitary gland.

9. Testosterone is a steroid hormone; its release from the testes in males is controlled by gonadotropin-releasing hormone (GnRH). Which of the following statements is true?
   A. Both of these hormones affect target cells using a second messenger system.
   B. An individual with McCune-Albright syndrome is likely to over-produce testosterone.
   C. An individual with McCune-Albright syndrome is likely to under-produce testosterone.
   D. An individual with McCune-Albright syndrome would produce normal amounts of testosterone, but the target cells responding to testosterone would overreact to the message.

10. Stanozolol is a synthetic steroid that resembles testosterone in its chemistry and action. It is often the drug athletes are ingesting when they are said to be on "steroids". Which of the following statements is true of Stanozolol?
   A. It is likely to exert its effects by binding to an external receptor and triggering a second messenger system.
   B. It is likely to be found in all cells of the body after being ingested.
   C. It will likely only be found in muscle cells after being ingested.
   D. It would affect an individual with McCune-Albright syndrome more drastically than a normal individual.

**Passage 3**

A hydrogen atom has energy levels given by

$$E_n = -(13.6 \text{ ev})\frac{1}{n^2}$$

where $n$ is a positive integer. A state in which the electron is infinitely far from the nucleus corresponds to 0 ev. (Note: 1 ev = $1.602 \times 10^{-19}$ J) The energy level diagram is shown in the figure below. An atom generally has an electron in a given energy level; however, a transition may occur if the atom absorbs or emits a photon, or particle of light.

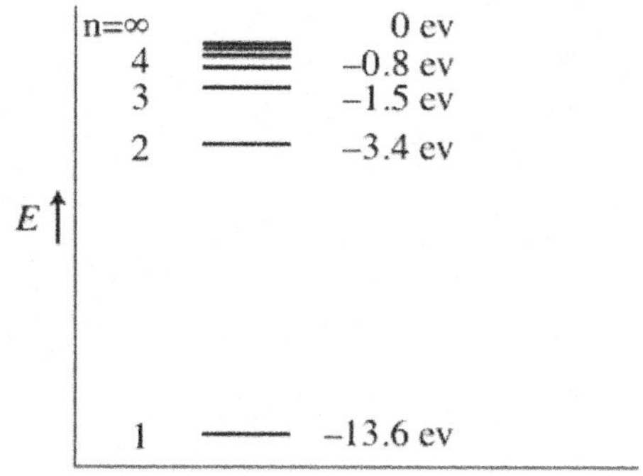


For the following questions, consider a hydrogen atom in an excited state initially at rest. It emits a photon of energy $E_{ph}$ which travels to the right. The hydrogen atom is now in its ground state.

The energy of a photon is given by $E_{photon} = hf$, where $f$ is the frequency and $h = 4.14 \times 10^{-15}$ ev s. The momentum of a photon is given by $p_{photon} = E/c$, where $c = 3.0 \times 10^8$ m/s, the speed of light. The mass of a hydrogen atom is $m_H = 1.7 \times 10^{-27}$ kg.

11. What is the smallest energy a photon must have to boost a hydrogen atom in its ground state to an excited state?
    A. 3.4 ev
    B. 10.2 ev
    C. 13.6 ev
    D. There is no smallest energy.

12. *Ionization energy* is the amount of energy required to remove an electron from an atom (or molecule) in its ground state. What is the ionization energy for a hydrogen atom?
    A. 1.5 ev
    B. 3.4 ev
    C. 10.2 ev
    D. 13.6 ev

13. A hydrogen atom in its ground state absorbs a photon and ends up in the excited state corresponding to $n = 3$. What is the most restrictive statement that can be concluded about the photon?
    A. The photon had an energy of 12.1 ev.
    B. The photon had an energy greater than or equal to 12.1 ev.
    C. The photon had an energy greater than 12.1 ev but less than 12.8 ev.
    D. The photon had an energy greater than 10.2 ev.

14. In the scenario mentioned in the second paragraph, what can be concluded?
    A. Momentum is conserved, but not energy.
    B. Energy is conserved, but not momentum.
    C. Both momentum and energy are conserved.
    D. Neither momentum nor energy is conserved.

15. What is the longest wavelength corresponding to a photon which could cause a transition in a hydrogen atom? Note: the frequency of a wave is given by $f = c/\lambda$, and the energy is given by $E = hf$.
    A. $9.1 \times 10^{-8}$ m
    B. $1.2 \times 10^{-7}$ m
    C. $3.6 \times 10^{-7}$ m
    D. There is no longest wavelength.

**Passage 4**

Considering the amazing quantities and diversity of pathogens that exist in the environment, vertebrates had to develop a complex defense system, or they would have perished. In humans, as in most animals, there are multiple lines of defense against potential pathogens. The integumentary membranes (skin and mucous membranes) provide a physical barrier to pathogens. If these are compromised, the inflammatory response is stimulated, during which neutrophils and macrophages phagocytize foreign invaders nonspecifically.

If an infection becomes systemic, the immune response is the last resort. Lymphocytes, specialized white blood cells that mainly reside in the lymph nodes, specifically attack pathogens in the blood and lymph. B lymphocytes are responsible for antibody-mediated immunity, which involves the reaction of protein antibody molecules with antigenic markers on the pathogen. T lymphocytes are responsible for cell-mediated immunity, during which antigens are presented to T cells by antigen presenting cells of the human body, often macrophages or virally infected cells.

Several diseases can cause the immune system to become weakened or almost completely destroyed. AIDS (Acquired Immune Deficiency Syndrome) is a disease caused by the HIV virus, which attacks helper T cells that normally produce chemicals (interleukins) that stimulate the entire immune system. While functional cytotoxic T cells and plasma B cells can be created, their general activity is low due to lack of interleukin stimulation. Type I diabetes mellitus is an example of an autoimmune disease, in which lymphocytes attack normal body cells, in this case the insulin producing cells of the pancreas. Perhaps the most devastating immune disease is called severe combined immune deficiency (SCID); individuals with this luckily rare condition have virtually no functioning lymphocytes, and often must live in vinyl bubbles to completely cut them off from any contact with pathogens.

16. What do the diseases AIDS and SCID have in common?
    A. In both cases, functional antibodies cannot be produced.
    B. In both cases, any secondary infection must cause death.
    C. In both cases, T cells are either absent or relatively inactive.
    D. Both diseases are caused by viral infection.

17. During HIV infection:
    A. Bacteria infect helper T cells, leading to compromised immunity.
    B. Functional antibodies to antigens on the HIV virus are produced.
    C. All T cells in the body are destroyed by the virus.
    D. HIV can be found infecting B cells.

18. Which of the following statements is true of the inflammatory response?
    A. B lymphocytes are important during inflammation, as they specifically phagocytize foreign invaders.
    B. Phagocytic white blood cells act nonspecifically to phagocytize invaders.
    C. It does nothing to actually destroy pathogens that enter the body; it only stops them from spreading to other regions of the body.
    D. Red blood cells must be involved in phagocytosis, since swelling and redness are caused by massive amounts of red blood cells entering the inflamed area.

19. Which physical barrier would be most susceptible to infection?
    A. Intact skin
    B. The respiratory mucous membrane
    C. Infection never occurs through skin or mucous membranes.
    D. They both present an equivalent barrier to pathogens.

20. Which of the following statements is true regarding the immune response?
    A. B cells can only be stimulated if antigen is presented to them by an antigen presenting cell.
    B. T cells can only be stimulated if antigen is presented to them by an antigen presenting cell.
    C. Macrophages exert their action by producing antibodies.
    D. Macrophages are very specific, and can only phagocytize pathogens of a single type.

**Passage 5**

Two charged balls which are near each other will exert a force on each other: attractive if they are oppositely charged, and repulsive if they are similarly charged. The magnitude of the force is given by

$$F = \frac{kq_1q_2}{r^2}$$

where $F$ is in N, $k$ is a constant $9 \times 10^9$ N m$^2$/C$^2$, $q_1$ and $q_2$ are the charges on the balls measured in C, and $r$ is the distance between the balls in m.

21. In a certain experiment, the separation between the balls is halved, while the charges on the balls are undisturbed. How would this affect the force between them?
 A. The force would decrease by a factor of 4.
 B. The force would decrease by a factor of 2.
 C. The force would increase by a factor of 2.
 D. The force would increase by a factor of 4.

22. In an experiment, the distance separating the balls is increased by 25%. How does this affect the force between the balls?
 A. It decreases by 50%.
 B. It decreases by 36%.
 C. It decreases by 25%.
 D. It increases by 25%.

23. If the force between the balls stays the same, but the charge $q_2$ is multiplied by 4, which is a possibility?
 A. The charge $q_1$ is also multiplied by 4, and all else is unchanged.
 B. The separation is decreased by a factor of 2, and all else is unchanged.
 C. The separation is increased by a factor of 2, and all else is unchanged.
 D. The separation is increased by a factor of 4, and all else is unchanged.

24. In a hypothetical situation, two balls of positive charge exert a force 12 N on each other. The charge on ball A is 2 C. If the charge on ball A is increased to 8 C, and all else unchanged, what would the force be?
 A. 8 N
 B. 16 N
 C. 18 N
 D. 48 N

25. Which graph best shows the relationship between the force between two balls $F$ and their separation $r$?

A.
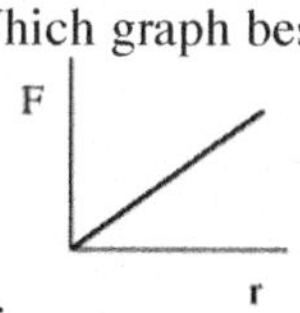


B.
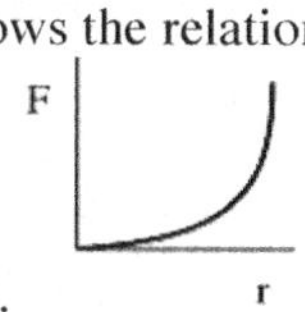


C.
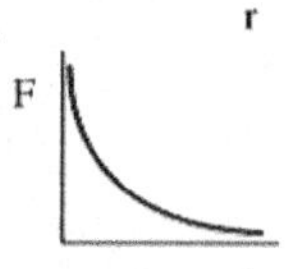


D.
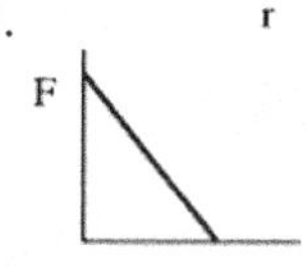

26. In a certain experiment two balls are both given a charge $q$, they are set a distance $r$ away from each other, and the force between them is recorded. Which graph best represents the relationship between $F$ and $q$?

A. 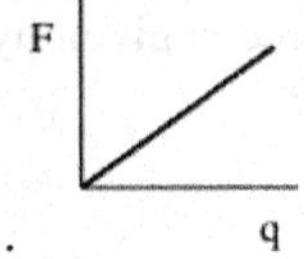


B. 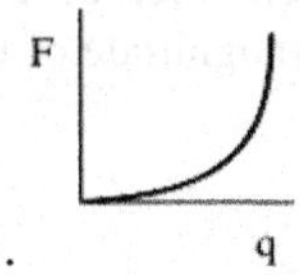


C. 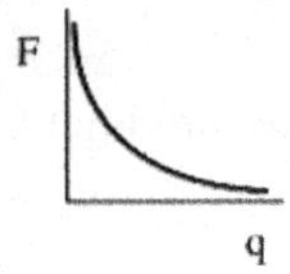


D. 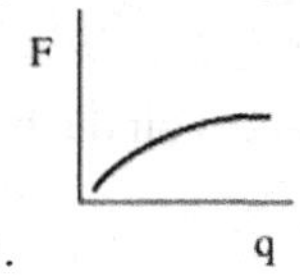

**Passage 6**

Some individuals do not produce sufficient amounts of the enzyme lactase to digest the sugar lactose, present in milk and dairy products, and are referred to as "lactose intolerant". Lactose is a disaccharide, which must be broken down by lactase into its constituent monosaccharides, glucose and galactose, before it can be absorbed. If not broken down, it remains in the small intestines and is passed along into the large intestine, a place it does not belong, with waste materials. The results can range from dehydration and diarrhea to intestinal cramps and bloating. The pain and bloating often experienced by the lactose intolerant after lactose ingestion are caused by intestinal bacteria that metabolize the lactose to which they are normally not exposed. Products of this fermentation include organic acids and various gases. Some products are available to allow lactose intolerant individuals to eat lactose containing products, and are meant to be ingested just before eating milk or dairy products.

27. Which of the following is a reasonable explanation of the observation that lactose intolerant individuals become dehydrated after lactose consumption?
    A. The presence of lactose in the large intestine inhibits the pituitary gland from releasing the hormone ADH.
    B. Large amounts of lactose in the large intestine increase the osmotic pressure, causing water to be retained there and eliminated with wastes, whereas it would normally be reabsorbed into the blood.
    C. Because they cannot drink milk, lactose intolerant individuals are deprived of an important water source.
    D. The presence of lactose in the large intestine causes more frequent urination.

28. What is the likely nature of the products that, if taken before lactose ingestion, will inhibit the symptoms of lactose intolerant individuals?
    A. They are likely antibiotics that kill the bacteria in the intestine so that they cannot metabolize the sugar.
    B. They probably contain the enzyme lactase, which will end up in the small intestine and allow lactose digestion.
    C. They are probably cocktails of various drugs that inhibit urination and help to ease the pain.
    D. They are likely drugs that allow the small intestine to absorb undigested lactose.

**Passage 7**

After glucose is absorbed by the small intestine, it is transported to the liver, where many "decisions" are made regarding its fate. These decisions are hormonally effected, and include release of some glucose into the blood, storage of some as glycogen, and conversion of some to fat.

The hormones insulin and glucagon are produced by the pancreas; insulin is produced in response to high blood sugar, as after a meal, and tells body cells to take up glucose from the blood, while notifying the liver to store or convert any excess. Glucagon targets the liver when blood sugar is low, telling it to break down some glycogen and release it into the blood as glucose. The interaction of these hormones acts like a thermostat to maintain a relatively constant blood glucose concentration.

In individuals with diabetes mellitus type I, insulin is no longer produced by the pancreas, and must be supplemented by intravenous injection, or the afflicted individual will eventually die.

29. Why must insulin be injected intravenously, while other enzymes such as lactase can be taken orally, by pill.
    A. Insulin is a protein; if it was taken by mouth, it would be broken down into its constituent amino acids, and would never enter the blood.
    B. Since insulin affects glucose concentrations, if it were ingested it would halt the absorption of glucose by the small intestine.
    C. If insulin was taken in pill form, it would interact with glucagon secreted by the pancreas in the small intestine, and the two would "neutralize" each other.
    D. None of the above answers is reasonable.

30. All of the following symptoms might be associated with untreated diabetes mellitus type I except:
    A. Extremely high blood glucose.
    B. Extremely low blood glucose.
    C. The excretion of glucose in the urine.
    D. Dehydration due to water loss in the urine.

31. Sometimes if the administration of glucose is not timed properly, insulin is injected but an individual does not eat for a prolonged period of time. This can lead to insulin shock, during which a person loses consciousness and can die. The probable cause of insulin shock is:
    A. The injected insulin causes the blood sugar to become so low that glucose is not available to fuel the body's needs.
    B. The injected insulin causes so much glucose to be released into the blood that the surge of energy can cause heart rhythm irregularities.
    C. The injected insulin suppresses hunger, so that the person does not realize they should eat.
    D. Injected insulin causes the inhibition of glucagon synthesis, so no glycogen can be broken down to glucose by the liver.

**Passage 8**

The space shuttle is a spaceship which was designed for transporting a payload to near-earth orbits and to be used many times. When it stands on the launching pad, it consists of the orbiter itself, the external tank, and two booster rockets. The orbiter is 7.32 X $10^4$ kg, while the whole assembly is 2.0 X $10^6$ kg. The thrust at liftoff is 2.86 X $10^7$ N, which is achieved by burning 3400 kg of fuel each second. The force provided by the engines is given by the product of the velocity of the exhaust gases relative to the ship and the rate (mass per time) at which fuel is burned.

As the space shuttle ascends, the rate of fuel burning is approximately constant, and so the mass of the shuttle decreases.

The following chart shows hypothetical data for the liftoff of a shuttle. We are assuming the shuttle moves in one dimension upward. Use 10 m/s$^2$ for the acceleration due to gravity. The force due to gravity is $F_{grav} = Mg$, where $M$ is the mass of the object, and $g$ is the acceleration due to gravity.

| $t$ (s) | $x$ (m) | $v$ (m/s) |
|---|---|---|
| 0 | 0.0 x $10^4$ | 0 |
| 90 | 2.0 x $10^4$ | 490 |
| 180 | 9.6 x $10^4$ | 1240 |
| 270 | 25.5 x $10^4$ | 2400 |

32. What is the approximate mass of the shuttle after 300 s?
 A. 1.0 x $10^6$ kg
 B. 1.2 x $10^6$ kg
 C. 1.5 x $10^6$ kg
 D. 2.0 x $10^6$ kg

33. Referring to the chart, what evidence is there that the acceleration is increasing?
 A. The velocity $v$ is linear with time.
 B. The velocity $v$ increases with time.
 C. The ratio of $\Delta v$ to $\Delta t$ increases with time.
 D. The ratio of $\Delta x$ to $\Delta v$ increases with time.

34. Which gives the best reason for the increase in acceleration?
 A. The first law of motion states that an unbalanced force implies a change in velocity.
 B. The second law of motion states that acceleration is proportional to force.
 C. The second law of motion states that acceleration is inversely proportional to mass.
 D. The third law of motion states that there must be a second force, equal in magnitude but of opposite direction to the force accelerating the shuttle.

35. Consider a situation when the shuttle is in space and fires its engines, creating a force accelerating the shuttle. Which of the following BEST describes the force accelerating the shuttle?
 A. The force of the shuttle on the air.
 B. The force of the exhaust gases on the shuttle.
 C. The force of the shuttle on the exhaust gases.
 D. If there is no air, then the shuttle cannot accelerate.

**Passage 9**

In the fruit fly, *Drosophila melanogaster*, many different genes influence eye color. Mutations in any one of these genes will change the color of the eyes. Normally, flies have brick red eyes (this is called the *wild type* condition). The inheritance patterns of these variations has been extensively studied. The results of some crosses are described below.

*Cross 1*:
Wild type flies were crossed with brown eyed flies. The $F_1$ progeny all had red eyes. When the $F_1$ were crossed with each other, 3/4 of the $F_2$ had red eyes and 1/4 had brown eyes.

*Cross 2*:
Wild type females were crossed with white eyed males. All of the $F_1$'s had red eyes. In the $F_2$ generation, all of the females had red eyes, while half of the males had red eyes and half had white eyes.

*Cross 3*.
Brown eyed flies were crossed with scarlet eyed flies (scarlet is known to be an autosomal recessive trait). All of the $F_1$ progeny had red eyes. The $F_2$ progeny showed a ratio of 9:3:3:1 red to brown to scarlet to white.

*Cross 4:*
Another mutation affecting the eye, called Bar, does not affect color. Instead, it affects the shape of the eye, changing it from the normal round phenotype to an elongated, oval shape. Wild type flies were crossed with Bar eyed flies. Half of the $F_1$ progeny had Bar shaped eyes and half had wild type eyes. When the wild type $F_1$ flies were crossed with the Bar eyed $F_1$ flies, the $F_2$ generation showed the same phenotypes as the $F_1$ generation: half were wild type, half had Bar eyes.

36. The inheritance pattern of brown eyes fits with which of the following mechanisms?
    A. autosomal recessive
    B. autosomal dominant
    C. sex-linked recessive
    D. sex-linked dominant

37. The inheritance pattern of white eyes fits with which of the following mechanisms?
    A. autosomal recessive
    B. autosomal dominant
    C. sex-linked recessive
    D. sex-linked dominant

38. The inheritance pattern of Bar eyes fits with which of the following mechanisms?
    A. autosomal recessive
    B. auttsomal dominant
    C. sex-linked recessive
    D. sex-linked dominant

39. What is the best way to describe the genotype of the Bar eyed mutants in the parental generation in cross 4?
    A. homozygous dominant
    B. homozygous recessive
    C. heterozygous
    D. hemizygous

40. The relationship between brown and scarlet eye color is an example of
    A. codominance.
    B. incomplete dominance.
    C. gene interaction.
    D. epistasis

# Test 4 Solutions

**Passage 1**

1. C.
There is an electric field $E_1$ between the plates, and the force on a charged particle is given by $F_{elec} = QE_1$. Since a helium nucleus has twice the charge of a proton, the force on it is double.

2. A.
In Experiment 2 the increase in distance results in a decrease in capacitance by a factor of 3 (see equation). Since $\Delta V$ is the same in Experiment 2, the charge ($Q = C\Delta V$) decreases by a factor of 3.

3. A.
The electric field is simply given by $E = \Delta V/d$, so an increase in $d$ by a factor of 3 results in a decrease in the electric field by a factor of 3.

4. B.
Once the wires are removed from the plates, no more charge may be transferred, so the charge must stay the same. This is a situation in which the question becomes easy if we visualize the charges and think about the experiment. The question is difficult or intractable if we rely on rote memorization of equations.

5. C.
The capacitance is decreased by a factor of 2 since the distance is increased by a factor of 2. The charge is still $Q_1$ since there is no way for it to change. Thus the potential between the plates *must* change. The only equation we have relating voltage and charge is $\Delta V = Q/C$, so $\Delta V$ increases by a factor of 2.

**Passage 2**

6. C.
Since testosterone is identified in the passage as a steroid hormone, it does not stimulate cells by using cAMP or a second messenger system, and thus should not affect cells abnormally in an individual with the disease. The other three hormones are all identified as proteins, which would use cAMP and thus affect cells abnormally in an individual with McCune-Albright syndrome.

7. C.
Hormones are chemical messengers that are produced by a gland, travel to their target cells, and send a message to those cells by interacting with a receptor. While neurotransmitters travel only a short distance, they are secreted by presynaptic dendrites, diffuse across the synaptic cleft, and interact with receptors on the postsynaptic axon, signaling that cell to initiate an action potential. The parallel with hormones is striking. Choices A and B refer to intermembrane proteins that allow transport through a plasma membrane. Choice D refers to proteins that would be more analogous to hormone receptors.

8. D.
The hypothalamus, which is composed mainly of nervous tissue and is considered a part of the brain, can also produce hormones, and controls the actions of the pituitary gland almost directly. The pituitary then controls the actions of all other glands in the body. Choices A and C are simply false statements, that may confuse you due to the presence of familiar sounding terminology. While choice B is a true statement, it does not demonstrate any general connection between the two systems, and certainly not one as important as the hypothalamus-pituitary connection.

9. B.
Since the release of testosterone is controlled by GnRH, which is identified in the passage as a protein hormone, the cells that secrete testosterone will overreact to the message and secrete excess quantities of testosterone. Choice D describes the opposite of the truth; we know from question one that once produced, testosterone will not have any abnormal effects on target cells. Its presence in large quantities, however, may cause abnormal effects. Choice A is not true because testosterone, being a steroid, does not function using a second-messenger system.

10. B.
Steroid hormones diffuse freely through plasma membranes and thus should be found in all cells. They only cause effects, however, in the cells that contain the appropriate intracellular receptor. A is incorrect because it implies a mode of action that would be employed by a protein hormone. D is also incorrect; for the same reason testosterone will not affect the cells of an individual with McCune-Albright syndrome abnormally, no steroid hormone will, including Stanozolol.

**Passage 3**

11. B.
We see from the energy level diagram that the smallest energy from ground state to excited state is from $n = 1$ to $n = 2$. The difference in energy is

$$E_{\text{photon}} = (-3.4 - [-13.6])\text{ ev} = 10.2\text{ ev}$$

12. D.
By definition, ionization energy is the amount of energy required to remove an electron from an atom (or molecule) in its ground state. The energy level diagram shows this to be 13.6 ev.

13. A.
The photon energy corresponds to the difference between the two energy states. Certainly B and C are incorrect. It is not the case that we must merely have enough energy to reach the next rung of the ladder. We have to hit it exactly. Compare this with problem 17, where we wanted to ionize the atom. In that case we can hit the atom with a photon of greater energy because the excess goes into kinetic energy. We have no way of soaking up the extra energy in this case.

14. C.
Both momentum and energy are conserved because there are no external forces in this scenario. Energy is converted from one form to another: from the energy of an excited state to photon energy and (much less) kinetic energy of the recoiling atom.

15. D.
We want the longest wavelength, which means the smallest frequency (since $f = c/\lambda$), which means the smallest energy (since $E = hf$). If we look at the energy level diagram for the differences between levels, we see there is a transition corresponding to a largest energy, but the energy difference can get as small as you like (at the upper end of the diagram).

**Passage 4**

16. C
Individuals afflicted by SCID have no T cells, while those with AIDS have T cells whose function has been compromised. Choice A is incorrect because the passage tells us that individuals with AIDS can in fact make antibodies (this is the basis for many common HIV tests). No information from the passage supports choice B or choice D.

17. B
As noted in question 1, individuals with AIDS can and do make antibodies to the HIV virus; they simply cannot do it efficiently enough to ward off the disease. A is incorrect because it is a viral infection that occurs, not a bacterial one. Choices C and D are incorrect because the passage does not support them; it implies that T cells are left alive, and it does not say that HIV infects any cell type except helper T cells.

18. B
The passage tells us that the inflammatory response acts only in a nonspecific fashion, which leaves out choice A as a possibility. The major blood cells involved in inflammation are the neutrophils and macrophages, which phagocytize foreign invaders. Choices C and D are not supported by the passage: it says that phagocytes do destroy invaders, and never mentions red blood cells. Red blood cells would never be involved in phagocytosis!

19. B
Intact skin, which contains an epithelial layer that is quite thick, is a much better barrier to infection than the respiratory mucous membrane, where most infections are acquired. While the passage does not state this directly, you are expected to be able to reason it through. Since gas exchange occurs across the surface of the respiratory membrane, it must be extremely thin, and could never be as good a barrier as intact skin. Choice C is not true and is not supported by the passage.

20. B
The passage states this specifically. B cells can react to circulating antigens, so choice A is incorrect. Choices C and D are incorrect, because the passage tells us that macrophages act by phagocytosis, and also that they are nonspecific.

**Passage 5**

21. D.
Since $r$ is in the denominator, if $r$ decreases, then the force increases. If $r$ changes by a factor of 2, then the force changes by a factor of $2^2 = 4$.

22. B.
The distance $r$ is multiplied by 1.25. Thus $F$ is multiplied by $1.25^{-2} = (5/4)^{-2} = (4/5)^2 = 0.8^2 = 0.64 = (1 - 36/100)$. The force decreases by 36%.

23. C.
Concerning choice A, a factor of 4 in both $q_1$ and $q_2$ will result in a factor of 16 in $F$, so this choice is incorrect. If $F$ is to stay the same, and $q_2$ increases, then the distance $r$ must increase, so choice B is incorrect. A factor of 4 in $q_2$ is equivalent to a factor of 2 in $r$, since $2^2 = 4$. Another way to see this is to solve for $r$, which you should do if this discussion was unclear.

24. D.
If the charge on one ball increases by a factor of 4 (from 2 C to 8 C), then the force must increase by a factor of 4.

25. C.
We can eliminate A and B immediately, since $F$ decreases as $r$ increases. We can eliminate D, since $F$ does not have a linear relationship with $r$ (that would look like $F = kr + c$). As $r$ approaches 0, the force becomes infinite, so C is a good choice. Also, as $r$ becomes large, $F$ approaches 0 but never reaches it.

26. B.
If both balls acquire a charge $q$, then the force between them is given by

$$F = \frac{kq^2}{r^2}$$

which is equivalent to $F = aq^2$, a quadratic equation. Thus the answer is B.

**Passage 6**

27. B.
The normal function of the large intestine is to reabsorb water from the digestive system into the blood; the presence of large amounts of any chemical not usually present will inhibit this function, and allow more water than normal to be eliminated with the feces. This also contributes to diarrhea. None of the other choices make sense, and none are supported by the passage.

28. B.
These supplements actually do contain the enzyme lactase; since it is a normal digestive enzyme, it will not be broken down by the proteases present in the digestive system as most other proteins would. Choice A would have far reaching consequences, as the bacteria of the colon are mutualistic and play an important role in homeostasis. Choice C is unlikely and unsupported by the passage. While choice D is tempting, only monosaccharides may ever be absorbed by the small intestines -- this has to do with their size and the lack of any machinery to transport them across the small intestinal wall.

**Passage 7**

29. A.
Since insulin is a protein and should never normally be found in the digestive tract, it would in fact be digested and useless if taken orally. Both choices B and C are unreasonable and unsupported by the passage; you should be aware that glucagon should never be present in the digestive tract. Thus choice D is also ruled out.

30. B.
If diabetes type I is untreated, no insulin exists to reduce the blood sugar, causing it to build up to extremely high levels, so much so that it upsets the osmotic balance between the cells and blood, causing water to leave the cells, enter the blood, and be excreted. The tremendous amounts of glucose also tax the kidneys' mechanism for reabsorbing it, so some is excreted in the urine.

31. A.
Since the job of insulin is to reduce the blood sugar, excess insulin will do just that, by any means it can. Choice B describes the opposite effect, and therefore must be discarded. Both choices C and D are unreasonable and unsupported by the passage.

**Passage 8**

32. A.
The passage states that 3400 kg of fuel are burned each second, and that this rate is fairly constant. The remaining mass after 300 s is $2.0 \times 10^6$ kg – (300 s) (3400 kg/s) = $1.0 \times 10^6$ kg.

33. C.
Choice A is not even a true statement. Choice B is a true statement, but $v$ increasing with time indicates only that the ship is speeding up, hence having positive acceleration. The ratio of $\Delta v$ to $\Delta t$ is the acceleration, so if this ratio is increasing, the acceleration is increasing. Choice C is correct. The ratio $\Delta x$ to $\Delta v$ does not have any obvious meaning.

34. C.
According to the passage, the mass of the ship decreases as it burns its fuel. In addition, there is reason to assume the force is constant, since the rate of fuel burning is approximately constant. The connection of force, mass, and acceleration is $a = F_{net}/M$, so if $M$ decreases, then $a$ increases. Thus C is correct.

35. B.
In order for the shuttle to accelerate, there must be a force on it by some agent. The third-law-pair force must be a force of the shuttle on that agent. This narrows the choices to A and C. Since there is little air in space, A is ruled out. The shuttle pushes off from the exhaust, and that is what creates the force which accelerates it forward, according to the third law of motion. (See figure.)

**Passage 9**

36. A.
Since the $F_1$ progeny all are phenotypically similar to one parent (they have red eyes), and the $F_2$ generation displays a phenotypic ratio of 3:1 red: brown, the brown trait must be recessive. Furthermore, since the pattern does not indicate a difference in heritability in males and females, it must also be an autosomal trait.

37. C.
The pattern seen is consistent with a sex-linked recessive trait. Genotypically, the $F_1$ generation consists of all females heterozygote for the trait (as they inherit one X from the male parent and one from the female parent) and all males will be hemizygous for the normal (wild type) allele (as they can only inherit the X from the female parent). In the $F_2$ generation, the $F_1$ females will contribute the wild type allele to half the male progeny and the white allele to the other half, resulting in the phenotypes reflecting the genotypes. However, the male $F_1$'s only have a wild type allele, which will be passed on to the females in the $F_2$ generation, so all these flies will have red eyes, regardless of what allele they inherit maternally. To extend your knowledge and understanding, determine what would happen if the reciprocal cross was done (white eyed females crossed with red eyed males in the parental generation).

38. B.
Since half of the progeny in a cross between affected and unaffected individuals display the mutant phenotype, this condition is autosomal dominant.

39. C.
As stated before, the mutation is dominant, but if one parent was homozygous for the allele, all the progeny would show the mutant phenotype. Only a heterozygote could produce unaffected progeny.

40. C.
Codominance and incomplete dominance (choices A and B) refer to interactions among alleles for one gene. In this situation, we are looking at the relationship between two genes, and how the alleles for one trait affect the expression of the other trait. Gene interaction and epistasis (choices C and D) describe relationships between genes; however, gene interactions result in new phenotypes (e.g. the white phenotype) whereas epistasis results in one phenotype expressed over another. If epistasis were taking place, the ratio of 9:3:3:1, the typical ratio expected in a dihybrid cross, would change to 9:4:3, if brown was epistatic to scarlet. The genotype representing the 1 in the above ratio is *bb/scsc* (both alleles at both genes recessive). If brown masks scarlet, as in epistasis, this genotype would phenotypically be brown. If a gene interaction occurs, a new phenotype would result (white).

# Test 4 Conversion Chart

**Scale Score Conversion Chart**

| Raw Score | Scale Score | Percentile Ranking |
|---|---|---|
| 40 | 36 | 99 |
| 39 | 35 | 99 |
| 38 | 34 | 99 |
| 37 | 33 | 99 |
| 36 | 32 | 98 |
| — | 31 | 98 |
| 35 | 30 | 97 |
| 34 | 29 | 95 |
| 33 | 28 | 94 |
| 31–32 | 27 | 91 |
| 30 | 26 | 88 |
| 29 | 25 | 84 |
| 27–28 | 24 | 77 |
| 25–26 | 23 | 71 |
| 24 | 22 | 64 |
| 22–23 | 21 | 56 |
| 20–21 | 20 | 48 |
| 19 | 19 | 38 |
| 17–18 | 18 | 30 |
| 15–16 | 17 | 23 |
| 14 | 16 | 18 |
| 13 | 15 | 14 |
| 11–12 | 14 | 10 |
| 10 | 13 | 7 |
| 9 | 12 | 5 |
| 7–8 | 11 | 3 |
| 6 | 10 | 1 |
| 5 | 9 | 1 |
| 4 | 8 | 1 |
| 3 | 7 | 1 |
| — | 6 | 1 |
| 2 | 5 | 1 |
| — | 4 | 1 |
| 1 | 3 | 1 |
| — | 2 | 1 |
| 0 | 1 | 1 |

# Test 5

(35 Minutes—40 Questions)

**Passage 1**

When a massive star uses up its nuclear fuel, there is no longer enough heat to hold up its core against gravitational forces, and the core collapses. The result is an explosion, called a supernova, which leaves behind a very dense core, called a neutron star.

A neutron star is composed mainly of neutrons. It has approximately the mass of the Sun, but the radius is about 14 km, that is, 50,000 times smaller than the Sun's. The structure inside a neutron star is different from anything known on Earth, and the gravity on the surface is strong enough to crush any ordinary material. The only thing which creates a stronger gravitational field is a black hole.

Away from the surface, the tidal forces near a neutron star are nevertheless prodigious. A man falling toward a neutron star would be stretched out as he fell. He would be killed when he was about 2000 km away, and as he got closer he would be drawn as thin as a wire. Finally he would land on the surface, creating a shower of X-rays.

For these problems you may want to use the following:

$G = 6.67 \times 10^{-11}$ N m$^2$/kg$^2$
$M_{Sun} = 2.0 \times 10^{30}$ kg
$V_{sphere} = 4/3\ \pi r^3$ (volume of a sphere)
$A_{sphere} = 4\pi r^2$ (surface area of a sphere)
$A_{circle} = \pi r^2$ (area of a circle)

1. What is the approximate density of a neutron star? Note: Density, $\rho$, is given by $\rho = m/v$, where $m$ is mass and $v$ is volume.
A. $2 \times 10^{17}$ kg/m$^3$
B. $8 \times 10^{20}$ kg/m$^3$
C. $1 \times 10^{24}$ kg/m$^3$
D. $1 \times 10^{27}$ kg/m$^3$

2. The acceleration due to gravity is given by

$$a_g = \frac{GM}{r^2}$$

where $M$ is the mass of the body, and $r$ is the radius. How does the acceleration due to gravity at the surface of a neutron star compare with that near the surface of the Sun?
A. $\sqrt{50,000}$ times stronger.
B. 50,000 times stronger.
C. $(50,000)^2$ times stronger.
D. $(50,000)^3$ times stronger.

3. The force due to gravity is

$$F_{grav} = \frac{GMm}{d^2}$$

where $G$ is the gravitational constant. The Earth is $1.5 \times 10^{11}$ meters away from the Sun. If there were a planet of the same mass which was $1.5 \times 10^{11}$ meters away from a neutron star, how would the neutron star's gravitational pull on that planet compare with the Sun's pull on the Earth?
A. It would be the same.
B. It would be 50,000 times stronger.
C. It would be $(50{,}000)^2$ times stronger.
D. It would be $1.5 \times 10^{11}/14$ times stronger.

4. What is the best explanation for the stretching of an object in free fall as it approaches a neutron star?
A. The density of the neutron star is huge.
B. The parts of an object which are nearer the neutron star are pulled more strongly than the parts which farther away from the star.
C. The strong surface gravity is due to the fact that surface gravity varies inversely as the radius.
D. The strong surface gravity is due to the fact that surface gravity varies inversely as the square of the radius.

**Passage 2**

In the skin, the outer epidermis consists of many layers of tightly packed dead cells containing large amounts of keratin.

Psoriasis is a chronic skin disease in which afflicted individuals develop red patches with a scale-like appearance. The cause of psoriasis is a relatively simple one, but it is difficult to treat. The epidermal cells of an individual with psoriasis are dividing 5-10 times more frequently than they should be! This accounts for the excessive accumulation of cells that leads to the symptoms, as they cannot be sloughed off quickly enough to prevent buildup.

5. Which of the following statements is true of the cells making up the scaly patches of a psoriasis sufferer?
   A. The cells are alive and contain large quantities of melanin.
   B. The cells are dead and highly keratinized.
   C. The dells are far from the dermis, but blood vessels from the dermis grow towards the surface of the skin and supply them with nutrients, or else they could not continue to divide.
   D. The cells are dead and contain large quantities of collagen.

6. What type of drugs might be effective in sever cases of psoriasis?
   A. Collagen creams designed to moisturize the skin.
   B. Antibiotics that can kill the bacteria responsible for the problem.
   C. Anticancer drugs that systemically inhibit mitosis.
   D. Drugs that inhibit the formation of sebum, and have also been used to effectively treat acne.

**Passage 3**

Most physical situations are quite complicated, involving a number of forces or interactions even in the simplest of cases. Much of the praxis of physics is breaking a problem into parts, treating some parts exactly and ignoring other parts. Once we have solved the idealized problem, we can use the solution to evaluate the appropriateness of the idealizing assumptions.

A simple example of this is the analysis of a tennis ball falling from a height at the surface of the Earth. The ball consists of many atoms, connected by chemical forces. In addition to the chemical forces, each atom is pulled by all the pieces of the Earth. The first idealization we make is that we can treat the ball as a point mass located at its center and the Earth as a point mass located at its center. Second, we ignore the gradient of the gravitational field, so that allows us to approximate the force of gravitation on the ball as $F_{grav} = mg$, where $g \approx 9.8$ m/s$^2$ is a constant.

The third effect we generally ignore is air resistance. If we ignore air resistance, we can calculate the idealized maximum velocity of the falling ball and then calculate the force of air drag. This is given by

$$F_{drag} \approx C\rho A v^2 \qquad (1)$$

where $C$ (= 0.2) is a constant, $\rho$ (= 1.3 kg/m$^3$) is the density of air, $A$ is the cross-sectional area of the ball, and $v$ is its velocity relative to the air. If the air resistance is small, then we were justified in ignoring it.

If air resistance is important, it is possible that we can still do the problem. If the ball falls far enough for there to be a force balance $F_{net} = 0$, then we can use equation (1) to solve the problem. (Actually we can only require that $F_{net}$ be small compared to the other forces in the problem.)

For the following problems, consider a ball of radius 0.03 m and mass 0.05 kg which is tossed upward at initial velocity 3 m/s.

7. What is the initial drag force on the ball?
   A. $3 \times 10^{-4}$ N
   B. $7 \times 10^{-3}$ N
   C. 0.7 N
   D. 50 N

8. If we are going to ignore air resistance, the drag force must be small compared to
   A. the normal force.
   B. the centripetal force.
   C. the gravitational force.
   D. the frictional force.

9. We have idealized the gravitational field as being uniform. If we remove that idealization, what happens to the force of gravity as the ball travels toward the top of its flight? Note: $F_{grav} = Gm_1m_2/r^2$.
   A. The force of gravity decreases.
   B. The force of gravity increases.
   C. The force of gravity decreases, then disappears at the top.
   D. The force of gravity decreases, then increases.

10. In the idealized problem, the ball attains a certain maximum height and afterward attains a final velocity just before it reaches the ground. If air resistance is included,
    A. the height is less, and the terminal velocity is less.
    B. the height is less, but the terminal velocity is greater.
    C. the height is greater, but the terminal velocity is less.
    D. the height is greater, and the terminal velocity is greater.

11. Cats falling from large heights often survive the fall. In fact, it has been found that a cat falling from a building at very great height (e.g., ten stories) has a better chance of surviving than a cat falling from a lesser height (five stories). Which, if true, could best explain this?
    A. The air is more dense near the ground.
    B. The force of gravity on the cat is greater near the ground.
    C. Cats falling for a while tend to stretch out their legs.
    D. Greater velocity leads to a greater force of drag.

**Passage 4**

A swiftly moving charged particle, called an incident particle, moving through a material composed of neutral atoms and molecules, such as gas or biological tissue, loses kinetic energy to the material. For instance, a massive positive particle, such as a bare nucleus, moving at a speed near that of light will ionize the atoms or molecules of the medium it is moving through. In the process, there is a drag force on the incident particle and hence a loss of energy.

The energy loss depends not on the mass of the incident particle, interestingly enough, but it does depend on its charge $z$, as well as on the average number of electrons $Z$ per atom or molecule in the material. The following equation gives energy loss per distance traveled:

$$\frac{\Delta E}{\Delta x} = 4\pi NZ \frac{z^2 e^4}{m_e c^2} \ln B \qquad (1)$$

where $N$ is the number of atoms per unit volume of material,

$Z$ is the average number of electrons per atom or molecule in the material,

$z$ is the charge (number of elementary charges) of the incident particle,

$e$ is the electron charge (in Coulombs),

$m_e$ is the electron mass,

$c$ is the speed of light,

$\ln B \approx 10$ (approximately constant).

Use this equation to answer the questions.

12. Consider the following statements:

I. Momentum is conserved.
II. Kinetic energy is conserved.
III. Total energy is conserved.

In an isolated collision between a fast charged particle and an electron, if the collision is elastic, then what can be definitely concluded?

A. I only
B. III only
C. I and III only
D. I, II and III

13. A hydrogen nucleus ($^1$H) and a helium nucleus ($^4$He) have the same initial kinetic energy, with near light speed. They travel through the air. Which tends to lose more energy in a given distance?
A. $^1$H, by a factor of 4
B. They lose approximately the same amount of energy per distance.
C. $^4$He, by a factor of 2
D. $^4$He, by a factor of 4

14. A hydrogen nucleus ($^1$H) and a tritium nucleus ($^3$H) have the same initial kinetic energy, with near light speed. They are traveling through water. Which tends to lose more energy in a given distance?
A. $^1$H, by a factor of 3
B. They lose approximately the same amount of energy per distance.
C. $^3$H, by a factor of 3
D. $^3$H, by a factor of 9

15. A relativistic proton beam is incident upon helium gas at STP. A second relativistic proton beam is incident upon neon gas at STP. Which beam loses more energy in a given distance?
    A. The beam in helium, by a factor of 5.
    B. They lose approximately the same in a given distance.
    C. The beam in neon, by a factor of 5.
    D. The beam in neon, by a factor of 25.

**Passage 5**

When the bombardier beetle is attacked or provoked, it sprays a jet of hot liquid toward its attacker. The jet comes out of an opening at the tip of its abdomen, which it can move in order to direct the stream.

The physical mechanism for creating the hot spray consists of two chambers in the abdomen: an inner chamber which stores hydrogen peroxide and hydroquinone in water solution, connected by a valve to an outer chamber which contains oxidative enzymes (peroxidase and catalase) adhered to the walls. The beetle squeezes the hydrogen peroxide and hydroquinone into the outer chamber where the following reaction takes place to form quinone (also called 1,4-benzoquinone):

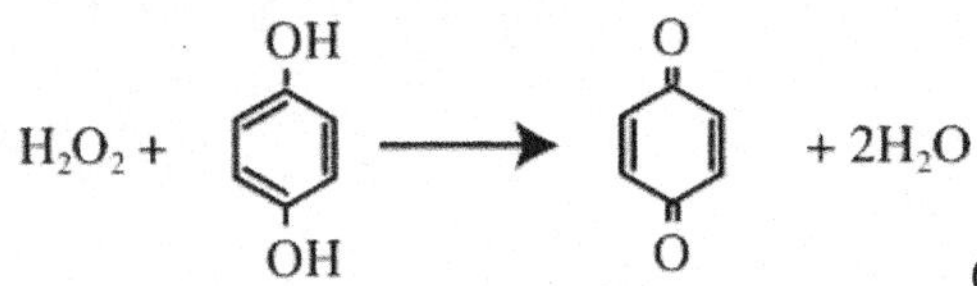


(1)

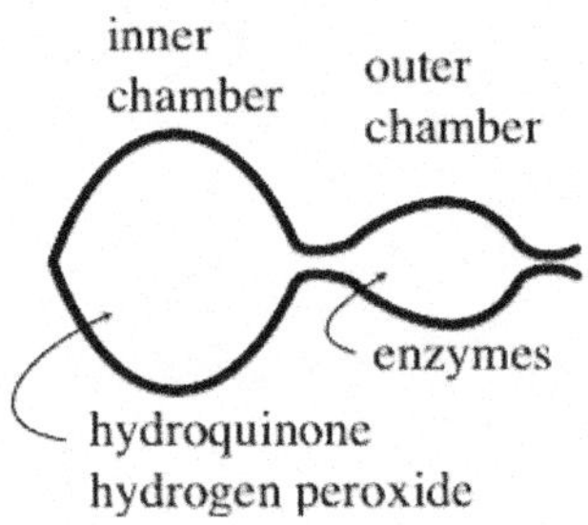


The heats of formation of the various compounds are shown in the table.

| species | $\Delta H^\circ\left(\frac{\text{kcal}}{\text{mole}}\right)$ |
|---|---|
| hydrogen peroxide | –45.68 |
| hydroquinone | –44.65 |
| quinone | –44.84 |
| water | –68.32 |

This reaction in the outer chamber creates a solution of the products of reaction (1) in water. This solution has enough heat and pressure to create a hot stream to shoot out of an opening (radius 0.1 mm) in the tip of the abdomen at a speed $v = 12$ m/s.

We can model this outer chamber as a volume (reaction flask) with a nozzle. The pressure in the chamber $P_{cham}$ is greater than the outside pressure $P_{atm}$, and this provides the force on the fluid in the nozzle. The nozzle has radius $r$ and cross-sectional area $A$, and the entire chamber has volume $V_{cham}$. The temperature of the fluid in the chamber is $T_{cham}$.

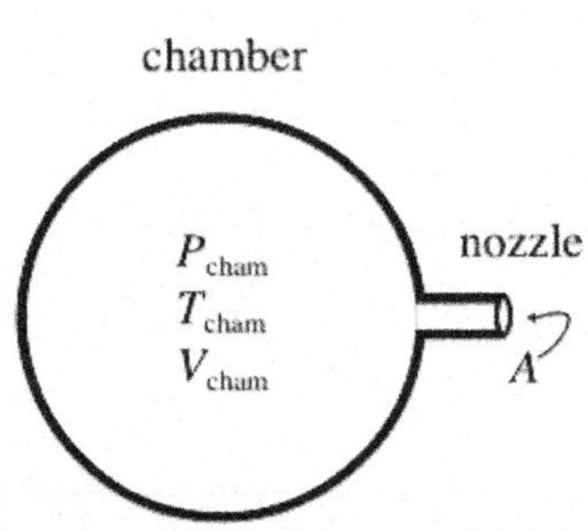

16. What is the primary reason that the reaction rate of the reaction (1) increases when the reactants go from the inner chamber to the outer chamber?
    A. The temperature is increased.
    B. The concentration of the reactants is increased.
    C. The surface area of reaction is increased.
    D. The activation energy is decreased.

17. How does the catalyst affect the heat of reaction for reaction (1)?
    A. The catalyst decreases the heat of reaction.
    B. The catalyst does not affect the heat of reaction.
    C. The catalyst increases the heat of reaction.
    D. The catalyst may decrease or increase the heat of reaction depending on the temperature.

18. The connection between force and pressure is $F = PA$, where $A$ stands for area. In reference to the model in the last paragraph, which gives the best expression for the total force on the part of the fluid in the nozzle? Note
    A. $(P_{\text{cham}} - P_{\text{atm}})A$
    B. $(P_{\text{cham}} + P_{\text{atm}})A$
    C. $\dfrac{(P_{\text{cham}} - P_{\text{atm}})V_{\text{cham}}}{r}$
    D. $\dfrac{(P_{\text{cham}} + P_{\text{atm}})V_{\text{cham}}}{r}$

19. Once reaction (1) takes place in the outer chamber, a scientist wants to determine whether there is enough heat to raise the temperature to the boiling point of the solution. In addition to the information given in the passage and the boiling point of the product solution, what information does she need to do the calculation?
    A. Heat capacity of the solution.
    B. Heat capacity of the solution and the concentration of quinone.
    C. Heat capacity of the solution, the concentration of quinone, and the heat of reaction (1).
    D. Heat capacity of the solution, the concentration of quinone, and volume of the outer chamber.

**Passage 6**

Sometimes we are concerned with what happens when a wave in one medium encounters a sharp boundary with a second medium. If the properties of the second medium are similar to those of the first medium, the wave passes from the first medium to the second, or is *transmitted* (see figure).

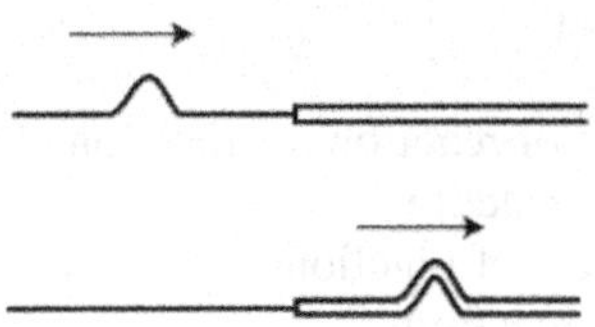

If the properties of the second medium are markedly different from the first, there is said to be an *impedance mismatch*, and much of the wave energy is reflected (see figure), while some is transmitted.

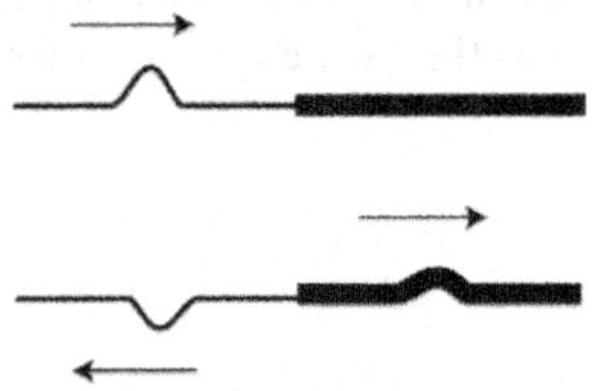

The situation is slightly more complicated if the boundary between the two media is gradual. In this case most of the energy of a wave may be transmitted to the second medium even if there is an impedance mismatch. This occurs if the length of the transition region is large compared with the wavelength of the wave (see figure). If not, then the wave "sees" the boundary as being sharp.

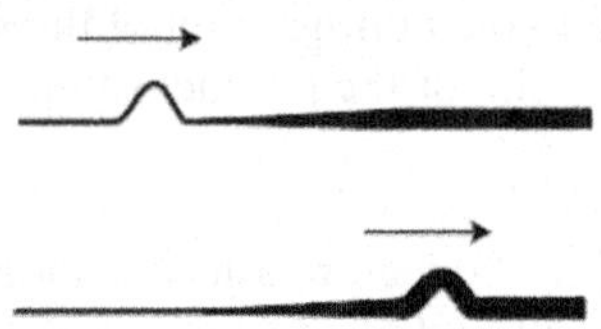

In addition to being reflected and transmitted, sometimes wave energy is *absorbed*. Certain media convert wave energy into heat energy. This can happen, for instance, for a wave traveling along a rope. The rope fibers rub against each other and the energy dissipates as heat (see figure).

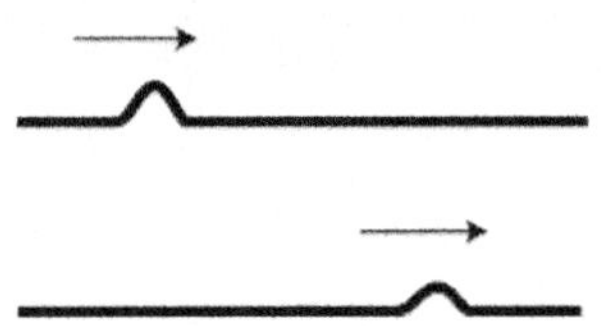

For the following questions, refer to the following chart, which shows the typical wavelengths of various waves.

| wave | typical wavelength |
|---|---|
| ocean wave | 10–100 m |
| swimming pool wave | 0.1 m |
| sound wave | 1 m |
| visible light | $10^{-7}$ m |

20. Water waves which strike the edge of a swimming pool are reflected, while ocean waves approaching the shore are generally not reflected back to sea. Which is a good explanation for this?
    A. Ocean waves have a longer wavelength than waves in pools.
    B. There is a gradual slope to shallower water at a shore.
    C. Ocean waves are generally parallel to the shore.
    D. Ocean waves are waves in a denser medium (salt water) than pool waves (fresh water).

21. What happens to the energy of ocean waves as they approach and break on the shore?
A. The energy is converted from kinetic to potential.
B. The energy is converted from potential to kinetic.
C. The energy is converted from mechanical to chemical.
D. The energy is converted from mechanical to heat.

22. Waves are approaching the ocean shore from the left in the following figure. According to the passage, which is true?

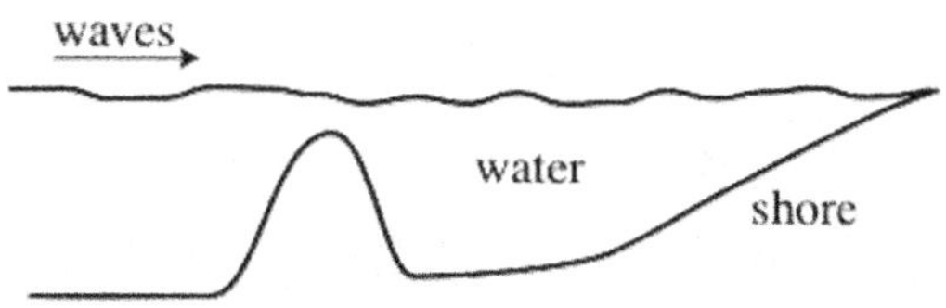


A. Waves with a long wavelength have the best chance of arriving at the shore.
B. Waves with a short wavelength have the best chance of arriving at the shore.
C. Waves of a large amplitude have the best chance of arriving at the shore.
D. Waves of a small amplitude have the best chance of arriving at the shore.

23. A photographer's lens often has a thin plastic coating on the surface of a glass lens. What is a likely purpose for that coating?
A. The coating increases the efficiency of converting light to heat.
B. The coating prevents absorption of light.
C. The coating increases the reflectivity of the lens.
D. The coating decreases the reflectivity of the lens.

24. One morning you go outside and find a blanket of newly fallen snow several centimeters thick. The outdoors seems very quiet. What is the best explanation for this?
A. The snow tends to reflect sound.
B. The snow tends to transmit sound energy to the ground.
C. The snow tends to absorb sound energy.
D. The cold temperature makes the air unable to carry sound.

25. When sound waves encounter a closed door, what is most likely to happen?
A. The sound energy is likely to be transmitted into the wood.
B. The sound energy is likely to be reflected.
C. The sound energy is likely to be absorbed.
D. The frequency of the sound is likely to be changed as it enters the door.

**Passage 7**

It is possible to construct a device for determining the speed of moving objects using sonar. The device consists of a sound emitter and a detector. The emitter creates a sound of a single frequency. The outgoing sound reflects from a moving target and is Doppler shifted. When the incoming signal arrives at the detector, the incoming signal and outgoing signal are combined, so that the detector actually detects the beat between them. The detector is a square-amplitude detector, which is thus able to pick up the beat (see figures below).

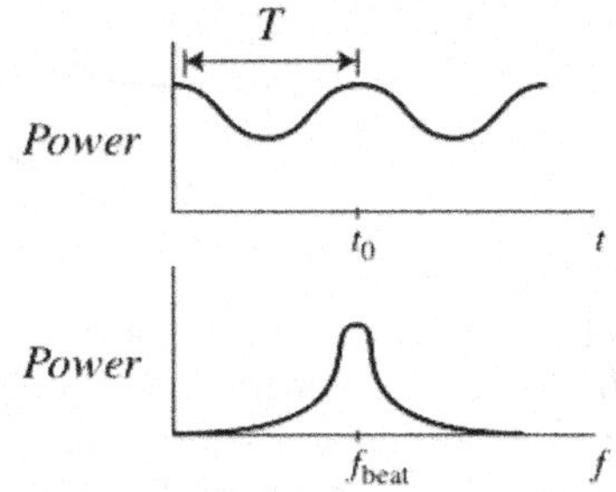


In one experiment with the device, the emitter created an outgoing signal of 80 kHz, which was reflected from a vehicle. The received signal was 70 kHz. The two graphs above were obtained. The top figure shows the output of the detector in terms of power versus time. The bottom figure shows power as a function of frequency (Fourier analyzed function of the top figure above).

For these questions, use 343 m/s for the speed of sound.

26. What is the wavelength, $\lambda$, corresponding to a 80 kHz wave? Note: Use $\lambda = v/f$.
    A. 4.3 mm
    B. 8.6 mm
    C. 27 mm
    D. 54 mm

27. What is the beat frequency referred to in this passage?
    A. 5 kHz
    B. 10 kHz
    C. 75 kHz
    D. 150 kHz

28. The *period* $T$ is the time it takes for the system to go through one cycle, and it has units [s]. The *frequency* $f$ is the number of cycles a system goes through in a unit of time, and it has units [1/s = Hertz = Hz]. We have $T = 1/f$. In the top figure above, what is $T$?
    A. $1.25 \times 10^{-5}$ s
    B. $1.33 \times 10^{-5}$ s
    C. $1.43 \times 10^{-5}$ s
    D. $10^{-4}$ s

29. Which can be concluded from the passage?
    A. The vehicle is approaching the detector directly.
    B. The vehicle is approaching the detector, but not necessarily directly.
    C. The vehicle is receding from the detector directly.
    D. The vehicle is receding from the detector, but not necessarily directly.

**Passage 8**

The mammalian eye is designed to collect light and focus it onto the retina. The retina consists of an array of cells, each having the ability to detect light falling on its surface. Most of the refraction (and thus focusing) of incoming light rays takes place at the interface between air and the cornea. The lens does the fine tuning, changing the focal length so the image lands exactly on the retina. The tuning is necessary since the eye must be able to bring into focus light from objects as close as 0.1 m as well as light from an infinitely distant source.

*Spatial resolution* is the ability of the eye to distinguish waves coming from different directions. For example, if a distant car is facing you at night with its headlights on, you can see two distinct headlights, since the light from the two headlights approaches your eye from two directions (see figure). If the car is far enough away, however, your eye lacks the resolution to distinguish the headlights, and you see only one light source.

Resolution is measured in degrees or radians. For instance, if your eye can just resolve two headlights which are 1.5 m apart on a car which is 1 km away, then the angular separation of the lights is approximately 1.5 m/1000 m = $1.5 \times 10^{-3}$ radians. The resolution of your eye is $1.5 \times 10^{-3}$ rad or 0.09 degrees (since 1 rad $\approx 57^\circ$) or 5 seconds of an arc. To a good approximation, the angular separation of two light sources (or features on any sort) is the ratio of spatial separation $\Delta x$ to distance from the point of reference $L$. (See figure.) Thus the better the resolution, the smaller the resolution angle.

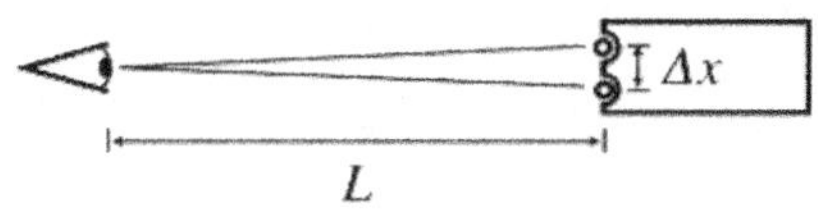


Ultimately the spatial resolution of any detector, including the eye, is limited by *diffraction*, which is the spreading of waves. When waves pass through an aperture, they spread on the other side subtending an angle given by

$$\theta_{\text{diff}} = \lambda/d \qquad (1)$$

where $\theta_{\text{diff}}$ is measured in radians, $\lambda$ is the wavelength of the wave involved, and $d$ is the diameter of the hole through which the waves must pass. Of course, diffraction is the physical limit of the resolution. The actual resolution of a detector may be much poorer than equation (1) would indicate if it is poorly designed. The human eye, when functioning properly, is essentially diffraction limited.

For the following problems use $c = 3.0 \times 10^8$ m/s. Green light has a wavelength of 520 nm in a vacuum. (1 nm = $10^{-9}$ m)

30. A certain eye does not focus correctly, being either near- or far-sighted. The front to back length of the eye is 0.025 m, but the focusing power of the resting eye is 35 diopters. What should be the approximate power of the appropriate corrective lens?
A. –5 diopters
B. –3 diopters
C. 3 diopters
D. 5 diopters

31. The cornea is made of a material which has a larger index of refraction for blue light than for red light. If the eye is focusing a beam of red light onto the retina, where would the focus for a beam of blue light fall?
A. In front of the retina.
B. On the retina.
C. Behind the retina.
D. Both in front of and behind the retina.

32. An engineer is working on a camera to photograph the distant landscape in foreign countries. He has designed a camera with a lens which focuses incoming light on a detector. The camera is essentially diffraction limited, but the resolution is not good enough. Which of the following could improve the resolution?
A. Increase the distance from the lens to the detector.
B. Improve the lens shape.
C. Change the material of the lens to be more transparent.
D. Increase the size of the whole camera.

33. A Seurat painting consists of many dots of paint about 0.002 m in diameter. If you view it from a great enough distance, the dots of color appear to blend together, and you see a coherent picture. The resolution of your diffraction-limited eye is $2 \times 10^{-4}$ radians. If you wanted to know how far away need you be for the dots to blur together, which paragraph in the passage gives the information to calculate this?
A. Paragraph 3, and you must stand 2 m away.
B. Paragraph 3, and you must stand 10 m away.
C. Paragraph 4, and you must stand 2 m away.
D. Paragraph 4, and you must stand 10 m away.

34. A cat's eye, adapted for seeing at night, has a larger pupil than a human eye and a much larger lens. The resolution for a cat's eye, however, is not better than that of a human eye, which is almost diffraction limited. Which of the following is a possible explanation for the lack of resolution in a cat's eye?
A. The larger pupil allows more light to enter the eye.
B. The larger pupil restricts the amount of directional information entering the eye.
C. The large lens introduces chromatic aberration.
D. The large lens introduces spherical aberration.

**Passage 9**

In a certain apparatus, a long wire along the $z$-axis carries a uniform charge on it. The resulting electric field outside the wire is directed away from the wire and has a magnitude given by

$$E = K/d$$

where

$$K = 30{,}000 \text{ Nm/C}$$

and $d$ is the perpendicular distance from the wire to the point in question.

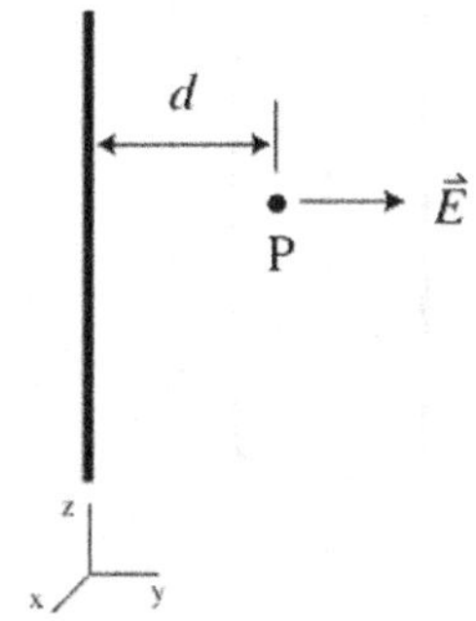


Charges near the wire experience forces due to the electric field. This is true of elementary particles, like electrons, as well as ions and molecules. Polar molecules have one end charged positively and the other end, negatively. The two sides of the molecule thus experience two forces.

If the electric field is great enough, the gas around the wire may undergo breakdown, that is, the molecules are ionized by the electric field.

For the following questions, use for the charge on an electron

$$q_{elec} = -1.6 \times 10^{-19} \text{ C}$$

35. What is the sign of the charge on the wire?
   A. Positive.
   B. Negative.
   C. Positive or negative.
   D. Positive and negative in equal amounts to make neutral.

36. Which of the following best depicts the electric field lines?

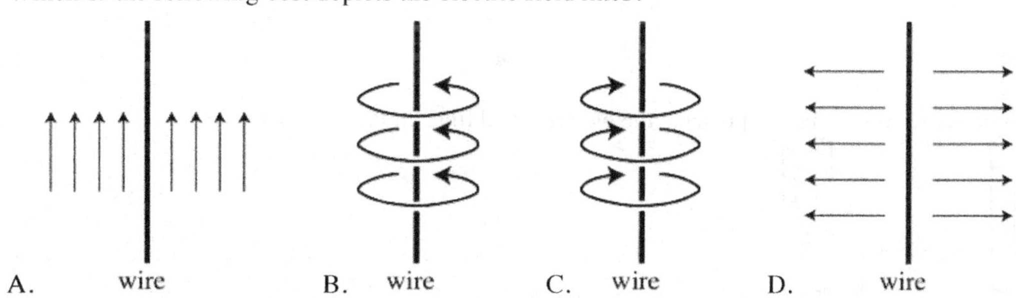


37. If a butanol molecule ($CH_3CH_2CH_2CH_2OH$) is near the wire, what is its most likely orientation?
   A. The methyl group will be near the wire.
   B. The methyl group will be away from the wire.
   C. The oxygen atom will be near the wire.
   D. The oxygen atom will be away from the wire.

**Passage 10**

An electrostatic precipitator is a device used in industry to remove pollution from exhaust gas. It consists of a long thin wire surrounded by a conducting cylinder, such that a potential about $5 \times 10^4$ volts is maintained between the negative wire and the positive cylinder. (See figure.)

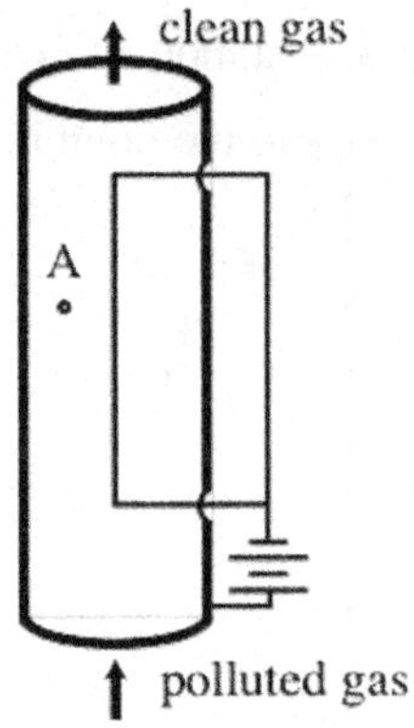


The resulting electric field inside the cylinder varies inversely with the distance from the center wire. Neutral particles of pollution are attracted to the center wire. The electric field near the wire is strong enough (greater than about $3 \times 10^6$ N/C) to ionize air, so the pollution particles are ionized negatively. They are attracted to the outer cylinder, where they collect and are eventually removed.

A typical energy requirement for such a device is about 300 Joules for each cubic meter of gas processed.

Note: The charge on an electron is $-1.6 \times 10^{-19}$ C. Point A is in the plane of the page.

38. If a sodium ion is at point A, in what direction would it experience a force?
 A. right
 B. left
 C. up
 D. down

39. If a $C_{10}H_{18}O_2$ molecule is at point A, in what direction would it experience a force?
 A. right
 B. left
 C. up
 D. down

40. Which of the following best represents the electric field lines in the cylinder?

A. 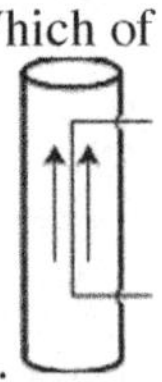

B. 

C. 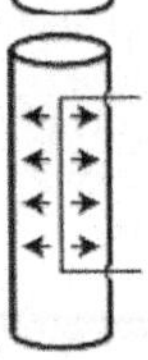

D. 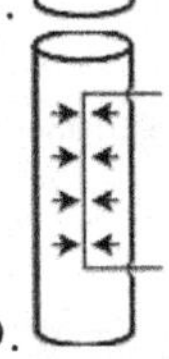

# Test 5 Solutions

**Passage 1**

1. A.
We estimate density as follows:

$$\rho = \frac{m}{v} = \frac{m}{\frac{4}{3}\pi r^3} = \frac{2.0\times 10^{30}\,\text{kg}}{4\left(1.4\times 10^{4}\,\text{m}\right)^3}$$

Here we have estimated $\pi = 3$. Next we estimate $1.4^3 = 1.4 \cdot 1.4 \cdot 1.4 = 2 \cdot 1.4 = 3$, so we have

$$\rho = \frac{10^{30}}{2\cdot 3\cdot 10^{12}}\frac{\text{kg}}{\text{m}^3} = \frac{10}{6}\cdot 10^{29-12}\frac{\text{kg}}{\text{m}^3} = 1.6\times 10^{17}\frac{\text{kg}}{\text{m}^3}$$

The arithmetic is spelled out so that you can see how you can do a lot of fast estimating and still get a reasonable answer. Even on questions which involve much less arithmetic, you should always be looking for shortcuts.

2. C.
The acceleration due to gravity is given by

$$a_g = \frac{GM}{r^2}$$

where $M$ is the mass of the body, and $r$ is the radius. The passage states that $M$ for a neutron star is the same as $M$ for the Sun, but $r$ is 50,000 times smaller. That means $a_g$ is greater by a factor of $(50{,}000)^2$.

3. A.
Considering the two situations in the problem, we know the mass of the neutron star is the same as the mass of the Sun. The mass of the two planets is the same. The distance $d$ is the same for the two situations. The force of gravity does not depend on the size of the bodies but on the distance from center to center.

4. B.
Choice A is not relevant to the things which happen 2000 km away from the neutron star. Concerning choices C and D, the surface gravity of the neutron star is also irrelevant. Choice B correctly states that the force of gravity between near things is greater than between far things. This difference is enough to pull a body apart.

**Passage 2**

5. B.
Like all outer epithelial cells, the cells of a psoriasis blemish are dead and highly keratinized, which is what gives them the scaly appearance. All of the other choices do not make sense, and are not supported by the passage.

6. C.
Since the problem here is excess cell division (which must be by mitosis), anti-cancer drugs that inhibit cell division should be effective (although there are side effects). While collagen is an important component of the dermis, it has nothing to do with psoriasis or the epidermis (choice A). Choice B would only be valid in the case of a bacterial infection, and none exists in this case. Sebum-inhibiting drugs do work against acne, but again, this is due to bacterial infection and is not applicable to psoriasis (choice D).

**Passage 3**

7. B.
We calculate the drag $F_{drag} = (0.2)\ (1.3\ kg/m^3)\ (\pi\ (0.03\ m)^2)\ (3\ m/s)^2 = 7 \times 10^{-3}$ N.

8. C.
According to the passage, the air resistance must be small compared to the other forces in the situation, but this is just the force of gravity.

9. A.
As the ball travels away from the center of the Earth, the force of gravity decreases slightly, since

$$F_{grav} = Gm_1m_2/r^2$$

10. A.
If air resistance is included, there is an initial force down while the ball is going up, so the maximum height is less. The ball loses energy to air resistance, so its speed just before hitting the ground will be less than in the idealized problem.

11. C.
Even if the density of air changed appreciably (it does not), it would not help cats survive a greater fall. This holds for choice B as well. Regarding choice C, if cats stretch their legs, then this increases their cross-sectional area, which would decrease their terminal velocity, so is a viable possibility for an answer. Regarding D, the statement is true, but the fact would not help a cat to have a lesser terminal velocity from a greater fall.

**Passage 4**

12. D.
If the collision is isolated, then there are no external forces and momentum is conserved, so I is true. The word "elastic" implies that kinetic energy is conserved, so II is true. The total energy is always conserved in an isolated collision, so III is true. Thus the answer is D.

13. D.
On the right-hand side of the equation, everything is constant except $N$, $Z$, and $z$. In this question we are comparing $^3$H and $^3$He, so the only difference is $z$. That is, $^3$H has $z = 1$ and $^3$He has $z = 2$. Since $z$ is doubled, $^3$He must lose 4 times as much energy per unit distance.

14. B.
The only difference in the two nuclei is the mass, which plays no role in equation (1).

15. C.
Since both gases are at STP, the ideal gas equation guarantees that the two gases have the same number of moles per unit volume, hence $N$ is the same. Neon has 10 electrons per gas particle, and helium has 2 per particle. So neon has a $Z$ which is 5 times larger and removes 5 times as much energy per unit distance.

**Passage 5**

16. D.
The main difference between the inner and outer chambers is the presence of enzymes in the outer chamber. Enzymes are catalysts which reduce the activation energy. Choice A may play a role, but a smaller one.

17. B.
A catalyst affects activation energy but has no effect on the heat of reaction.

18. A.
The connection between force and pressure is $F = PA$, so the answer is A or B. The pressure inside the chamber is pushing in the opposite direction as the atmospheric pressure outside the chamber, so A is correct.

19. B.
The table in the passage allows us to calculate the heat released per mole of quinone produced. We need to know the concentration of quinone to know how much heat per quantity of solution is produced. Thus A is incorrect: not enough information. Once we know the heat available per quantity of solution, then we can use the heat capacity to get the temperature change. B is correct. As for choice C, the heat of reaction is necessary but it is derivable from information in the passage. As for choice D, the volume of the chamber is unnecessary information, since the question is about intensive properties of the solution and not about extensive properties. That is, the question is not about how much stuff there is, but about quantities that do not depend on volume (such as temperature).

**Passage 6**

20. B.
Choice A is true but does not do a good job of explaining anything. Choice B refers to the transition from deep to shallow water, which happens over length scales very long compared to the 10-m ocean waves. The pool waves hit the sharp boundary of the side of the pool and are reflected. Choices C and D are true but irrelevant.

21. D.
As waves, the energy is initially both kinetic and potential, which is refer3d to as mechanical. Choices A and B do not express what happens when a wave breaks, and the mechanical energy goes into another form. There are no chemical changes, so C is incorrect. Choice D is reasonable. The source of friction is the sandy beach.

22. A.
According to the second paragraph, a wave has its best chance of being transmitted if it has short wavelength.

23. D.
Choice A is incorrect, since we certainly do not want to convert light energy to anything else. Choice B is incorrect, since the situation more closely resembles paragraphs 1 and 2, the coating providing a gradual transition. Visible light waves have a small wavelength, so a thin coating would suffice. This would decrease the reflectivity.

24. C.
Choice A is incorrect, since the point is that the snow is *not* reflecting sound, resulting in your not hearing it. Choice B would be a good explanation, if the coat of snow were thick compared with the wavelength of sound (paragraph 2). This is not the case. It must be the case that the snow absorbs the sound energy, turning it to heat.

25. B
Choice A is incorrect, because the transition from air to wood is abrupt. Choice B seems likely. There is nothing to indicate C is correct. Choice D is definitely incorrect, since the frequency of a wave stays constant as it travels from place to place. There is nothing in the passage indicating a change of frequency.

**Passage 7**

26. A.
$\lambda = v/f = 343/80 = 4.2875 \approx 4.3$.

27. B.
The beat frequency is merely the difference between the outgoing and returning sound frequencies, as implied in the first paragraph: $f_{beat} = f_1 - f_2 = 80\text{ kHz} - 70\text{ kHz} = 10\text{ kHz}$.

28. D.
In the upper figure we see the beat, so $T$ refers to the beat period, which is $1/f_{beat}$. In problem, we calculated the beat to be 10 kHz, so $T = 1/f = 1/10\text{ kHz} = 1/10^4 = 10^{-4}$.

The reason that the power never reaches zero is that the outgoing wave has a larger amplitude than the incoming wave.

29. D.
Because the detected frequency is lower, the vehicle is receding, but there is no way to know from the passage whether it is doing so directly away or at an angle. The Doppler effect happens in three dimensions, and we have discussed only the one dimension parallel to the motion (which is the important one).

**Passage 8**

30. D.
The desired power of the combination of corrective lens plus eye lens is 1/0.025 m = 40 D. Since $P_{combo} = P_{eye} + P_{correct}$, we must have $P_{correct} = 5$ D.

31. A.

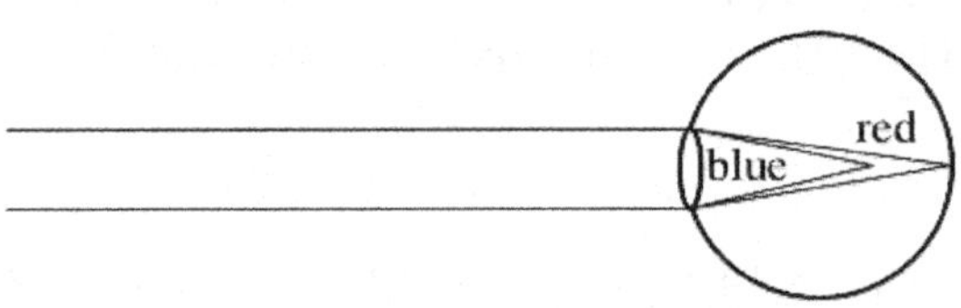


The figure shows the red light focused on the retina. This lens refracts the blue light more than red light and hence focuses blue rays in front of the red focus. The figure exaggerates the case.

32. D.
According to the passage, if the camera is diffraction limited, then the resolution depends on the size of the lens (that is, light gathering hole) and the wavelength of the light used. Choices A, B, and C address neither of these issues. Increasing the size of the entire camera would increase the size of the lens, which would increase the resolution, by decreasing the resolution angle.

33. B.
Trying to determine if two dots are separate or blurred is analogous to trying to distinguish two headlights, so the information is in paragraph 3. The resolution of the eye is the ratio of dot separation to standing distance, so we write

$2 \times 10^{-4} = 0.002\text{m}/D$

$D = 10$ m.

You must stand 10 m away.

34. D.
Concerning choice A, a larger pupil does allow in more light, but this does not explain a decrease in resolution. For choice B, the opposite is true: a larger pupil allows more directional information to enter the eye, which should improve resolution if there were not another factor present. For choice C, both cats and humans contend with chromatic aberration. The large lens introduces spherical aberration, so D is correct.

**Passage 9**

35. A.
Since the electric field is directed away from the wire, this indicates that the charge on the wire is positive.

36. D.
The electric field lines point in the same direction that the field points, that is, away from the wire.

37. C.
In an butanol molecule, the oxygen atom is the most electronegative atom, so it is negatively charged relative to the rest of the molecule. Since the electric field points away from the wire, the negative oxygen experiences a force toward the wire.

**Passage 10**

38. A.
A sodium ion ($Na^+$) is positively charged, and the negatively charged central wire will attract it.

39. A.
It does not matter what kind of molecule this is. The only thing that matters is its neutral charge. The passage states that it will be attracted to the wire.

40. D.
The electric field lines point away from the positive charge and toward the negative charge.

# Test 5 Conversion Chart

**Scale Score Conversion Chart**

| Raw Score | Scale Score | Percentile Ranking |
|---|---|---|
| 40 | 36 | 99 |
| 39 | 35 | 99 |
| 38 | 34 | 99 |
| 37 | 33 | 99 |
| 36 | 32 | 98 |
| — | 31 | 98 |
| 35 | 30 | 97 |
| 34 | 29 | 95 |
| 33 | 28 | 94 |
| 31–32 | 27 | 91 |
| 30 | 26 | 88 |
| 29 | 25 | 84 |
| 27–28 | 24 | 77 |
| 25–26 | 23 | 71 |
| 24 | 22 | 64 |
| 22–23 | 21 | 56 |
| 20–21 | 20 | 48 |
| 19 | 19 | 38 |
| 17–18 | 18 | 30 |
| 15–16 | 17 | 23 |
| 14 | 16 | 18 |
| 13 | 15 | 14 |
| 11–12 | 14 | 10 |
| 10 | 13 | 7 |
| 9 | 12 | 5 |
| 7–8 | 11 | 3 |
| 6 | 10 | 1 |
| 5 | 9 | 1 |
| 4 | 8 | 1 |
| 3 | 7 | 1 |
| — | 6 | 1 |
| 2 | 5 | 1 |
| — | 4 | 1 |
| 1 | 3 | 1 |
| — | 2 | 1 |
| 0 | 1 | 1 |

# Test 6

(35 Minutes—40 Questions)

**Passage 1**

Certain toys found in cereal boxes display the phenomenon of phosphorescence. If a child exposes the toy to a bright light (for example, the Sun), then the toy will glow with a characteristic color when the child takes it into a dark closet.

The phenomenon of phosphorescence requires the interaction of light with three energy levels in an atom (or molecule). The figure below (not to scale) shows a hypothetical energy diagram in which the lowest state shown is the ground state ($E = 0$). When the toy is taken into the Sun, a photon causes a transition from the ground state to state 2. Almost immediately, the atom emits a photon, making a fast transition to state 1.

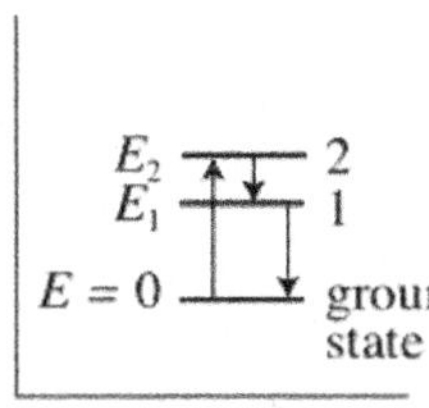


The atom makes a transition to the ground state only slowly. Thus when the child takes the toy into a closet, he is able to observe the phosphorescent photons from this final transition. The transition from state 1 to the ground state is said to be *forbidden*, and it takes place by different (and thus slower) processes than the other transitions.

Let $f_{abs}$ be the frequency of the absorbed photons, $f_{f1}$ be the frequency of the photons emitted while the object is still in the sunlight, and $f_{ph}$ be the frequency of the photons released in the closet. (Planck's constant is $h = 4.14 \times 10^{-15}$ ev s.)

1. How does $f_{abs}$ compare to the $f_{ph}$?
   A. $f_{abs} < f_{ph}$
   B. $f_{abs} = f_{ph}$
   C. $f_{abs} > f_{ph}$
   D. It depends on the specific atoms used in the material.

2. How does $f_{f1}$ compare to the $f_{ph}$?
   A. $f_{f1} < f_{ph}$
   B. $f_{f1} = f_{ph}$
   C. $f_{f1} > f_{ph}$
   D. It depends on the specific atoms used in the material.

3. The frequency of the photons is given by $f = E_{photon}/h$. Which of the following gives a correct expression for $f_{ph}$?
   A. $\frac{E_1}{h}$  B. $\frac{E_2}{h}$
   C. $\frac{(E_1 + E_2)}{h}$  D. $\frac{(E_2 - E_1)}{h}$

4. After the child brings the object into the closet, which atomic state is the most populated? That is, a number of atoms have electrons in the ground state, a number in state 1, and a number in state 2. Which number is greatest?
   A. The ground state is the most populated.
   B. State 1 is most populated.
   C. State 2 is most populated.
   D. Either the ground state or state 1 is the most populated.

5. Which is correct?
   A. The Sun emits photons of many frequencies, but only photons corresponding to the absorption frequency will be absorbed by the object.
   B. The Sun emits photons of many frequencies, but only photons corresponding to the absorption frequency make it through Earth's atmosphere.
   C. The Sun emits photons of one frequency, and the toy is constructed to absorb that frequency.
   D. The Sun emits photons of one frequency, and the toy converts the photons to the desired frequency.

**Passage 2**

A cannon is a device for imparting a large velocity to a mass of iron, generally for the purpose of warfare. Over the centuries, the manufacture of cannons has taken many forms, but the basic construction has remained the same.

A cylinder (or *barrel*) is closed at one end (the *breech*) and open at the other (the *muzzle*). An explosive (or *charge*) is placed in the cylinder at the breech, and a ball is placed in the cylinder on top of the charge. The explosive is ignited and the reaction produces hot gases which increase the pressure. Thus the gases push the ball along the cylinder and out the muzzle at great velocity.

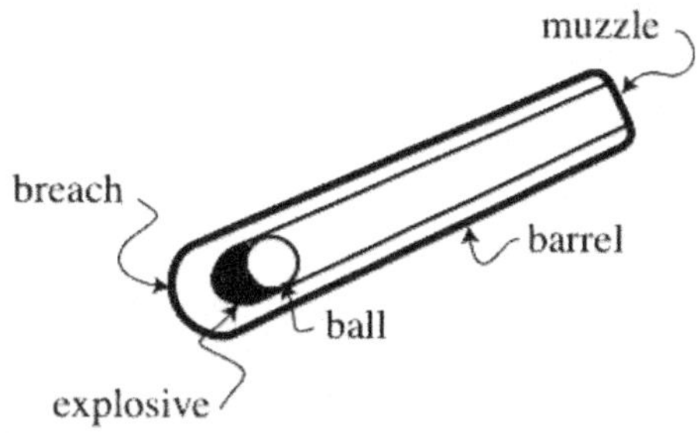


Beginning in the 1500s, gunners began using large-grained explosive in order to decrease the rate of burning. Faster burning charge creates a large pressure very quickly, thus creating stress on the cannon and creating a risk of failure. The slower burning charge ensures that the pressure behind the ball stays more nearly constant as the ball travels the length of the barrel.

For the following questions, consider a cannon which is 2.3 meters long with a bore (hole) of radius 5 centimeters. Assume the pressure inside the cannon after the explosive has been set off is constant. While the ball is in the cannon, the forces due to the gases are so much greater than the force of gravity that the force of gravity can be ignored.

6. What additional piece of information would be sufficient to allow the calculation of the kinetic energy of the ball upon leaving the cannon?
   A. The mass of the cannonball.
   B. The final velocity of the cannonball.
   C. The force the cannonball experiences in the barrel.
   D. None of the above.

7. Pressure and force are related by the formula $P = F/A$, where $A$ stands for area. What additional piece of information would be sufficient to allow the calculation of the pressure in the barrel while the cannonball is still inside?
   A. The temperature of the gas in the barrel.
   B. The final velocity of the cannonball.
   C. The force the cannonball experiences in the barrel.
   D. None of the above.

8. Which of the following would best explain why a large-grained charge would burn more slowly than a small-grained charge?
   A. The rate of reaction depends on the surface area.
   B. The rate of reaction depends on the concentration of reactants.
   C. The rate of reaction depends on the temperature of reactants.
   D. The activation energy is reduced for smaller-grained charge.

9. Which of the following is necessarily true?
   A. The entropy change during burning is zero.
   B. The free energy change during burning is negative.
   C. The free energy change during burning is zero.
   D. The free energy change during burning is positive.

10. Which of the following best describes the energy flow in the passage?
   A. chemical to kinetic to heat
   B. chemical to heat to kinetic
   C. chemical to potential to heat
   D. chemical to potential to kinetic

**Passage 3**

A recent innovation in automobile safety is the airbag. In the event of a collision, an airbag in front of the driver inflates in about 0.02 seconds. As the driver's body comes to a rapid stop, the airbag provides a soft cushion, softer than the steering wheel in any case.

The essential mechanism of the airbag consists of a sealed combustion chamber containing iron(III) oxide and sodium azide, which react to form the nitrogen gas which fills the airbag:

$$Fe_2O_{3(s)} + 6NaN_{3(s)} \rightarrow 9N_{2(g)} + 3Na_2O_{(s)} + 2Fe_{(s)} \qquad (1)$$

In the same chamber is an igniter, consisting of an electric coil, zirconium and potassium hyperchlorate, which react somewhat as follows:

$$4Zr_{(s)} + KClO_{4(s)} \rightarrow 4ZrO_{(s)} + KCl_{(s)} \qquad (2)$$

The main active ingredient in the airbag is the sodium azide. This ingredient is prepared using the reaction of sodium amide and nitrous oxide:

$$NaNH_{2(s)} + N_2O_{(g)} \xrightarrow[180°C]{} NaN_{3(s)} + H_2O_{(g)} \qquad (3)$$

During a collision, an electric coil starts reaction (2). This provides the heat which ignites reaction (1). After 1–5 ms, the seal of the combustion chamber bursts and releases the products (see figure). A filter screen removes the solid products of reaction (1) and other trace reaction products (some possibly noxious) while passing the nitrogen gas.

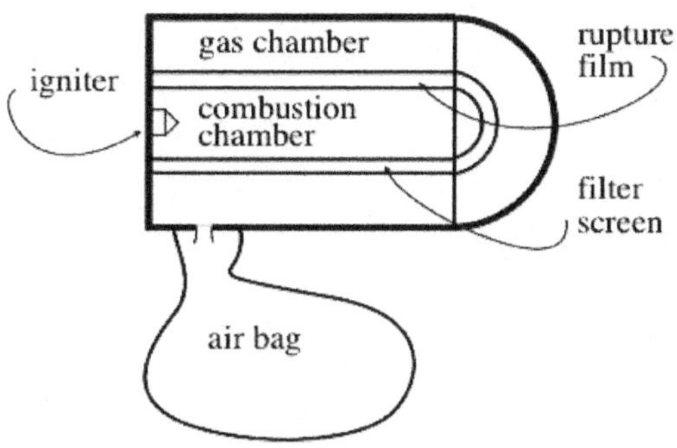


The nitrogen gas fills the bag in about 20 ms, and the whole reaction (1) continues for about 50 ms. The bag is porous, so as the driver presses against it, the bag deflates. Thus, it is almost completely deflated by the time the accident is over. After that, it is hoped that the driver can walk away from the collision with no injuries worse than abrasions from the inflating bag.

11. What is a possible purpose for keeping the combustion chamber sealed during the first 1–5 ms of reaction?
   A. This ensures that poisonous byproducts are prevented from entering the bag.
   B. The concentration of reactants is kept high to speed the reaction.
   C. The reactants are protected from water vapor, which could contaminate the reaction.
   D. This keeps the reactants near the catalyst.

12. Of reactions (1), (2), and (3), which has the greatest entropy (the degree of disorder or randomness in a system) increase?
   A. (1)
   B. (2)
   C. (3)
   D. All three have zero entropy increase.

13. What is the best possible purpose for having the bag deflate as the driver presses into it?
    A. As the bag deflates, the temperature of the gas decreases to safe levels.
    B. Allowing the gas to flow out of the bag ensures that the bag does not burst.
    C. The bag's deflation decreases the energy of the collision.
    D. Increasing the distance over which the deceleration occurs decreases the required force.

14. What is the advantage of having an airbag with larger area (as one views it from the driver seat)?
    A. Increasing the area increases the gas flow rate.
    B. Increasing the area decreases the gas flow rate.
    C. Increasing the area decreases the pressure on any given body part.
    D. Increasing the area decreases the density of the gas inside the bag.

15. Which of the following is the best description of the energy flow for the airbag inflation?
    A. chemical to heat
    B. chemical to gravitational potential.
    C. chemical to kinetic to heat.
    D. chemical to heat to work against the atmosphere.

16. Which of the following is the *least* likely byproduct for reaction (1) referred to in Paragraph 3?
    A. $NaO_2$
    B. Na
    C. FeO
    D. $Na_3N$

**Passage 4**

The human ear can hear sounds with frequencies from 20 Hz to 20,000 Hz. Frequencies of sound higher than this are called ultrasound. Although they cannot be heard, they are used in the technique of ultrasound imaging, for example, to take the image of a fetus in the womb. The sound waves are reflected off the interface between the fetus and the surrounding fluid. In order for this to provide information, the wavelength of the sound has to be smaller than the object being observed. Otherwise the wave passes right around the object.

At the other end of the sound spectrum there are very low frequency sounds. These can be highly injurious to humans if they have sufficient intensity. The sounds can cause internal organs to vibrate and eventually rupture, tearing the connective tissue holding the organ in place. For this reason, there are limits in the workplace as to how intense low frequency sounds can be.

For the following, use 1500 m/s for the speed of sound in biological tissue. Use 343 m/s for the speed of sound in air. A mass *m* on a spring (constant *k*) has a frequency given by

$$f = \frac{1}{2\pi}\sqrt{\frac{k}{m}}$$

17. If a doctor wanted to take the image of a fetus and wanted to resolve features of size on the order of one millimeter, what frequency sound could she use?
    A. Any frequency less than about $3 \times 10^5$ Hz.
    B. Any frequency greater than about $3 \times 10^5$ Hz.
    C. Any frequency less than about $1.5 \times 10^6$ Hz.
    D. Any frequency greater than about $1.5 \times 10^6$ Hz.

18. What period, $T$, corresponds to the lowest frequency, $f$, a human ear can hear? Use $T = 1/f$.
    A. 0.05 seconds
    B. 0.05 meters
    C. 17 seconds
    D. 17 meters

19. Which best describes the flow of energy in paragraph 2?
    A. sound to potential
    B. sound to heat
    C. sound to gravitational
    D. sound to chemical and heat

20. In paragraph 2, which of the following would result in an increase in intensity?
    A. An increase in frequency.
    B. A decrease in frequency.
    C. An increase in the mass of the internal organ.
    D. None of the above.

21. Which of the following would be part of an explanation of why low frequency sounds are injurious?
    A. The wavelength of the sound is smaller than the size of the organ.
    B. The wavelength of the sound is larger than the size of the organ.
    C. The sound is reflected by the organ.
    D. The frequency of the natural oscillation of the organ is similar to that of the sound.

**Passage 5**

If two parallel metal plates separated by a small gap are charged, one positively and one negatively, then there will be an electric field between the plates. If the electric field is larger than a certain threshold, then a spark may jump across the gap, partially discharging the plates. The threshold electric field depends on the type and pressure of the gas between the plates.

To see how a spark occurs, imagine a lone electron between the plates. It experiences a force and an acceleration, so it will move in the direction of the positive plate. Before it gets there, it is likely to collide with a gas particle. The average distance an electron travels before encountering a gas particle is the *mean free path*. The mean free path depends only on the number density of the gas, that is, the number of gas particles per unit volume. If the electron gains enough energy before colliding with the gas particle to ionize it, then after the collision there are more electrons to continue the process. The original electron loses much of its kinetic energy but is still available to accelerate and ionize other gas particles. (See figure.)

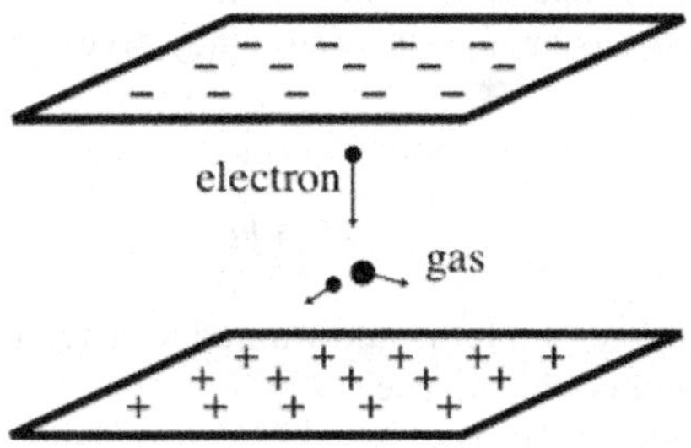


A lone electron thus undergoes acceleration and energy gain, followed by a collision, depositing energy into the gas particle and debris. Each collision releases several electrons, so the phenomenon grows exponentially. The process described is a run-away chain reaction. The result is a transfer of charge from one plate to the other. In addition, the discharge leaves many molecules in an excited state, which releases photons as they decay to the ground state.

For air at atmospheric pressure the threshold electric field is about

$$E_{th} = 3 \times 10^6 \text{ N/C}$$

You may also use the following:

$$k = 9 \times 10^9 \frac{\text{Nm}^2}{\text{C}^2}$$

$$Q_{elementary} = 1.6 \times 10^{-19} \text{ C}$$

$r_0$ = the average atomic radius = $6 \times 10^{-11}$ meters
$l_{mfp}$ = the mean free path = $8 \times 10^{-9}$ meters

22. The force on a charged particle in an electric field is $F = QE$. If an oxygen ion ($O^{2-}$) and an electron are both between the plates, which experiences the greater force?
A. The oxygen ion by a factor of 8.
B. The oxygen ion by a factor of 2.
C. They experience the same magnitude of force.
D. The electron experiences a much greater force because of its small mass.

23. How much force would a calcium atom experience if it were between the plates? (Assume the electric field is the threshold electric field for air.)
    A. 0 N
    B. 4.8 x $10^{-13}$ N
    C. 9.6 x $10^{-13}$ N
    D. 9.6 x $10^{-12}$ N

24. How would the threshold electric field change if the pressure of the gas were increased?
    A. The threshold would decrease because the mean free path would decrease.
    B. The threshold would decrease because the mean free path would increase.
    C. The threshold would increase because the mean free path would decrease.
    D. The threshold would increase because the mean free path would increase.

25. Sulfur hexafluoride is a dense gas which absorbs electrons. What would be the effect of replacing air with sulfur hexafluoride between the plates?
    A. The threshold would decrease because the density would be greater.
    B. The threshold would increase because the density would be greater.
    C. The increased density would increase the threshold, while the lone electrons would be absorbed, increasing the threshold even more.
    D. The lone electrons would be absorbed, increasing the threshold. (Density is not a factor.)

26. Which graph could be a typical graph of the electron described in paragraph 2?

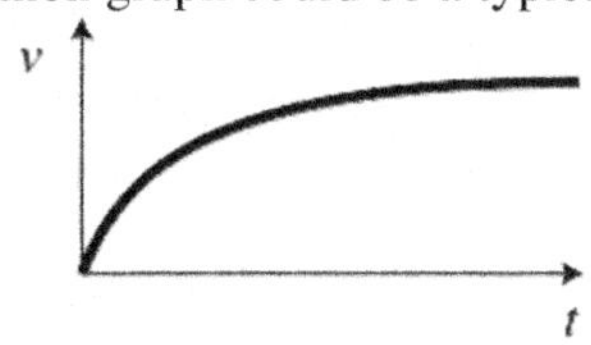


A.

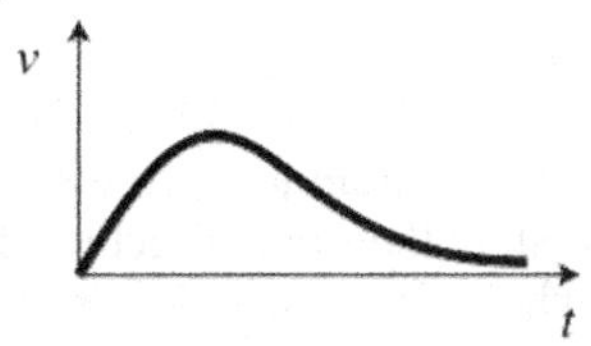


B.

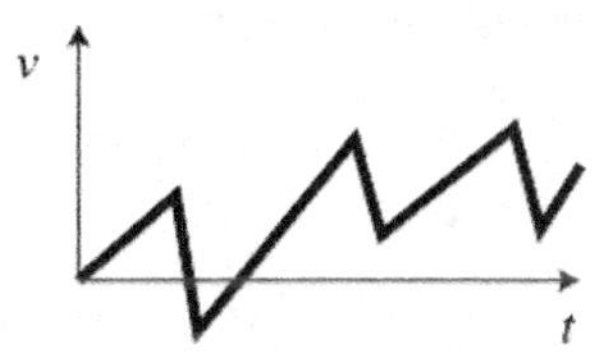


C.

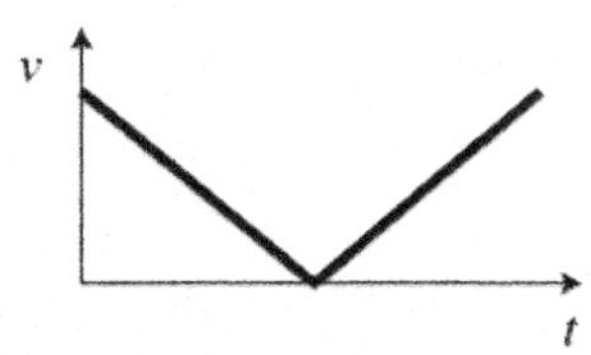


D.

**Passage 6**

A number of aquatic species have evolved the capability to produce sizable electric fields. For some species this helps them to detect other organisms, either predator or prey. Other species, such as the electric eel, use this capability to stun or kill a predator or prey.

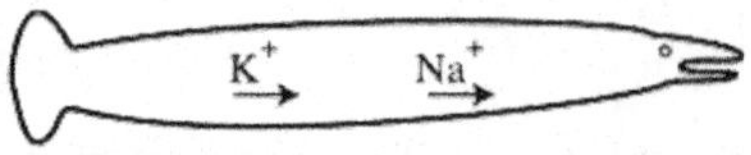


The organism generates the electric field in an "electric organ" which can take up most of the body cavity. The following description refers to a hypothetical electric organ, which includes many features by which these organs operate.

Electrocytes are flattened cells (like disks) which are stacked in a series, as shown in the figure below. During the equilibrium state, the cells actively exclude sodium ions ($Na^+$), creating a potential difference between the inside of the cell and the outside of the cell. The cell membrane is permeable to potassium ions ($K^+$), so the interior of the cell becomes enriched with $K^+$ ions, partially but not fully compensating for the potential difference due to the imbalance of $Na^+$ ions. The magnitude of the potential difference between the inside and outside is of the cell is about 0.1 V.

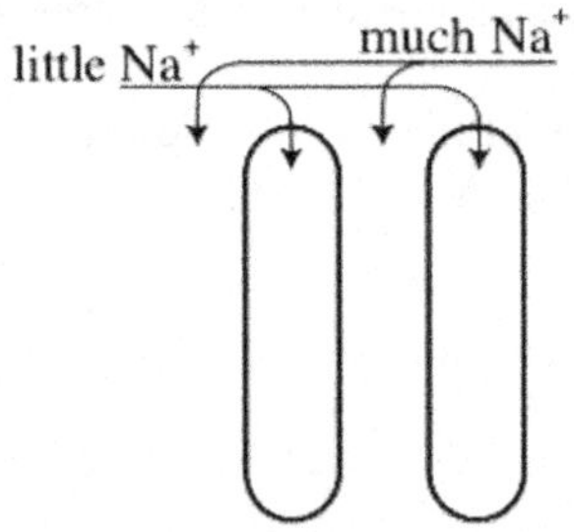


*Equilibrium State*

During the activated state, the posterior side of the cells becomes permeable to sodium ions, so they rush in through the posterior face due to the potential difference. Potassium ions rush in the same direction through the anterior face of the cell. The result is a potential difference across the whole organ, which acts, in effect, as a battery. (See the figure below.) The circuit is completed in the surrounding water, as shown in the second figure below.

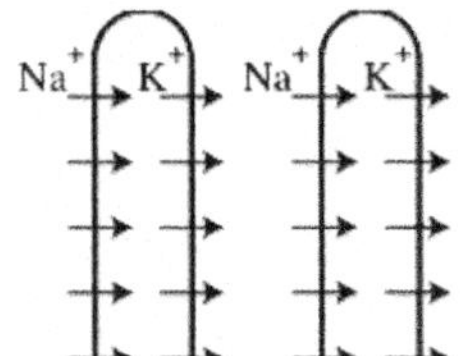


*Activated State*
*(arrows show current flow)*

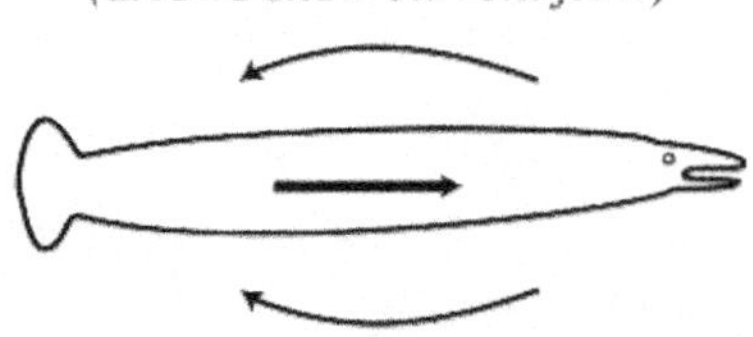

*arrows show current flow*

During the activated state, the current across the anterior and posterior faces of a cell in this hypothetical electric organ is 30 milliamps. The duration of a pulse is 2 milliseconds.

The charge on an electron is $-1.6 \times 10^{-19}$ C.

27. Consider $x_0$ and $x_2$ to be points in the extracellular medium and $x_1$ a point in the interior of the cell. During the equilibrium state, which graph best shows the electric potential as a function of x?

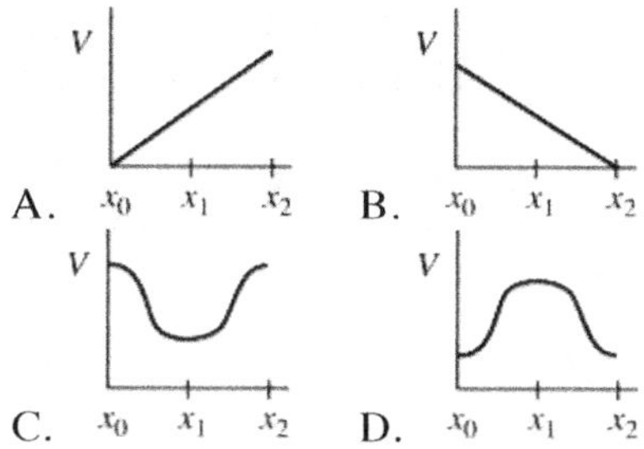

28. If the electric organ of a fish consists of 5000 electrocytes stacked in a series, what is the approximate total current through the fish during the activated state?
    A. 30 milliamps
    B. 75 amps
    C. 150 amps
    D. 300 amps

29. Which of the following best shows the electric field due to the fish during the activated state?

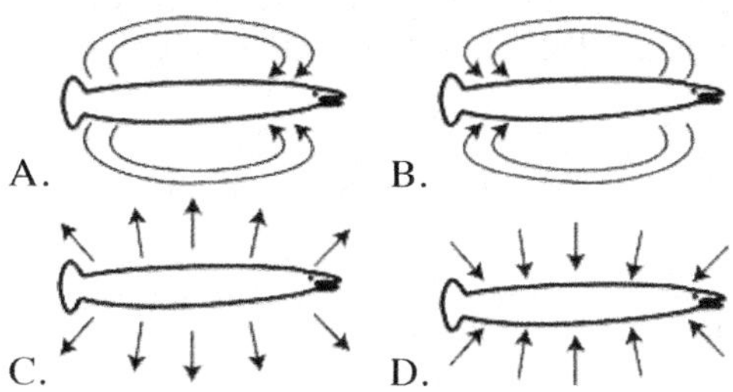

30. Which of the following best shows the magnetic field due to the fish during the activated state?

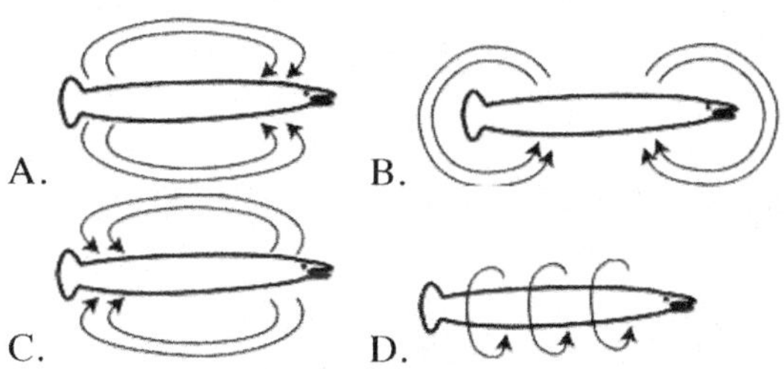

31. Which is a better conductor of electricity, fresh water or seawater?
    A. Fresh water, because seawater is slightly more dense.
    B. Sea water, because seawater is slightly more dense.
    C. Fresh water, because ions impede current in seawater.
    D. Sea water, because ions carry current in seawater.

**Passage 7**

One strategy in the fight against cancer involves the interaction of neutrons with a particular isotope of boron, called boron neutron capture therapy (BNCT). In BNCT, boron-10 is introduced into tumor cells, and the area is irradiated with slow neutrons. The interaction of a boron-10 nucleus and a neutron results in the production of an alpha particle with 2.3 Mev of kinetic energy.

A charged particle with large kinetic energy will lose that energy as it ionizes molecules that it passes by. In this case the alpha particles disrupt the DNA (among other molecules) of the tumor cells, effectively destroying cancerous tissue.

There are several problems with this method. It is important not to use a neutron flux so high that neutrons react substantially with oxygen and hydrogen in normal tissue, producing gamma rays. This sets an upper limit on the flux of neutrons and thus a lower limit on the necessary concentration of $^{10}$B in tumor cells.

It has proven difficult to concentrate $^{10}$B in tumor cells. One idea involves attaching boron to certain monoclonal antibodies which would recognize antigens on cancer cells. Another idea involves attaching boron to nucleosides, which tend to be taken up by dividing cells.

Another problem is finding a significant source of slow neutrons. Nuclear reactors are few, expensive, and relatively immobile. The isotope $^{252}$Cf has been suggested as a source of neutrons. Although its main channel of decay is alpha, with a 2.6-year half-life, a significant fraction of it spontaneously fissions, yielding neutrons. Thus $^{252}$Cf can be taken to the site where it is needed. The neutrons from the fission can be moderated and directed to the area to be irradiated.

32. Why do the neutrons used in irradiation not cause significant ionization and thus radiation damage of normal tissue?
    A. The neutrons do not contain any orbitals.
    B. The neutrons get captured before they have a chance to ionize anything.
    C. The neutrons have overall zero charge.
    D. The neutrons are not an elementary particle.

33. Paragraph 3 alludes to nuclear reactions which may be dangerous to normal tissue. Which of the following is a possibility, according to this paragraph?
    A. $^{16}\text{O} + {}^{1}\text{n} \rightarrow {}^{13}\text{C} + {}^{4}\text{He}$
    B. $^{16}\text{O} + {}^{1}\text{n} \rightarrow {}^{17}\text{O} + \gamma$
    C. $^{16}\text{O} + {}^{1}\text{n} \rightarrow {}^{17}\text{F} + \gamma$
    D. $^{1}\text{H} + {}^{1}\text{n} \rightarrow {}^{2}\text{He} + \gamma$

34. A sample of $^{252}$Cf is shipped to a hospital, and 7.8 years later there are 0.01 moles of $^{252}$Cf left. How much was shipped to the hospital in the first place?
    A. 0.01 moles
    B. 0.03 moles
    C. 0.04 moles
    D. 0.08 moles

35. Why is it dangerous to have too large a neutron flux?
    A. The neutrons will ionize molecules in biological tissue.
    B. The neutrons will convert hydrogen and oxygen to other elements.
    C. The neutrons will react with hydrogen and oxygen to produce ionizing radiation.
    D. The neutrons will react with boron in normal cells, not just tumor cells.

**Passage 8**

An electrostatic precipitator is a device used in industry to remove pollution from exhaust gas. It consists of a long thin wire surrounded by a conducting cylinder, such that a potential about $5 \times 10^4$ volts is maintained between the negative wire and the positive cylinder. (See figure.)

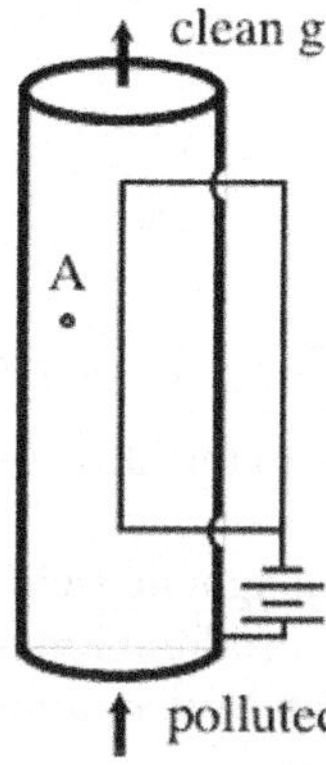


The resulting electric field inside the cylinder varies inversely with the distance from the center wire. Neutral particles of pollution are attracted to the center wire. The electric field near the wire is strong enough (greater than about $3 \times 10^6$ N/C) to ionize air, so the pollution particles are ionized negatively. They are attracted to the outer cylinder, where they collect and are eventually removed.

A typical energy requirement for such a device is about 300 Joules for each cubic meter of gas processed.

Note: The charge on an electron is $-1.6 \times 10^{-19}$ C. Point A is in the plane of the page.

36. If a sodium ion is at point A, in what direction would it experience a force?
    A. right
    B. left
    C. up
    D. down

37. If a $C_{10}H_{18}O_2$ molecule is at point A, in what direction would it experience a force?
    A. right
    B. left
    C. up
    D. down

38. If the potential maintained across the wire/cylinder were increased, how would the capacitance of the device be affected?
    A. The capacitance would decrease.
    B. The capacitance would stay the same.
    C. The capacitance would increase.
    D. The capacitance could decrease or increase depending on the gas in the cylinder.

39. Which of the following best represents the electric field lines in the cylinder?

A. 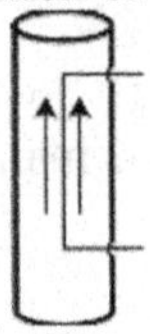

B. 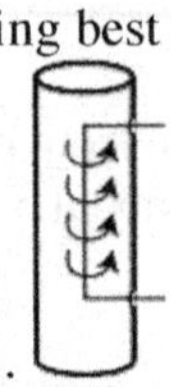

C. 

D. 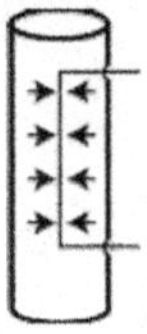

40. Which of the following is the closest analogy to the attraction of neutral particles of pollutants to the center wire?

A. The aligning of a small magnet to Earth's magnetic field.
B. The attraction between a chloride ion and a sodium ion in a salt crystal.
C. The van der Waals attraction between two nitrogen molecules in air.
D. A charged comb picking up pieces of paper.

# Test 6 Solutions

**Passage 1**

1. C.
The energy level diagram makes clear that the largest transition is the transition of absorption, from the ground state to the highest energy level. Thus those photons have the most energy and the highest frequency.

2. D.
On the other hand, nothing in the passage indicates that the energy difference between levels 1 and 2 should be less than (or greater than) the energy difference between level 1 and the ground state.

3. A.
The question refers to the transition from state 1 ($E_1$) to the ground state ($E = 0$), so the frequency is

$$f = (E_1 - 0)/h = E_1/h$$

4. D.
Clearly state 2 is not very populated, since transitions out of state 2 to state 1 (or to the ground state) occur very quickly. Otherwise, the passage is silent on this issue, so D is the answer.

5. A.
The Sun emits photons of all sorts of frequencies; that is what makes the colors of the rainbow. And many photons make it through the atmosphere.

**Passage 2**

6. C.
The definition of kinetic energy involves mass and velocity, and we know neither, so A and B are not right. Choice C mentions force, and there is a connection between force and energy, the equation $W = F\Delta x\cos\phi$. Does this equation apply in this case? The change in kinetic energy is given by the total work done on an object, that is $F_{net}\Delta x \cos\phi$, so the equation does apply. The quantity $\Delta x$ is the length of the barrel, and we have $\cos\phi = 1$.

7. C.
Pressure and temperature go together in the ideal gas equation, and though we may have the volume of the "reaction flask", we do not have the number of moles, so A is not the answer. It seems difficult to connect velocity with pressure, so let us look for a better answer than B. Pressure and force together remind us of the definition of pressure $P = F/A$. Since we have the cross-sectional area of the cannon, this is a connection between force and pressure. C is the answer.

8. A.
The rate of reaction does depend on surface area, concentration, and temperature, but grain size affects only the surface area. So A is the answer. Only a catalyst could reduce the activation energy.

9. B.
Since the reaction is irreversible, the entropy change is positive, so A is false. The reaction is spontaneous, so the free energy change must be negative.

10. B.
The energy starts as chemical energy and turns to heat after burning. The whole point of a cannon is to convert energy to kinetic energy of the ball. Note that the conversion of heat to kinetic energy is inefficient. Among the choices given, however, B is the best answer.

**Passage 3**

11. B.
Option A seems good, except it is the purpose of the *filter screen* to filter out byproducts, so this is not a likely possibility. The rupture film would certainly keep in the reactants, increasing the concentration and speeding the reaction. B is a likely possibility. The rupture film does keep the reactants dry, but that is important during the days and years before the accident, not the 1–5 ms during the reaction, so C is incorrect. There is no catalyst involved, so D is incorrect.

12. A.
In the absence of other information about entropy, it is reasonable to assume that the reaction which produces the most gas has the largest entropy increase. At a given temperature gases tend to have much more entropy than liquids and solids.

13. D.
Having the bag deflate will not affect the temperature appreciably, so A is incorrect. There are a number of ways to ensure that a bag will not burst (safety valve, and so on) that are better than having it deflate, so B is incorrect. The energy of collision or, rather, the work that must be done on the driver to stop him is a constant (equal to the negative of his kinetic energy before the collision), so C is incorrect. The language of choice D reminds us of $W_{tot} = F\Delta x$. And, indeed, the presence of an airbag increases the distance over which the decelerating force acts: not all at once at the steering wheel, but "gradually" over 0.3 meters. This decreases the force and reduces grievous bodily harm.

14. C.
It is not clear how a larger area of the bag could affect the flow rate. A larger area of the nozzle could increase the flow rate, but the question is asking about the airbag, so A and B are incorrect. Option C reminds us of the formula $P = F/A$. If $A$ is increased, and $F$ is the same, then pressure is reduced, hence the cushioning effect. Thus C is correct. D makes a correct statement, but does not lend advantage to the driver, so it is incorrect.

15. D.
Option A, chemical to heat, misses the point. If this were a reaction that generated a lot of heat that then dissipated, then A would be a possibility. The point of energy release in an airbag is that the airbag is actually inflated. So A is incorrect, as well as B. Option C seems like a possibility, since the bag seems to gain kinetic energy, which might later be turned to heat. Option D is a better description; however, since the true energy flow involves pushing back the atmosphere and not in moving the bag (which has little mass).

16. A.
All the options are reasonable byproducts except A, which involves Na or O in bizarre valence states.

**Passage 4**

17. D.
According to the first paragraph, the wavelength must be smaller than the observed object. A wavelength of $10^{-3}$ m corresponds to a frequency of about 1.5 x $10^6$ Hz in biological tissue, where we use $v = 1500$ m/s.

18. A.
The lowest frequency is 20 Hz, so $T = 1/f = 1/(20\text{ Hz}) = 0.05$ s.

19. D.
The two real choices are B and D. Since there is tearing and rupturing involved, there must be chemical bonds broken, so D is it.

20. D.
Frequency is not directly connected to intensity. Amplitude and intensity are connected, but amplitude is not mentioned among the choices.

21. D.
Choices A and B are irrelevant. Concerning choice C, the problem is not energy reflecting off the organ but being absorbed by it. This is an example of resonance, and there must be a weak coupling between two oscillators: the sound is one and the oscillating organ is the other.

**Passage 5**

22. B.
The force on a charged particle in an electric field is $F = QE$, so the oxygen ion ($O^{2-}$) experiences twice the force.

23. A.
A calcium atom has no charge, so the force on it would be zero.

24. C.
Increasing the pressure would put the gas particles closer to each other, so an electron could not travel as far before colliding with a particle. Thus the mean free path would decrease. If the electron is not able to travel as far, then the energy it gains is less. (The kinetic energy it gains is about $F_{elec}\, l_{mfp}$, from the standard formula for work.) Thus the electron is not likely to create a spark. A greater electric field is required, so the threshold is higher.

25. D.
The mass density does not have anything to do with this phenomenon, so choices A, B, and C are out. It is true that an electron absorber will inhibit the phenomenon like killing a baby in its crib.

26. C.
This graph shows the constant acceleration (due to the constant force) of the electron in the parts of the graph which have a small positive slope. The rapid decreases in velocity are the times of collision where the electron loses energy.

**Passage 6**

27. C.
According to the third paragraph, the positive sodium ions are transported to the outside of the cell. The outside of the cell acquires a positive charge and a higher potential. So C is the correct choice. The diffusion of potassium ions lessens the effect, but the qualitative picture is the same.

28. A.
The cells are stacked in series. Thus the current through all the cells is the same, and the total current is simply 30 milliamps. Remember, the currents add up if the components of the current are in parallel. The voltages add up in series.

29. B.
The electric field points away from positive charges and toward negative charges. So where are the charges during the activated state? A glance at Figures 1 and 3 shows that the fish pushes positive charges from its tail to its head, so a positive charge collects at the head of the fish, and a negative charge at its tail. Thus the answer is B.

30. D.
To obtain the magnetic field, we need to consider the current. The main current during the activated state is along the fish from tail to head. The right-hand rule applied to this implies a magnetic field circling the body in the direction shown in choice D. The secondary currents outside the fish in the fourth figure reinforce this magnetic field. On problems like this one, if you are in doubt, choose one typical point and apply the right-hand rule. For instance, choose the point under the fishes belly. If the magnetic field there is to the right, choose A; up, B; left, C; or into the page, D.

31. D.
Seawater has the charge carriers which make it more conductive. You knew this.

**Passage 7**

32. C.
Choice A is nonsensical. Choice B misses the point, since the question asks *why* these neutrons do not tend to ionize the tissue. The passage indicates (paragraph 2) that a charged particle with large energy is ideal for ionizing tissue, so a neutron with no charge and little energy will do little harm. Choice C is a likely possibility. Choice D is excluded since neutrons are an elementary particle.

33. B.
Choices C and D do not have the superscripts and subscripts balanced. Choice A does not satisfy the description in the passage, in that there is no gamma ray in the products.

34. D.
The 7.8 years represents 3 half-lives, so the original sample decreased by a factor of 2 three times. The original sample must have been 0.08 moles.

35. C.
This is a paraphrase of the second sentence of paragraph 3.

**Passage 8**

36. A.
A sodium ion ($Na^+$) is positively charged, and the negatively charged central wire will attract it.

37. A.
It does not matter what kind of molecule this is. The only thing that matters is its neutral charge. The passage states that it will be attracted to the wire.

38. B.
The main formula we have for capacitance is $C = Q/\Delta V$, which may tempt you to think that an increase in potential would lead to a decrease in capacitance, but this is not so. The capacitance of a device is fixed by the construction of that device, and the amount of stored charge goes up proportionally as the potential. An increase in potential would result in an increase in the charge on the wire $Q_{wire}$, but the capacitance remains the same.

39. D.
The electric field lines point away from the positive charge and toward the negative charge.

40. D.
The attraction of neutral particles to a charged wire is like a charged comb attracting neutral pieces of paper. Choice A does not refer to charges at all. Choice B involves two charged particles, and choice C involves two neutral species. Choice D refers to a charged and a neutral species, just like the wire and the pollutants. A charge is induced on the pollutants, and the resulting net force is attractive.

# Test 6 Conversion Chart

**Scale Score Conversion Chart**

| Raw Score | Scale Score | Percentile Ranking |
|---|---|---|
| 40 | 36 | 99 |
| 39 | 35 | 99 |
| 38 | 34 | 99 |
| 37 | 33 | 99 |
| 36 | 32 | 98 |
| — | 31 | 98 |
| 35 | 30 | 97 |
| 34 | 29 | 95 |
| 33 | 28 | 94 |
| 31–32 | 27 | 91 |
| 30 | 26 | 88 |
| 29 | 25 | 84 |
| 27–28 | 24 | 77 |
| 25–26 | 23 | 71 |
| 24 | 22 | 64 |
| 22–23 | 21 | 56 |
| 20–21 | 20 | 48 |
| 19 | 19 | 38 |
| 17–18 | 18 | 30 |
| 15–16 | 17 | 23 |
| 14 | 16 | 18 |
| 13 | 15 | 14 |
| 11–12 | 14 | 10 |
| 10 | 13 | 7 |
| 9 | 12 | 5 |
| 7–8 | 11 | 3 |
| 6 | 10 | 1 |
| 5 | 9 | 1 |
| 4 | 8 | 1 |
| 3 | 7 | 1 |
| — | 6 | 1 |
| 2 | 5 | 1 |
| — | 4 | 1 |
| 1 | 3 | 1 |
| — | 2 | 1 |
| 0 | 1 | 1 |

Made in the USA
Monee, IL
13 September 2026

63177768R00201